D1064929

Practical Handbook of
Environmental
Control

Edited by
Conrad P. Straub, M.C.E., Ph.D.
Emeritus Professor
Environmental Health
School of Public Health
University of Minnesota
Minneapolis, Minnesota

CRC Press, Inc.
Boca Raton, Florida

Library of Congress Cataloging-in-Publication Data

Practical handbook of environmental control / editor, Conrad P.
  Straub.
    p. cm.
  Includes bibliographies and index.
  ISBN 0-8493-3707-0
  1. Pollution—Handbooks, manuals, etc. 2. Sewage disposal-
-Environmental aspects—Handbooks, manuals, etc. I. Straub, Conrad
P.
  [DNLM: 1. Air Pollution--prevention. 2. Refuse Disposal.
3. Sewage. 4. Water Pollution—prevention. WA 754 P895]
TD176.4.P73 1989
363.7'37—dc19
DNLM/DLC   88-22218
for Library of Congress

Direct all inquiries to CRC Press, Inc., 2000 Corporate Blvd., N.W., Boca Raton, Florida, 33431.

International Standard Book Number 0-8493-3707-0

Library of Congress Card Number 88-22218
Printed in the United States

PREFACE

Much of the material in this volume was culled from the previously published five-volume *Handbook of Environmental Control*. Whereas each of the previous volumes dealt with a particular aspect of the macroenvironment (air pollution, solid wastes, water supply and treatment, wastewater treatment and disposal) and the microenvironment (hospital and health care facilities), this volume includes important aspects of the earlier five volumes in the context of the total environment.

Our environment is complex. Any consideration of the effects of present-day contaminants on the ecological system and the potential health effects on human and animal populations must concern itself with considerations of the total environment, examining the sources of these contaminants, their pathways — physical and biological—and the effect each route of exposure has on the ultimate exposure of the individual.

The most pressing problems relate to our handling and disposal of solid, potentially hazardous, waste; atmospheric contamination from energy sources, motor vehicles, and industry; contamination of surface water supplies, and, in particular, contamination of ground water supplies by organic and inorganic substances.

This volume should prove useful to students in a variety of disciplines since it provides basic information in an interdisciplinary area and encourages students, through the references supplied, to seek out and to read the original articles cited. It should also be of interest to professionals in academia, in consulting offices, and government agencies, and to others responsible for the investigation, evaluation, and solution of the complicated environmental issues.

My deepest appreciation is expressed to the authors, journals, and book publishers for permitting the use of copyrighted materials, and to Ms. Amy G. Skallerup, Associate Managing Editor, CRC Handbooks, CRC Press, Inc., for her patience, suggestions, and recommendations which helped make this volume a reality.

Conrad P. Straub
Columbia Heights, MN
September, 1988

THE EDITOR

Conrad P. Straub, editor of this volume, received his B.S. in C.E. and C.E. degrees from the Newark College of Engineering and his M.C.E. and Ph.D. degrees in sanitary engineering and public health from Cornell University.

He joined the U.S. Public Health Service in 1941, served in various capacities in the U.S., China, and Poland, and retired as Director, Robert A. Taft Sanitary Engineering Center, Cincinnati, in 1966. He joined the School of Public Health, University of Minnesota, as Professor of Environmental Health, served as Director, Environmental Health Research and Training Center, Department Head, and Director, Midwest Occupational Health and Safety Center, and retired in 1981.

Dr. Straub has been on many committees and has been a consultant to the World Health Organization, and to other national bodies, governmental agencies, and private organizations. He has published a book *Low Level Radioactive Wastes,* has contributed chapters to several books, has published over 170 papers in various professional journals, and has co-edited or edited the five-volume *CRC Handbook of Environmental Control.* Currently, he edits *Critical Reviews in Environmental Control,* CRC Press.

ACKNOWLEDGMENTS

With exceptions as noted, much of the material in this volume was derived from the five-volume *CRC Handbook of Environmental Control* (of which Volumes I through IV were edited by Richard G. Bond and Dr. Conrad P. Straub; and Volume V was edited by Conrad P. Straub), as follows:

Section 1: from Volume I, *Air Pollution;* except Table 1.2-16.

Section 2: from Volume III, *Water Supply and Treatment;* except new Tables 2.4-9, 2.5-6 through 2.5-9, 2.6-1, 2.6-12 through 2.6-15, and 2.6-19.

Section 3: from Volume IV, *Wastewater Treatment and Disposal;* except new tables 3.1-7, 3.2-3, 3.2-4, 3.2-7, 3.2-9, 3.4-2, 3.4-3, 3.4-8, 3.4-23, 3.4-25, 3.4-27, 3.4-30, 3.4-32, 3.4-34, 3.4-39, 3.4-46, 3.4-52, 3.4-54, 3.4-55, 3.4-67, 3.4-68, 3.4-69, 3.4-75, 3.5-1, 3.6-3, and 3.7-7; from Volume II, 3.5-5, 3.5-6, and 3.5-7; from Volume III, 3.2-18, 3.2-44, 3.3-18, 3.5-2, 3.5-3, 3.6-1, and 3.6-4.

Section 4: from Volume II, *Solid Waste;* except new Table 4.3-5; and Table 4.2-24 from Volume IV.

Section 5: from Volume V, *Hospital and Health Care Facilities,* except new or revised tables (figures) 5.1-1, 5.1-7, 5.1-13 through 5.1-15, 5.1-17, 5.1-18, 5.2-1, 5.2-4, 5.2-8, 5.2-13, 5.2-16, 5.2-17, 5.2-19, 5.3-2, 5.4-5, 5.4-16, and 5.4-19; from Volume III, 5.4-14; and from Volume IV, 5.4-12 and 5.4-13.

TABLE OF CONTENTS

Section 1
Air

1.1   THE ATMOSPHERE AND AIR POLLUTANTS

1.1.1   ATMOSPHERIC DATA

1.1–1A   AVERAGE COMPOSITION OF DRY AIR

For most applications the following accepted values for the average composition of the atmosphere are adequate. These values are for sea level or any land elevation. Proportions remain essentially constant to 50 000 ft altitude.

Gas	Molecular weight	Percentage by volume, mol fraction	Percentage by weight
Nitrogen	$N_2 = 28.016$	78.09	75.55
Oxygen	$O_2 = 32.000$	20.95	23.13
Argon	$Ar = 39.944$	0.93	1.27
Carbon dioxide	$CO_2 = 44.010$	0.03	0.05
		100.00	100.00

For many purposes the percentages 79% N_2 –21% O_2 by volume and 77% N_2 –23% O_2 by weight are sufficiently accurate, the argon being considered as nitrogen with an adjustment of molecular weight to 28.16.

Other gases in the atmosphere constitute less than 0.003% (actually 27.99 parts per million by volume), as given in the following table.

1.1–1B   MINOR CONSTITUENTS OF DRY AIR

Gas	Molecular weight	Parts per million By volume	By weight
Neon	$Ne = 20.183$	18.	12.9
Helium	$He = 4.003$	5.2	0.74
Methane	$CH_4 = 16.04$	2.2	1.3
Krypton	$Kr = 83.8$	1.	3.0
Nitrous oxide	$N_2O = 44.01$	1.	1.6
Hydrogen	$H_2 = 2.0160$	0.5	0.03
Xenon	$Xe = 131.3$	0.08	0.37
Ozone	$O_3 = 48.000$	0.01	0.02
Radon	$Rn = 222.$	(0.06×10^{-12})	

Minor constituents may also include dust, pollen, bacteria, spores, smoke particles, SO_2, H_2S, hydrocarbons, and larger amounts of CO_2 and ozone, depending on weather, volcanic activity, local industrial activity, and concentration of human, animal, and vehicle population. In certain enclosed spaces the minor constituents will vary considerably with industrial operations and with occupancy by humans, plants, or animals.

The above data do not include water vapor, which is an important constituent in all normal atmospheres.

Source: *Handbook of Tables for Applied Engineering Science*, R.E. Bolz and G. L. Tuve, Eds., The Chemical Rubber Co., Cleveland, O., 1970, p. 533.

1.1.2   AIR POLLUTANT PROPERTIES

Air pollution is the presence in the ambient air of one or more contaminants, which can be naturally occurring or man-made. The quantities, characteristics, and duration of these contaminants are, or may tend to be, injurious to human, plant, and animal life or may interfere with the enjoyment of life or use of property.

1.1-2   CLASSIFICATION OF AIR POLLUTANTS

Major classes	Subclasses	Typical members
Inorganic gases	Oxides of nitrogen (NO_X)	Nitrogen dioxide, nitric oxide
	Oxides of sulfur (SO_X)	Sulfur dioxide, sulfuric acid
	Other inorganics	Ammonia, carbon monoxide, chlorine, hydrogen fluoride, hydrogen sulfide, ozone
Organic gases	Hydrocarbons	Benzene, butadiene, butene, ethylene, isooctane, methane
	Aldehydes, ketones	Acetone, formaldehyde
	Other organics	Acids, alcohols, chlorinated hydrocarbons, peroxyacyl nitrates, polynuclear aromatics
Aerosols	Solid particulate matter	Dusts, smoke
	Liquid particulates	Fumes, oil mists, polymeric reaction-products

Source: *Environmental Biology,* P.L. Altman and D.S. Dittmer, Eds., Federation of American Societies for Experimental Biology, Bethesda, Md., 1966. With permission.

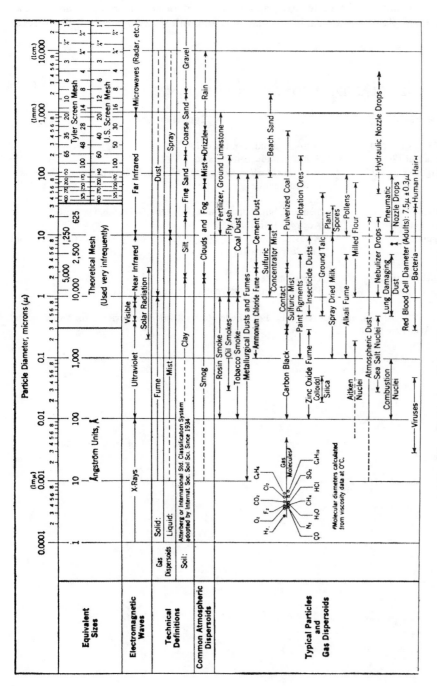

1.1-3   Characteristics of particles and particle dispersoids.

1.1-3 Characteristics of particles and particle dispersoids (continued).

Source: C.E. Lapple, *Stanford Res. Inst. J.*, *5*:95, 1961. Reprinted by permission of the author and the publisher.

1.1-4 MOBILITIES OF ATMOSPHERIC IONS

Type of ions	Average mobility, cm^2/volt sec	Approximate diameter, μm
Small +	1.4[a], 1.1[b]	0.001–0.005
Small –	1.9[a], 1.2[b]	0.001–0.005
Intermediate	0.05	0.005–0.015
Large	0.000 4	0.015–0.10
Fog droplets	0.5 x 10^{-6} n[c]	2–70

[a]In dry air.
[b]In moist air.
[c]With n charges.

GENERAL REFERENCES

1.   H. Neuberger, "Introduction to Physical Meteorology," Pennsylvania State University Press, University Park, Pa., 1951.
2.   V.A. Gordieyeff, *Arch. Ind. Health*, 14:471, 1956.

1.1-5 PROPERTIES OF SOME TYPICAL AEROSOLS

Type of dispersal system	Size range of particles diameter, μm	Terminal velocity (cm/sec) due to gravity settling in air, 20°C and 1 atm
Raindrops	5 000–500	Turbulent motion $v_t = k_1 pD$ for particles down to about 1 000 μm
Natural mist	500–40	p = particle density
Natural fog	40–1.0	D = particle diameter, μm
Dust		
Foundry sand	2 000–200	
Ground limestone, fertilizers	800–30	Intermediate region for particles between 1 000–100 μm
Sand tailings from flotation	400–20	$v_i = k_2 p^{2/3} D$
Pulverized coal	400–10	Stokes' law for streamline region, applicable to particles from 100-1.0 μm
Ground sulfide ore for flotation	200–4	
Foundry shake-out dust	200–1	$v_s = KpD^2/n$
Cement	150–10	n = coefficient of viscosity of air, poises
Fly ash	80–3	
Silica dust in silicosis	10–0.5	$v_s = k_3 pD^2$
Pigments	8–1.0	
Pollens	60–20	
Plant spores	30–10	
Bacteria	15–1.0	

1.1-5 PROPERTIES OF SOME TYPICAL AEROSOLS (continued)

Type of dispersal system	Size range of particles diameter, μm	Terminal velocity (cm/sec) due to gravity settling in air, 20°C and 1 atm
Fumes and Mist		
Metallurgical fumes	100−0.1	Cunningham's correction for particles in range 1.0− 0.10 μm
H$_2$SO$_4$ concentrator mist	10−1.0	
Alkali fume	2−0.1	$v_c = v_s \left(\dfrac{1 + 1.72\ 1}{D}\right)$
SO$_3$ mist	3−0.5	1 = mean free path of gas molecules, μm
NH$_4$Cl fume	2−0.1	
Zinc oxide fume	0.3−0.03	$v_c = v_s \left(\dfrac{1 + 0.172}{D}\right)$ in air
Smokes		
Oil smoke	1.0−0.03	Velocity due to Brownian Motion exceeds velocity of gravity settling for particles less than 0.1 μm
Rosin smoke	1.0−0.01	
Tobacco smoke	0.15−0.01	
		Einstein equation
Carbon smoke	0.2−0.01	
		$x = k_4\ t/D$
Normal impurities in quiet atmosphere	1.0−0.01	x = average displacement in cm of spherical particle in air in time, t sec

	Spheres	Irregular shape
$k_1 =$	24	16
$k_2 =$	0.41	0.26
$k_3 =$	0.003 0	0.002
$k_4 =$	0.000 68	−

Sources: C.E. Miller, "Pointers on Selecting Equipment for Industrial Gas Cleaning," *Chem. Metallurg. Eng., 45*:132, 1938; and L. Silverman, "What Process Wastes Cause Air Pollution?" *Chem. Eng., 5*(58):132, 1951.

1.1-6   AIRBORNE PARTICLE SHAPES

Shape	Examples
Spherical	Smoke, pollen, fly ash
Irregular	Mineral
Cubical	Cinder
Flakes	Mineral, epidermis
Fibrous	Lint, plant fiber
Condensation flocs	Carbon, smoke, fume
Loose flocs	Magnesium oxide
Platelet	Mica
Solids of revolution	
Ellipse	
Short right circular cylinder	Asbestos
Needles	Talc
Chains	Carbon black, smokes

GENERAL REFERENCES

1. *Air Pollution Manual*, Part II, Control Equipment, American Industrial Hygiene Association Detroit, Mich., 1968.
2. K.T. Whitby *et al.*, *Heat., Piping, Air Cond.*, 29:185, 1957.

1.1-7   PARTICLE DENSITIES FOR AGGLOMERATES

Material	Floc density, g/cc	Normal density, g/cc
Aluminum oxide	1.18	3.70
Antimony trioxide	0.63	5.57
Arsenic trioxide	0.91	3.7
Cadmium oxide	0.51	6.5
Lead monoxide	0.62	9.36
Magnesium oxide	0.35	3.65
Mercury	1.70	13.6
Silver	0.94	10.5
Stannic oxide	0.25	6.71

REFERENCE

1. G.R. Whytlaw and H.H. Patterson, *Smoke*, Edward Arnold and Co., London, England, 1932.

1.1-8   FREE NONVOLATILE FATTY ACIDS AS DETERMINED IN PARTICULATE MATTER OF AIR POLLUTANT SAMPLES[a]

	Sampled in Detroit, Mich., on freeway-interchange, October–November 1963		Sampled in New York, N.Y., at high-traffic city location, February 1964	
	Per cent in acidic portion[b]	μg per 1 000 m^3 air	Per cent in acidic portion[b]	μg per 1 000 m^3 air
Saturated Acids				
Lauric	0.29	13.7	0.43	39.4
Myristic	0.47	17.5	0.53	48.6
Palmitic	3.10	146.8	2.40	220.0
Stearic	2.42	113.7	1.50	137.5
Behenic	1.31	61.6	1.30	119.2
Unsaturated Acids				
Oleic	0.68	32.2	1.34	122.8
Linolenic	0.56	26.5	Trace	Trace

[a]Values represent averages from three analyses.
[b]As defined in separation scheme for organic particulate matter, E.L. Wynder and D. Hoffmann, *J. Air Pollut. Contr. Assn., 15:*155, 1965.

Source: D. Hoffman and E.L. Wynder, in *Air Pollution,* A.C. Stern, Ed., Vol. 2, Academic Press, New York, N.Y., 1968, p. 221. With permission.

1.1-9   CHEMICAL ANALYSIS AND PHYSICAL PROPERTIES OF FLY ASH

Per Cent by Weight as Collected Dried

Constituent	Utilities	Industrials
Carbon, as C	0.37–36.2	2.4–64
Iron, Fe_2O_3 or Fe_3O_4	2.0–26.8	10.6–43.6
Magnesium, as MgO	0.06–4.77	0.10–1.49
Calcium, as CaO	0.12–14.73	1.92–11.8
Aluminum, as Al_2O_3	9.81–58.4	10.0–21.0
Sulfur, as SO_3	0.12–24.33	0.66–28.0
Titanium, as TiO_2	0.50–2.8	0.22–3.81
Carbonate, as CO_3	0.05–2.6	0.05
Silica, as SiO_2	17.3–63.6	25.0–46.0
Phosphorus, as P_2O_5	0.07–47.2	0.05–1.93
Undetermined	0.08–18.9	–
Other Tests		
Hydrogen ion concentration, pH	4.7–11.7	6.3–12
Apparent specific gravity	0.8–3.0	0.77–1.45
Retained on 325-mesh screen (44 μm), %	8–95	29–84
Through 325-mesh screen (44 μm), %	5–92	16–71
Fusion temperature of ash, °F	1 910–3 000	1 980–2 900

REFERENCE

1. *Air Pollution Abatement Manual,* Manufacturing Chemists' Association, Inc. Washington, D.C.

1.1-10   MINERAL ASSEMBLAGES IN ATMOSPHERIC DUSTS

SYMBOLS:

1–Talc	4–Mica
2–Quartz	5–Amphibole
3–Plagioclase	6–Chlorite

Sample	Location	Collection date	Solid phase concentration, mg/l	Mineral found
Rain	San Diego, Calif.	7 Nov. 1966	1.9	1, 2, 3
	32° 50′ N	7 Nov. 1966		1, 2, 3, 4, 5
	117° 16′ W	5 Dec. 1966		1, 2, 3, 4, 5, 6
		13 March 1967	2.5	1, 2, 3, 4, 5
		13 March 1967	0.4	2, 3
Atmospheric dust	San Diego, Calif.	17 Nov. 1966		1, 2, 3, 4, 5, 6
		15 Feb. 1967		1?, 2, 3, 4, 5, 6
		16 Feb. 1967		2, 3, 4
		17 Feb. 1967		2, 3, 4, 5
		21 Feb. 1967		2, 3, 4, 5
		22 Feb. 1967		2, 3, 4, 5
		23 Feb. 1967		1, 2, 3, 4, 5
		24 Feb. 1967		2, 3, 4
		27 Feb. 1967		1, 2, 3, 4, 5
		28 Feb. 1967		1, 2, 3, 4, 5
		6 March 1967		2, 3, 4, 5?
		7 March 1967		1, 2, 3, 4, 5, 6
		8 March 1967		1, 2, 3, 4, 5, 6
		9 March 1967		2, 3, 4, 5
	Scotts Bluff, Neb.			
	41° 52′ N	28 April 1966		1, 2, 3, 4
	103° 40′ W	28 April 1966		1, 2, 3, 4
		10 May 1966		1, 2, 3, 4
	Minicoy Island, India	1963		1, 2, 3, 4, 6
	8° 20′ N	1965		1, 2, 3, 4, 6
	73° 01′ E			
Dust storm	Bagdad, Iraq	17 Feb. 1966		1?, 2, 3, 4, 6
	33° 20′ N			
	44° 26′ E			
Glacier[a]	Yukon	1967	0.8	1, 2, 3, 4, 5, 6
	60° 45′ N			
	139° 30′ W	1961	0.5	1?, 2, 3, 4, 5, 6
		1955	0.5	1?, 2, 3, 4, 5, 6
		1947	0.3	2, 3, 4, 6
		1936	0.5	1?, 2, 3, 4, 6
	Orizaba, Mexico	1962–67	89	1, 2, 3, 4, 5, 6
	19° 01′ N	1957–62	734	1, 2, 3, 4, 5, 6
	97° 15′ W	1952–57	17	1, 2, 3, 5, 6
		1942–47	19	1, 2, 3, 4, 5
		1937–42	49	1, 2, 3, 4, 5, 6
	Popocatepetl, Mexico	1960–67	302	1, 2, 3, 4, 5
	19° 01′ N	1949–53	38	1, 2?, 3
	98° 37′ W	1943–46	159	1, 2, 3, 4, 5

1.1-10 MINERAL ASSEMBLAGES IN ATMOSPHERIC DUSTS (continued)

Sample	Location	Collection date	Solid phase concentration, mg/1	Mineral found
Glacier	Washington	1967	1.8	1, 2, 3, 4, 5, 6
(continued)	47° 52′ N	1962	1.9	1, 2, 3, 4, 5, 6
	123° 36′ W	1950	2.4	1, 2, 3, 4, 5, 6
		1940	1.8	1, 2, 3, 4, 5, 6
		1932	9.9	2, 3, 4, 5, 6
		1925	3.7	2, 3, 4, 5, 6
		1916	10.1	2, 3, 4, 5, 6
	Palomar	1965	2.2	1, 2, 3, 4, 5, 6
	33° 24′ N		1.4	1, 2, 3, 4, 5, 6
	116° 53′ W			
Snow samples	Julian	1965	2.9	1, 2, 3, 4, 5, 6
	33° 04′ N			
	116° 36′ W			
	Mount Rainier	1965	30	1, 2, 3, 4, 6
	46° 51′ N			
	121° 45′ W			
	Canadian Rockies	1967	2.7	2, 3, 4, 5, 6
	52° 55′ N			
	118° 10′ W			

[a]For the glacial materials the collection date is the time of accumulation of the strata sampled.

Source: Reprinted with permission from H. Griffin, *Environ. Sci. Technol.*, *1*(11):925, November 1967. Copyright 1967, American Chemical Society.

1.1-11 NATURAL CHARGES ON REPRESENTATIVE PARTICLE DISPERSOIDS

Dispersoid	Method of dispersal	Charge distribution, %			Specific charge, esu/g	
		Positive	Negative	Neutral	Positive	Negative
Raw cement mix	Agitation in air stream	35	35	30	0.7 x 10⁴	0.7 x 10⁴
Gypsum dust (Schumacher Plant, L.A.)	Grinding, drying in flash dryer	44	50	6	1.6	1.6
Copper smelter dust (Tooele, Utah)		40	50	10	0.2	0.4
Fly ash (Stateline, Chicago)		31	26	43	1.9	2.1
Fly ash (Rochester Electric)		40	44	16	4.8	4.2
Gypsum dust (U.S. Gypsum, Phila., Pa.)	Grinding and drying in rotary kiln				0.2	0.2
Lead fume (Tooele, Utah)	Dwight-Lloyd sintering machine	25	25	50	0.003	0.003
Laboratory oil fume	Condensation from vapor	0	0	100	0	0

REFERENCE

1. H.J. White, "Report on Electrical Characteristics of Industrial Dispersoids," Unpublished report, Research Corporation, 1941.

1.1-12   MOST COMMON AEROALLERGENIC FUNGI[1,2]

Alternaria	Hormodendrum
Aspergillus	Macrosporium
Botrytis	Penicillium
Cladosporium	Phoma
Curvularia	Pullalaria
Epicoccum	Spondylocladium
Fusarium	Stemphyllum
Helminthosporium	

REFERENCES

1. R.H. Criep, R.A. Teufel, and C.S. Miller, "Fungicidal Agents in the Treatment of Allergy to Molds," *J. Allergy, 29:*258, 1968.
2. M.C. Harris and N. Shure, *Sensitivity Chest Diseases*, F.A. Davis Company, Philadelphia, Pa., 1964.

1.1-13   RATE CONSTANTS FOR ATOMIC OXYGEN AND ATMOSPHERIC POLLUTANTS

Reaction	Rate constant at 300°K	Reference
$O + O_2 + M \rightarrow O_3 + M$	$1.5 \times 10^8 \ l^2 \ mol^{-2} \ sec^{-1}$	1–4
$O + NO_2 \rightarrow NO + O_2$	$5.3 \times 10^9 \ l \ mol^{-1} \ sec^{-1}$	5
$O + NO_2 + M \rightarrow NO_3 + M$	$4.2 \times 10^{10} \ l^2 \ mol^{-2} \ sec^{-1}$	5
$O + NO + M \rightarrow NO_2 + M$	$2.3 \times 10^{10} \ l^2 \ mol^{-2} \ sec^{-1}$	5
$O + SO_2 + O_2 \rightarrow SO_3 + O_2$	$2.7 \times 10^9 \ l^2 \ mol^{-2} \ sec^{-1}$	6

Source: Reprinted with permission from A.P. Altshuller and J.P. Bufalini, "Photochemical Aspects of Air Pollution: A Review," *Environ. Sci. Technol.*, 5(1):39, 1971. Copyright 1971, American Chemical Society.

REFERENCES

1. S.W. Benson and A.E. Axworthy, *J. Chem. Phys., 26:*1718, 1957.
2. W.M. Jones and N. Davidson, *J. Amer. Chem. Soc., 84:*2868, 1962.
3. F. Kaufman and J.R. Kelso, *J. Chem. Phys., 40:*1162, 1964.
4. J.A. Zaslowsky *et al., J. Amer. Chem. Soc., 82:*2682, 1960.
5. E.A. Schuck, E.R. Stephens, and R.R. Schrock, presented at the 59th Annual Meeting of Air Pollution Control Association, San Francisco, Calif., June 20–24, 1966.
6. M.F.R. Mulcahy, J.R. Steven, and J.C. Ward, *J. Phys. Chem., 71:*2124, 1967.

1.1-14 AIR POLLUTANT REACTIONS

Reactant	Reaction	Product
General Reactions		
Sulfur dioxide, oxygen (+ catalysts)	$SO_2 + O \rightarrow SO_3 \rightarrow H_2SO_4$	Sulfuric acid, sulfates, aerosols
Olefins, sulfur dioxide, oxides of nitrogen, oxygen, and sunlight	$SO_2 + ROO\cdot \rightarrow RO\cdot + SO_3$	Sulfuric acid, aerosols
Styrene, halogens, sunlight	$C_6H_5CH = CH_2 + Cl_2$	Eye irritant
Nitric oxide, oxygen	$2NO + O_2 \rightarrow 2NO_2$	Nitrogen dioxide (slow reaction)
Photolysis[a]		
Nitrogen dioxide	$NO_2 \rightarrow NO + O$	Nitric oxide, atomic oxygen (main primary reaction)
Aldehydes	$R - COH \rightarrow R\cdot + HCO$	Alkyl, formyl
Ketones	$R_1R_2CO \rightarrow R\cdot + RCO$	Alkyl, acyl
Alkyl nitrites	$RONO \rightarrow RO\cdot + NO \rightarrow R\cdot + NO_2$	Alkyl, alkoxyl, nitric oxide, and nitrogen dioxide
Nitrous acid	$HNO_2 \rightarrow HO\cdot + NO \rightarrow H + NO_2$	Hydroxyl radical, atomic hydrogen, nitric oxide, and nitrogen dioxide
Thermal Reactions[a]		
Ozone, olefins	$O_3 + R_2 C = CR_2 \rightarrow R\cdot, RO\cdot, ROO\cdot$	Alkyl, alkoxyl, formyl
Atomic oxygen, hydrocarbon	$O + RH \rightarrow R\cdot + HO\cdot$	Alkyl, hydroxyl
Atomic oxygen, aldehydes	$O + RCO\cdot H \rightarrow RCO + HO\cdot$	Acyl, hydroxyl
Organic Chain Reactions		
Alkyl, oxygen	$R\cdot + O_2 \rightarrow ROO\cdot$	Peroxylalkyl
Peroxyalkyl, oxygen	$ROO\cdot + O_2 \rightarrow RO\cdot O_3$	Alkoxyl, ozone
Organic Chain Reactions		
Alkoxyl, hydrocarbon	$RO\cdot + RH \rightarrow ROH + R\cdot$	Alkyl alcohol
Peroxyalkyl, hydrocarbon	$ROO\cdot + RH \rightarrow ROOH + R\cdot$	Alkyl hydroperoxide
Hydroxyl, hydrocarbon	$HO\cdot + RH \rightarrow R\cdot + H_2O$	Alkyl, water
Consumption of Free Radicals and Reactive Intermediates		
Peroxyalkyl, nitric oxide	$RO\dot{O} + NO \rightarrow ROONO \rightarrow RO + NO_2$	Alkoxyl, nitrogen dioxide
Peroxyalkyl, olefin	$RO\dot{O} + : C = C : \rightarrow ROO - C - C\cdot$	Polymers
Peroxyalkyl, nitrogen dioxide	$RO\dot{O} + NO_2 \rightarrow ROONO_2$	Alkyl nitrate
Peroxyalkyl, sulfur dioxide	$RO\dot{O} + SO_2 \rightarrow SO_3 + RO\cdot$	Sulfur trioxide
Peroxyacyl, nitrogen dioxide	$R(CO)O\dot{O} + NO_2 \rightarrow R(CO)OONO_2$	Peroxyacyl nitrate
Alkoxyl	$2RCH_2 O \rightarrow RCH_2 OH + RCOH$	Aldehyde, alkoxyl
Alkoxyl, nitric oxide	$R\dot{O} + NO \rightarrow RONO$	Alkyl nitrite
Alkyl, hydroxyl	$R\cdot + H\dot{O} \rightarrow ROH$	Alcohol
Atomic oxygen, oxygen	$O + O_2 \rightarrow O_3$	Ozone
Atomic oxygen, sulfur dioxide	$O + SO_2 \rightarrow SO_3$	Sulfur trioxide
Ozone, olefins	$O_3 + : C = C : \rightarrow R - COH$	Aldehydes, ketones, ozonides
Ozone, nitric oxide	$O_3 + NO \rightarrow NO_2 + O_2$	Nitrogen dioxide
Ozone, nitrogen dioxide	$O_3 + NO_2 \rightarrow N_2O_5 \rightarrow HNO_3$	Nitric acid

[a]Generation of free radicals and reactive intermediates.

Source: P.L. Altman and D.S. Dittmer, Eds., *Environmental Biology,* Federation of American Societies for Experimental Biology, Bethesda, Md., 1966, p. 270. With permission.

1.1-14   AIR POLLUTANT REACTIONS (continued)

GENERAL REFERENCES

1.   A.J. Haagen-Smit, in *Air Pollution,* A.C. Stern, Ed., Vol. 1, Academic Press, New York, N.Y., 1962, p. 41.
2.   H.F. Johnstone and D.R. Coughanowr, *Ind. Eng. Chem., 50:* 1169, 1958.
3.   M. Katz, *World Health Organ. Monogr. Ser., 46:* 97, 1961.
4.   P.A. Leighton, *Photochemistry of Air Pollution,* Academic Press, New York, N.Y., 1961.
5.   N.A. Renzetti and G.J. Doyle, *J. Air Pollut. Contr. Assn., 8:* 293, 1959.
6.   N.A. Renzetti and G.J. Doyle, *Int. J. Air Pollut., 2:* 327, 1960.
7.   B.E. Saltzman, *Ind. Eng. Chem., 50*(4):677, 1958.
8.   E.A. Shuck, G.J. Doyle, and N.A. Endow, *Air Pollution Found.,* Los Angeles, Report 31, 1960.
9.   L.G. Wayne, Los Angeles County Air Pollution Control District Technical Progress Report 3, 1962.

1.1-15   RATE CONSTANT FOR ATOMIC OXYGEN
AND HYDROXYL RADICALS
WITH SOME ORGANIC COMPOUNDS

Reaction	Rate constant at 300°K	Reference
$O + CH_3CHO$	4.6×10^8 l mol^{-1} sec^{-1}	1,2
	2.4×10^8 l mol^{-1} sec^{-1}	3
$O + CH_4$	3.3×10^6 l mol^{-1} sec^{-1} in O_2 at 295°K	4
	0.085×10^6 l mol^{-1} sec^{-1} in N_2 at 295°K	4
$O + CH_4$	0.7×10^4 l mol^{-1} sec^{-1} at 298°K	5
$O + C_2H_6$	5.6×10^5 l mol^{-1} sec^{-1} at 298°K	5
$O + n\text{-}C_4H_{10}$	3.3×10^7 l mol^{-1} sec^{-1} with O_2 present	4
	2.5×10^7 l mol^{-1} sec^{-1} without O_2	
	$1.44-1.59 \times 10^7$ l mol^{-1} sec^{-1} at 307°K	6
$O + n\text{-}C_5H_{12}$	$4.02-4.72 \times 10^7$ l mol^{-1} sec^{-1} at 307°K	6
$O + n\text{-}C_6H_{14}$	$6.19-6.87 \times 10^7$ l mol^{-1} x sec^{-1} at 307°K	6
$O + HCHO$	2.1×10^7 l mol^{-1} sec^{-1}	7
	9×10^7 l mol^{-1} sec^{-1}	8
	2.1×10^8 l mol^{-1} sec^{-1}	9
$OH + HCHO$	$\geqslant 4 \times 10^9$ l mol^{-1} sec^{-1}	8
	6×10^7 l mol^{-1} sec^{-1}	10
$OH + C_2H_6$	1.76×10^8 l mol^{-1} sec^{-1}	11
$OH + C_3H_8$	8.2×10^8 l mol^{-1} sec^{-1}	11
$OH + iso\text{-}C_4H_{10}$	1.28×10^9 l mol^{-1} sec^{-1}	11

1.1-15   RATE CONSTANT FOR ATOMIC OXYGEN
AND HYDROXYL RADICALS
WITH SOME ORGANIC COMPOUNDS
(continued)

REFERENCES

1. R.J. Cvetanovic, *Can. J. Chem., 34:*775, 1956.
2. L. Elias and H.I. Schiff, *Can. J. Chem., 38:*1657, 1960.
3. R.D. Cadle and J.W. Powers, *J. Phys. Chem., 71:*1702, 1967.
4. R.D. Cadle and E.R. Allen, *J. Phys. Chem., 69:*1611, 1965.
5. A.A. Westenberg and N. deHaas, *J. Chem. Phys., 46:*490, 1967.
6. J.T. Herron and R.E. Huie, *J. Chem. Phys., 73:*3327, 1969.
7. H. Niki, *J. Chem. Phys., 45:*2330, 1966.
8. J.T. Herron and R.D. Penzhorn, *J. Phys. Chem., 73:*191, 1969.
9. J.J. Bufalini and K.L. Brubaker, "Symposium on Chemical Reactions in Urban Atmospheres," Research Laboratories, General Motors Corp., Warren, Mich., October 6–7, 1969.
10. L.I. Avramenko and R.V. Lorentzo, *Dokl. Akad. Nauk SSSR, 69:*205, 1949.
11. N.R. Greiner, *J. Chem. Phys., 46:*3389, 1967.

1.1-16   OZONE OR OXIDANT YIELDS FROM PHOTOOXIDATION
OF ORGANIC SUBSTANCE–NITROGEN OXIDE MIXTURES

Organic substance	Static irradiations			Dynamic irradiations, Altshuller *et al.* (1966),[4] ppm by volume
	Haagen-Smit and Fox (1956),[1] cracking depth, mm	Schuck and Doyle (1959),[2] ppm by volume	Altshuller and Cohen (1963),[3] ppm by volume	
1,3-Butadiene	12	0.65		0.72
2-Alkenes	8	0.55–0.73		
1,3,5-Trimethylbenzene	7		1.1	0.87
Xylenes	6–7	0.18	0.65–1.0	
1-Alkenes	5	0.58–1.00		0.4
Methanol, ethanol	5			
Formaldehyde	4			1.05
Propionaldehyde	4		1.0	0.80
3-Methylheptane	3			
n-Nonane	3		0.2	
Ethylene	2	1.1		0.69
Hexanes, heptanes	1	~0.2		0.0
Toluene	0.6		0.5	0.36
Acetylene	0.5		0.0	
C_1–C_5 paraffins	<0.2	0.0–0.2		

Source: Reprinted with permission from A.P. Altshuller and J.P. Bufalini, "Photochemical Aspects of Air Pollution: A Review," *Environ. Sci. Technol.*, 5(1):39, 1971. Copyright 1971, American Chemical Society.

REFERENCES

1. A.J. Haagen-Smit and M.M. Fox, *Ind. Eng. Chem., 48:*1484, 1956.
2. E.A. Schuck and G.J. Doyle, "Photooxidation of Hydrocarbon in Mixtures Containing Oxides of Nitrogen and Sulfur Dioxide," Report No. 29, Air Pollution Foundation, San Marino, Calif., 1959.
3. A.P. Altshuller and I.R. Cohen, *Int. J. Air Water Pollut., 7:*787, 1963.
4. A.P. Altshuller *et al., Int. J. Air Water Pollut., 10:*81, 1966.

1.1-17   PRIMARY REACTIONS OF SULFUR DIOXIDE

Reaction No.	Reaction[a]	Rate constants[b]
1	$SO_2 + hv \rightarrow {}^1SO_2$	$\Phi I_{absorbed}$
2	${}^1SO_2 + SO_2 \rightarrow (2SO_2)$	$2.0 \pm 0.1 \times 10^{10}$ 1 mol^{-1} sec^{-1}
3	${}^1SO_2 + SO_2 \rightarrow {}^3SO_2 + SO_2$	$0.18 \pm 0.08 \times 10^{10}$ 1 mol^{-1} sec^{-1}
4	${}^1SO_2 \rightarrow SO_2 + hv_f$	$5.1 \pm 4.0 \times 10^3$ sec^{-1}
5	${}^1SO_2 \rightarrow SO_2$	$1.7 \pm 0.4 \times 10^4$ sec^{-1}
6	${}^1SO_2 \rightarrow {}^3SO_2$	$1.5 \pm 0.8 \times 10^3$ sec^{-1}
7	${}^3SO_2 \rightarrow SO_2 + hv_p$	$0.10 \pm 0.06 \times 10^2$ sec^{-1}
8	${}^3SO_2 \rightarrow SO_2$	$1.3 \pm 0.2 \times 10^2$ sec^{-1}
9	${}^3SO_2 + SO_2 \rightarrow (2SO_2)$	$2.5 \pm 0.5 \times 10^7$ 1 mol^{-1} sec^{-1}

[a]Parentheses were used in Reactions 2 and 9 to designate that either chemical change or relaxation to ground-state molecules was occurring.
[b]The rate constant for Reaction 9 was calculated by Rao *et al.* from the data of S.J. Strickler and D.B. Howell, *J. Chem. Phys., 49:*1947, 1968.

Source: Reprinted with permission from T.N. Rao, S.S. Collier, and J.G. Calvert, *J. Amer. Chem. Soc., 91:*1609, 1969. Copyright 1969, American Chemical Society.

1.1-18   COMPARISON OF ARRHENIUS PARAMETERS, VELOCITY CONSTANTS, AND RATES OF FORMATION OF *RX* AT 25°C FOR THE REACTION:$R + X(+M) \rightarrow RX(+M)$[1]

R	X	log A mol^{-1} cm^3 sec^{-1}	E kcal mol^{-1}	log $k(25°C)$ mol^{-1} cm^3 sec^{-1}	log R_{RX}	Reference
CH_3	SO_2	10.82	1.5	9.7	−11.3	1
C_2H_5	SO_2	11.0	3.1	8.6	−12.4	1
CH_3	O_2	16.18[a]	0	16.18[a]	−10.8	2
C_2H_5	O_2	10.7	0	10.7	−10.4	3

Source: Reprinted with permission from M. Bufalini, "Oxidation of Sulfur Dioxide in Polluted Atmospheres — A Review," *Environ. Sci. Technol., 5*(8):694, 1971. Copyright 1969, American Chemical Society.

REFERENCES

1.   A. Good and J.C.J. Thynne, *Trans. Faraday Soc., 63:*2708, 2720, 1967.
2.   D.E. Hoare and A.D. Walsh, *Trans. Faraday Soc., 53:*1102, 1957.
3.   J.E. Jolley, *J. Amer. Chem. Soc., 79:*1537, 1957.

1.1.3 AIR POLLUTION VARIABLES

Meteorological Factors

Studies of air pollution have implicated several factors of meteorological origin that are generally associated with intense episodes. These factors tend to limit the movement of air so that rapid accumulation of pollutants and reactants in the photochemical process can occur. The major factors involved are reduced wind speeds and the presence of a layer of warm air blanketing an area. Under these conditions pollutants emitted into the atmosphere are not dissipated by dilution or by being blown away, and gradually they build up pollution concentration.

Topographic Factors

Coupled with stagnation of the air mass and presence of an inversion lid is the topographic structure of many areas. Features generally include mountain barriers creating huge bowls within which the accumulated pollutants tend to stagnate. The net effect of this combination of low-wind speeds, inversion layer, high temperature, and natural bowl is to create a huge chemical reaction cell in which the photo and other chemical processes may occur.

Radiant Energy

The prime mover under the conditions outlined above is the energy derived from the sun that initiates the atmospheric reactions resulting in smog. The intensity of sunlight and the general absence of cloud cover and ultraviolet absorbing layers tend to allow maximal intensities of energy, in the 3 000 Å range, to reach the surface of the earth.

Source: M. Feldstein, in *Handbook of Analytical Toxicology,* I. Sunshine, Ed., The Chemical Rubber Co., Cleveland, O., 1969, p. 694.

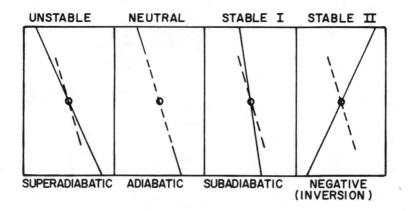

1.1-19   Stability of an air parcel, determined by environmental lapse rate.

Note: The approximate value of 10°C/km (5.5°F/1000 ft) is commonly used for the adiabatic process lapse rate.

Source: R.C. Wanta, in *Air Pollution,* A.C. Stern, Ed., Vol. 1, Academic Press, New York, N.Y., 1968, p. 189. With permission.

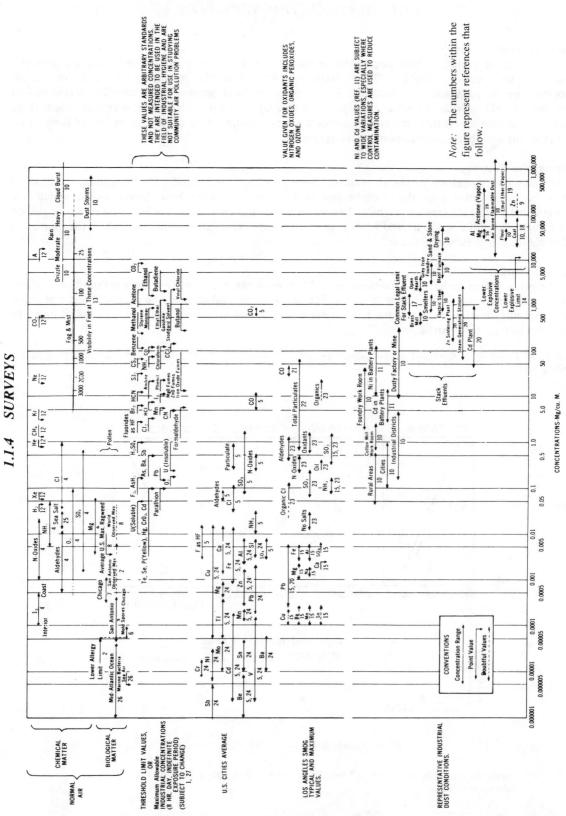

1.1-20   Concentrations of materials in the air.

1.1-20   CONCENTRATIONS OF MATERIALS IN THE AIR   (continued)

REFERENCES

1. F.J. Barkley, *Accepted Limit Values of Air Pollutants,* Information Circular 7682, U.S. Bureau of Mines, 1954.
2. F.W. Bieberdorf, Private Communication.
3. H.R. Brown, *Dust-exposion Hazards in Plants Producing or Handling Aluminum, Magnesium, or Zinc Powders,* Information Circular 7183, U.S. Bureau of Mines, 1941.
4. H. Cauer, "Some Problems of Atmospheric Chemistry," *Compendium of Meteorology,* American Meteorological Society, Boston, Mass., 1951, p. 1126.
5. J. Cholak, "The Nature of Atmospheric Pollution in a Number of Industrial Communities," *Proceedings of the National Air Pollution Symposium,* Vol. 2, Stanford Research Institute, Los Angeles, Calif., 1952, p. 6.
6. O.C. Durham, "Airborne Fungus Spores as Allergenes," *Aerobiology,* Publication No. 17, American Association for the Advancement of Science, 1942, p. 43.
7. O.C. Durham, "The Volumetric Incidence of Atmospheric Allergenes, IV, A Proposed Standard Method of Gravity Sampling, Counting, and Volumetric Interpolation of Results," *J. Allergy, 17:*79, 1946.
8. O.C. Durham, "The Volumetric Incidence of Atmospheric Allergenes, V, Spot Testing in the Evaluation of Species," *J. Allergy, 18:*231, 1947.
9. P.W. Edwards and L.R. Leinbach, *Explosibility of Agricultural and Other Dusts as Indicated by Maximum Pressure and Rates of Pressure Rise,* Technical Bulletin 490, U.S. Department of Agriculture, 1935.
10. M.W. First and P. Drinker, "Concentrations of Particulates Found in Air," *Arch. Ind. Hyg. Occup. Med., 5:*387, 1952.
11. L.T. Friberg, "Proteinuria and Emphysema among Workers Exposed to Cadmium and Nickel Dust in a Storage-settling Plant," *Proceedings of the Ninth International Congress of Industrial Medicine,* London, 1949, p. 641.
12. E. Glueckauf, "The Composition of Atmospheric Air," *Compendium of Meteorology,* American Meteorological Society, Boston, Mass., 1951, p. 3.
13. G. Houghton and W.H. Radford, "On the Measurement of Drop Size and Liquid Water Content in Fogs and Clouds," *Papers Phys. Ocean. Meteor., 6*(4):1, 1938.
14. C.E. Lapple, "Dust and Mist Collection," *Chemical Engineer's Handbook,* 3rd ed., McGraw-Hill Book Company, New York, N.Y., 1950, p. 1016.
15. G.P. Larson, "Recent Identification of Atmospheric Contaminants in the Los Angeles Area," *Proceedings of the Air Pollution Smoke Prevention Association of America, 44:*127, 1951.
16. R.B. Mason and C.S. Taylor, "Explosion of Aluminum Powder Dust Clouds," *Ind. Eng. Chem., 29:*626, 1937.
17. C.E. Miller, *What Can the Small Plant Do about Fly Ash?* Bituminous Coal Research, Inc., Pittsburgh, Pa., 1949.
18. G.S. Rice and H.P. Greenwald, *Experiment to Determine the Minimum Amount of Coal Dust Required for Propagation of a Mine Explosion,* Report of Investigation 3132, U.S. Bureau of Mines, 1931.
19. N.I. Sax, *Handbook of Dangerous Materials,* Reinhold Publishing Corporation, New York, N.Y., 1951.
20. H.H. Schrenk *et al., Air Pollution in Donora, Pa.,* Public Health Bulletin No. 306, U.S. Public Health Service, 1949.
21. M. Shepherd *et al.,* "Isolation, Identification, and Estimation of Gaseous Pollutants of Air," *Anal. Chem., 23:*1431, 1951.
22. Stanford Research Institute, *The Smog Problem in Los Angeles County,* Western Oil and Gas Association, Los Angeles, Calif., 1948.
23. *Third Interim Report on the Smog Problem in Los Angeles County,* Stanford Research Institute, Western Oil and Gas Association, Los Angeles, Calif., 1950.
24. H.E. Stokinger, "Toxicologic Perspective in Planning Air Pollution Studies," *Amer. J. Pub. Health, 43:*742, 1953.
25. *Threshold Limit Values for 1955,* American Conference of Governmental Industrial Hygienists, April 1955.
26. A. Woodcock and M.M. Gifford, "Sampling Atmospheric Sea-salt Nuclei over the Ocean," *J. Marine Research, 8:*177, 1949.
27. C. Zobell, "Microorganisms in Marine Air," *Aerobiology,* Publication No. 17, American Association for the Advancement of Science, 1942, p. 55.

1.1-21   DUSTFALL VALUES FOR A NUMBER OF CITIES

City	Year	Dustfall, metric ton/ km² month	City	Year	Dustfall, metric ton/ km² month
North America			**Australia**		
Detroit	1956		Melbourne, Victoria		
Area 1		25.7	Commercial	1965	21.1
Area 2		27.5	Commercial		20.1
Area 3		17.2	Industrial		17.7
Windsor, industrial	1955	29.1	Residential		6.9
Residential–commercial		16.6	**South Australia**		
Residential–semirural		12.5	Adelaide		
Toronto, industrial[a]	1956	29.8	Industrial	1965	24.9
Industrial–residential		15.4	Commercial		17.9
Residential–semirural		7.6	Residential		9.2
New York[a]	1956	30.0	**Japan**		
	1955	27.4	Nagoya		
Chicago[a]	1947	26.3	Residential	1964	8.75
Pittsburgh[a]	1951	19.7			12.13
Cincinnati[a]	1946	14.6	Semiindustrial		–
Los Angeles[a]	1948	14.4	Industrial		15.75
Rochester[a]	1942	11.3	Yokohama		
Great Britain			Residential	1964	6.46
Birmingham[a]	1954	32.7	Semiindustrial		9.03
Bristol[a]	1954	19.2	Industrial		13.35
Glasgow (East)[a]	1954	25.5			
Leeds (Park Square)[a]	1954	32.2			
London (Westminster)[a]	1954	39.0			
Manchester (Phillips Park)[a]	1954	34.2			

[a]Dustfall values for these cities have been corrected to correspond roughly to the data obtained with the type of collector used in Detroit-Windsor.

Sources: "Australasia," *J. Air Pollut. Contr. Assn.*, *16*(11):581, 1966; and "Asia," *J. Air Pollut. Contr. Assn.*, *16*(11):578, 1966.

1.1-22   CONCENTRATION AND PARTICLE SIZE OF CHROMIUM PARTICULATES

Location	Concentration, $\mu g/m^3$				Mass median diameter, μm			
	max	min	avg	max/min ratio	max	min	avg	max/min ratio
Cincinnati[a]	0.91	<0.01	0.16	>91.0	>10	<0.1	1.7	>100
Cincinnati[b]			0.31				1.5	
Fairfax[c]			0.28				1.9	
Indian Creek[d]	0.39	<0.01	0.11	>39	>10	<0.1	2.0	>100

[a]Samples collected in downtown Cincinnati, Ohio, May 24–June 3, 1967.
[b]Samples collected in downtown Cincinnati, Ohio, Sept. 8–23, 1966.
[c]Samples collected in Fairfax, Ohio, a suburb of Cincinnati, Ohio, Feb. 3–16, 1967.
[d]Samples collected at the Indian Creek Wild Life Preserve, Ohio, May 24–June 3, 1967.

Sources: R.E. Lee, Jr., R.K. Patterson, and J. Wagman, "Particle-size Distribution of Metal Components in Urban Air," *Environ. Sci. Technol.,* 2(4):288, 1968; and R.E. Lee, Jr., R.K. Patterson, and J. Wagman, *Concentration and Particle Size of Metals in Urban and Rural Air,* Preprint, U.S. Department of Health, Education, and Welfare, National Air Pollution Control Administration, Cincinnati, O., 1969.

1.1-23   CONCENTRATION OF LEAD IN ATMOSPHERE
$\mu g/m^3$

Circumstance	Cincinnati	Los Angeles	Philadelphia
Annual average value			
Downtown	2	3	3
Outlying area	1	2	1
All stations	1.4	2.5	1.6
Seasonal distribution (all stations)			
Summer	1.3	1.9	1.4
Fall	1.7	2.8	1.9
Winter	1.3	3.1	1.9
Spring	1.3	2.1	1.4
Diurnal distribution (annual – all stations stated as a fraction of annual mean)			
2 300–0 300	1.3	1.0	0.9
0 300–0 700	1.1	1.1	0.8
0 700–1 100	1.4	1.2	1.4
1 100–1 500	0.8	0.7	0.8
1 500–1 900	0.9	0.8	1.1
1 900–2 300	1.0	1.1	1.1

Source: *Survey of Lead in the Atmosphere of Three Urban Communities,* U.S. Department of Health, Education, and Welfare, Public Health Service Publication No. 999-AP-12, 1965.

1.1-24   CONCENTRATIONS OF LARGE ORGANIC COMPOUNDS
IN THE AVERAGE U.S. URBAN ATMOSPHERE

Compound	Airborne particulate, μg/g	μg/1 000 m³ air	Compound	Airborne particulate, μg/g	μg/1 000 m³ air
Benzo(*f*)quinoline	2	0.2	Coronene	15.	2.
Benzo(*h*)quinoline	3	0.3	*n*-Heptadecane	20.	2.5
Benzo(*a*)acridine	2	0.2	*n*-Octadecane	110.	14.
Benzo(*c*)acridine	4	0.6	*n*-Nonadecane	160.	20.
11*H*-Indeno(1,2-*b*)quinoline	1	0.1	*n*-Eicosane	180.	23.
Dibenz(*a,h*)acridine	0.6	0.08	*n*-Heneicosane	320.	40.
Dibenz(*a,j*)acridine	0.3	0.04	*n*-Docosane	480.	60.
Benz(*a*)anthracene	~30.	~4.	*n*-Tricosane	620.	77.
Fluoranthene	~30.	~4.	*n*-Tetracosane	480.	60.
Pyrene	42.	5.	*n*-Tentacosane	480.	60.
Benzo(*a*)pyrene	46.	5.7	*n*-Hexacosane	85.	11.
Benzo(*e*)pyrene	42.	5.	*n*-Heptacosane	260.	32.
Perylene	5.5	0.7	*n*-Octacosane	340.	43.
Benzo(*ghi*)perylene	63.	8.	**Total**	**3 800**	**480.**
Anthanthrene	2.3	0.26			

REFERENCE
1. E. Sawicki *et al., Int. J. Air Water Pollut.*, 9:515, 1965.

1.1-25   HIGHEST VANADIUM CONCENTRATIONS, U.S. CITIES

Quarterly Composite Values, 1967

Community	Concentrations, μg/m³ Avg	Max	Min	Rank Avg	Max	Min
New York, N.Y.	.905	1.40	.34	1	1	1
Paterson, N.J.	.565	1.20	.14	2	2	2
New Haven, Conn.	.490	0.74	.10	2	6	10
Jersey City, N.J.	.487	1.10	.18	4	3	3
Bayonne, N.J.	.445	0.99	.22	5	4	2
Perth Amboy, N.J.	.390	0.86	.092	6	5	11
Newark, N.J.	.345	0.62	.16	7	7	4
Providence, R.I.	.271	0.35	.076	8	10	12
Philadelphia, Pa.	.264	0.43	.076	9	9	13
Concord, N.H.	.258	0.51	.072	10	8	14
Baltimore, Md.	.200	0.35	.13	11	11	6
Wilmington, Del.	.190	0.24	.13	12	14	7
Washington, D.C.	.165	0.23	.10	13		9
Hartford, Conn.	.160	0.21	.11	14		8
Bayamon, P.R.	.132	0.31		15	13	
Scranton, Pa.		0.32			12	
Marlton, N.J.		0.24			14	
Warminster, Pa.		0.24			15	
East Providence, R.I.			.062			15

Source: *Raw Data Tabulations of the Measurements for Metals in Quarterly Composites of 1966 and 1967,* National Air Sampling Network, High-volume Samples of Urban and Nonurban Sites, Air Quality and Emission Data Division, National Air Pollution Control Administration, Cincinnati, O.

1.1-26   TYPICAL ATMOSPHERIC CARBON MONOXIDE CONCENTRATION
FOUND IN NONURBAN AREAS

Place	Date	Local time	Wind direction and velocity, mph	CO concentration, ppm
Camp Century, Greenland	7/3/65	1245	SE-8	0.90
	7/3/65	1245	SE-8	0.85
	7/5/65	0800	SE-15	0.24
	7/5/65	0805	SE-15	0.32
North coast, Calif.	6/23/65	1400	W-8	0.85
	6/24/65	1400	W-10	0.80
Coastal forest, Calif.	6/24/65	1130	Calm	0.80
Crater Lake, Ore.				
(700 ft elev)	9/27/65	0905	NE-22	0.30
	9/28/65	0835	Lt and Var	0.08
	9/28/65	1650	W-4	0.06
	9/29/65	0910	Lt and Var	0.03
	9/29/65	1700	Lt and Var	0.05
	10/2/65	1630	S-5	0.04
	10/3/65	1800	W-10	0.04
	10/4/65	1145	W-10	0.04
	10/5/65	0840	E-2	0.80
	10/5/65	1320	W-5	0.06
	10/6/65	1320	Lt and Var	0.34
	10/6/65	1650	W-8	0.06

Source: R. Robbins *et al., J. Air Pollut. Contr. Assn., 18*(2):110, 1969. With permission.

1.1-27   CARBON MONOXIDE AVERAGES IN URBAN AREAS

	Carbon monoxide, ppm					
Site	Spring	Summer	Autumn	Winter	Weekday	Weekend
New York area						
Herald Square	8.5	11.5	12	10.5	10.5	5.5
Columbus Circle	6	9.5	8.5	4	7.5	7
Queens Expressway Interchange	2.5	3	7	2.5	3.5	4.5
Queens residential	3.5	3	4	3	3	–
Scarsdale, Westchester County	1.5	2	2.5	2.5	2	–
New York area average	4	5.5	7	4	5	–
Los Angeles area						
Pico Boulevard	7	5.5	9	11	9	6.5
Harbor-Santa Monica						
Freeway Interchange	5.5	7	9	11	8	8
Santa Monica	4	4.5	4	6.5	5	4.5
Monrovia	4.5	4.5	8.5	5.5	5.5	4
Los Angeles area average	5.5	5.5	8	10.5	7.5	6

Source: Reprinted with permission from J. Calucci *et al., Environ. Sci. Technol., 3*(1):43, 1969. Copyright 1969, American Chemical Society.

1.1-28   HYDROCARBON CONCENTRATIONS IN URBAN AIR SAMPLES

Hydrocarbon	Expected concentration range, ppm	Hydrocarbon	Expected concentration range, ppm
Methane	1.6–10.0	*trans*-2-Butene	0.000 5–0.01
Ethane	0.05–0.50	*cis*-2-Butene	0.000 5–0.01
Propane	0.05–0.40	1,3-Butadiene	0.000 5–0.01
Isobutane	0.05–0.30	Acetylene	0.015–0.25
n-Butane	0.05–0.45	Propylene	0.005–0.05
Isopentane	0.05–0.35	Benzene	0.01–0.05
n-Pentane	0.05–0.35	Toluene	0.01–0.05
Ethylene	0.012–0.25	Ethyl benzene	0.002–0.02
Propene	0.001–0.10	Xylenes	0.01–0.1
Butene-1	0.001–0.02	C_9 + aromatics	0.01–0.2
Isobutylene	0.001–0.01		

Source: A.P. Altshuller, *Atmosphere Analysis by Gas Chromatography*, U.S. Department of Health, Education, and Welfare, Public Health Service, Division of Air Pollution, 1969; and California Department of Public Health, SDPH1-SPDH1-50, August, 1966.

1.1-29   OXIDANT CONCENTRATIONS IN SELECTED CITIES, 1964–67

Station	Days of observation	Percentage of total days with maximum hourly average equal to or greater than			Maximum hourly average, ppm	Peak concentration, ppm
		0.15 ppm	0.10 ppm	0.05 ppm		
Pasadena	728	41.1	55.1	75.0	0.46	0.67
Los Angeles	730	30.1	48.5	74.0	0.58	0.65
San Diego	623	5.6	20.9	70.6	0.38	0.46
Denver[a]	285	4.9	17.9	79.3	0.25	0.31
St. Louis	582	2.4	10.1	62.2	0.35	0.85
Philadelphia	556	2.3	10.9	41.9	0.21	0.25
Sacramento	711	2.3	14.6	62.3	0.26	0.45
Cincinnati	613	1.6	9.0	52.0	0.26	0.32
Santa Barbara	723	1.5	10.5	70.5	0.25	0.28
Washington, D.C.	577	1.2	11.3	54.2	0.21	0.24
San Francisco	647	0.9	4.5	28.6	0.18	0.22
Chicago	530	0	4.5	50.8	0.13	0.19

Note: A maximum hourly average concentration of *ozone* in the range 0.15–0.25 ppm has been reported for several U.S. cities.

[a]Eleven months of data beginning February 1965.

Source: *Air Quality Criteria for Photochemical Oxidants*, AP-63, National Air Pollution Control Administration, March 1970.

1.1-30 MAXIMUM PESTICIDE LEVELS FOUND IN AIR SAMPLES

Levels in $\mu g/m^3$

Numbers in parentheses within the table denote the number of samples that contained detectable pesticides. The values in parentheses in the column heads indicate the total number of samples taken for the city cited.

Pesticide	Baltimore, Md. (123)	Buffalo, Md. N.Y. (57)	Dothan, Md. Ala. (90)	Fresno, Calif. (120)	Iowa City, Md. Iowa (94)	Orlando, Md. Fla. (99)	Riverside, Calif. (94)	Salt Lake City, Utah (100)	Stoneville, Md. Miss. (98)	Industrial TLV,c $\mu g/m^3$
p,p'-DDT[e]	.019 5 (89)a	.011 0 (40)a	.177 0 (88)a	.011 2 (62)a	.002 7 (56)	1.560 (99)	.024 4 (85)	.008 6 (62)	0.950 (98)	1 000d
o,p'-DDT[f]	.003 0 (59)	.002 9 (24)	.088 0 (72)	.005 5 (28)	.002 1 (21)	0.500 (95)	.006 2 (44)	.001 4 (29)	0.250 (98)	1 000d
p,p'-DDE[g]	.002 4 (4)		.013 2 (32)	.006 4 (3)	.003 7 (10)	0.131 (29)	.011 3 (6)		0.047 (76)	
o,p'-DDE[h]			.003 9 (13)			.009 6 (7)			0.001 9 (25)	
α-BHC[i]	.004 5 (27)			.004 5 (4)	.004 4 (9)			.009 9 (30)		500d
Lindane[j]	.002 6 (4)				.000 1 (1)			.007 0 (24)		
β-BHC[i]	.002 2 (4)							.001 8 (3)		
δ-BHC[i]								.009 9 (5)		
Heptachlor[k]					.019 2 (37)	0.002 3 (7)				500d
Aldrin[l]					.008 0 (1)					250d
Toxaphene[m]			.068 0 (11)			2.520 (9)				
2,4-D[n]								.004 0 (1)	1.340 (55)	10 000
Dieldrin[o]										250d
Endrin[p]						0.029 7 (50)			0.058 5 (25)	100d
Parathion[q]						0.465 (37)				100d
Methyl parathion[r]			.029 6 (9)			0.005 4 (3)			0.129 (40)	
Malathion[s]						0.002 0 (4)				15 000d

a Numbers in parentheses denote the number of samples containing detectable amounts of the pesticide.
b Rural sampling sites located as much as 20 miles from the community indicated.
c Threshold limit values (TLV's) refer to airborne concentrations of substances and represent conditions under which nearly all workers may be exposed daily without adverse effects. The TLV's should be used as guides in the control of health hazards and should not be regarded as fine lines between safe and dangerous concentrations.
d Tentative value for exposure through the skin.
e 1,1,1-Trichloro-2,2-*bis*(*p*-chlorophenyl)-ethane.
f 1,1,1-Trichloro-2-(*o*-chlorophenyl)-2-(*p*-chlorophenyl) ethane.
g 1,1-Dichloro-2-*bis*-(*p*-chlorophenyl)-ethylene.
h 1,1-Dichloro-2-(*o*-chlorophenyl)-2-(*p*-chlorophenyl)-ethylene.
i 1,2,3,4,5,6-Hexachlorocyclohexane (benzene hexachloride), α-,β-,γ-, and δ-isomers.
j γ-1,2,3,4,5,6-hexachlorocyclohexane.
k 1,4,5,6,7,8,8-Heptachloro-3a,4,7,7a-tetrahydro-4,7-methanoindene.
l 1,2,3,4,10,10-Hexachloro-1,4,4a,5,8,8a-hexahydro-*endo-exo*-5-8-di-methanonaphthalene.
m 2,2,-Dimethyl-3-methylenenorbornane octachloro derivative.

1.1-30   MAXIMUM PESTICIDE LEVEL FOUND IN AIR SAMPLES (continued)

[n]2,4-Dichlorophenoxyacetic acid (including esters and salts).
[o]1,2,3,4,10,10-Hexachloro-6,7-epoxy-1,4,4a,5,6,7,8,8a-octahydro-1,4-*endo-exo*-5-8-di-methanonaphthalene.
[p]1,2,3,4,10,10-Hexachloro-6,7-epoxy-1,4,4a,5,6,7,8,8a-octahydro-1,4-*endo-endo*-5,8-dimethanonaphthalene.
[q]*O,O*-Diethyl *O-p*-nitrophenyl phosphorothioate.
[r]*O,O*-Dimethyl *O-p*-nitrophenyl phosphorothioate.
[s]Diethyl mercaptosuccinate, *S*-ester with *O,O*-dimethyl-phosphorodithioate.

Source: Reprinted with permission from C.W. Stanley *et al.,* "Measurement of Atmospheric Levels of Pesticides," *Environ. Sci. Technol.,* 5:430, 1971. Copyright 1971, American Chemical Society.

GENERAL REFERENCE

1.   *Threshold Limit Values,* American Conference of Governmental and Industrial Hygienists, Cincinnati, O., 1967.

1.1-31 PARTICULATE AND BENZO (a) PYRENE EMISSIONS[a]

Location or source	mg benzene sol per 1000 m³	mg particulates per 1000 m³	μg BaP per g benzene sol	μg BaP per g particulate	μg BaP per 1000 m³	Reference
SOURCE EFFLUENT						
Coal-tar pitch kettle 310°C, 20 cm from surface	420 000	420 000	14 000	14 000	6 000 000	1
Gasworks, retort fumes	—	—	—	—	2 300 000	2
Stack—home heating, coal	150 000	470 000	9 400	3 100	1 400 000	3
Incinerator, garbage	98 000	820 000	14 000	1 700	1 400 000	3
Outdoor burning—floor mats, auto seats, etc.	40 000	460 000	4 300	380	170 000	3
Incinerator, auto parts	300 000	1 300 000	580	130	170 000	3
Incinerator, vegetable matter	3 100	200 000	4 500	70	14 000	3
Outdoor burning—grass, leaves	93 000	140 000	29	45	4 200	3
Municipal incinerator, refuse	63 000	91 000	29	41	2 600	3
Power plant stack	1 500	620 000	220	0.51	320	3
Motel, space heater (gas)	1 400	—	51	—	70	3
AIR POLLUTED MAINLY BY ONE TYPE OF POLLUTANT						
Sidewalk tarring	17 300	17 300	4 500	4 500	78 000	3
Coal tar pitch working area	50 000	50 000	5 200	5 200	75 000	1
Roof tarring area	740	740	19 000	19 000	14 000	3
Blackwell Tunnel—mainly auto traffic		930-2 400	—	91-151	120-290	3
Boston Sumner Tunnel						
Outgoing air	240	610	300	110	69	3
Incoming air	5.4	100	500	26	2.6	3
Bus garage	—	1 400	—	18	80	4
INDOOR AND PERSONAL POLLUTION						
Hut I in mountain of Kenya	2 580	2 700	33	31.5	85	5
Hut II in mountain of Kenya	2 750	3 625	106	71	291	5
Room polluted by tobacco smoke	—	—	—	—	28-87	6
Cigarette smoke[b]	82 000 000	95 000 000	1.15	1.0	95 000	7

1.1-31 PARTICULATE AND BENZO (a) PYRENE EMISSIONS (continued)

Location or source	mg benzene sol per 1000 m³	mg particulates per 1000 m³	µg BaP per g benzene sol	µg BaP per g particulate	µg BaP per 1000 m³	Reference
URBAN ATMOSPHERE						
London[c]	–	440	–	154	68	4
Birmingham, Ala.[c]	41	320	1 800	230	74	3
St. Louis, Mo.[c]	–	–	1 800	230	54	3
Richmond, Va.[c]	24	110	1 900	410	45	3
Hammond, Ind.[c]	–	–	2 600	280	39	3
Detroit, Mich.[c]	49	–	344	–	17	8
Scarsdale, Westchester County, N.Y. (average for four seasons)	2-12	–	25-50	–	0.1-0.6	8
U.S. cities, 100 average–1958	6.5	92	550	–	6	3
NONURBAN ATMOSPHERE						
Coconino County, Ariz., Grand Canyon Park	1.3	14	31	3	0.04	3
Curry County, Ore., near Cape Blanco	1.0	67	9.3	0.15	0.01	3

[a]Most values were taken from E. Sawicki, *Proc. Arch. Environ. Health, 14*:46, Copyright 1967, American Medical Association.

[b]An 85-mm U. S. cigarette without filter tip was tested and smoked once a minute with a 35-ml puff of 2 seconds duration. To reach a 23-mm butt an average of 10.5 puffs had to be taken.

[c]Highest value found.

Source: *Air Pollution*, A. C. Stern, Ed., 2nd ed., Vol. 2, Academic Press, 1968, pp. 210-211. With permission.

REFERENCES

1. J. Bonnet, *Natl. Cancer Inst. Monograph, 9*:221, 1962.
2. P. J. Lawther, B. T. Commins, and R. E. Waller, *Brit. J. Ind. Med., 22*:13, 1965.
3. E. Sawicki, *Proc. Arch. Environ. Health, 14*:46, 1967.
4. R. E. Waller, B. T. Commins, and P. J. Lawther, *Brit. J. Ind. Med., 22*:128, 1965.
5. S. S. Epstein, M. Small, E. Sawicki, and H. L. Falk, *J. Air Pollut. Contr. Assn, 15*:174, 1965.
6. V. Galuskinova, *Neoplasma, 11*:465, 1964.
7. E. L. Wynder and D. Hoffmann, *New Engl. J. Med., 262*:540, 1960.
8. D. Hoffmann and E. L. Wynder, cited by J. M. Colucci and C. R. Begeman, *J. Air Pollut. Contr. Assn., 15*:113, 1965.

1.1.5   SAMPLING AND ANALYSIS

1.1-32   RECOMMENDED UNITS FOR AIR SAMPLING AND ANALYSIS

Item	Recommended units	Alternative or derived units
Particulate contaminants (liquid or solid) of known composition	milligrams per cubic meter (mg/m³)	micrograms per cubic meter (μg/m³)
Suspended or airborne particulate matter	milligrams per cubic meter (mg/m³)	micrograms per cubic meter (μg/m³)
Gases or vapors	milligrams per cubic meter (mg/m³)	micrograms per cubic meter (μg/m³)[a]
Gas volumes	cubic meters (m³) at standard conditions[a]	
Volume emission rates	cubic meters per second (m³/sec)	
Mass emission rates	kilograms per second (kg/sec)	grams per second (g/sec)
Velocity	meters per second (m/sec)	
Air sampling rates	cubic meters per minute (m³/min) or cubic centimeters per minute (cm³/min)	liters per minute (l/min)
Temperature	degrees Celsius (°C)	
Pressure	millibars (mbar) or newtons per square meter (N/m²)	
Visibility	kilometers (km)	
Light transmission	percentage transmittance (%T)	
Light reflection	percentage reflectance (%R)	
Particle size	micrometers (μm) (10^{-6} m)	
Wavelength of light	nanometers (nm) (10^{-9} m)	

Note: Time of day should be specified in terms of the 24-hour clock, e.g., 15.00 hours, not 3 p.m.

[a]"Standard conditions" means 0°C and standard pressure, i.e., 760 mm Hg or 1 013.25 millibars.

Source: Based on M. Katz, *Measurement of Air Pollutants,* Geneva, 1969, with the inclusion of more recent information supplied by the World Health Organization.

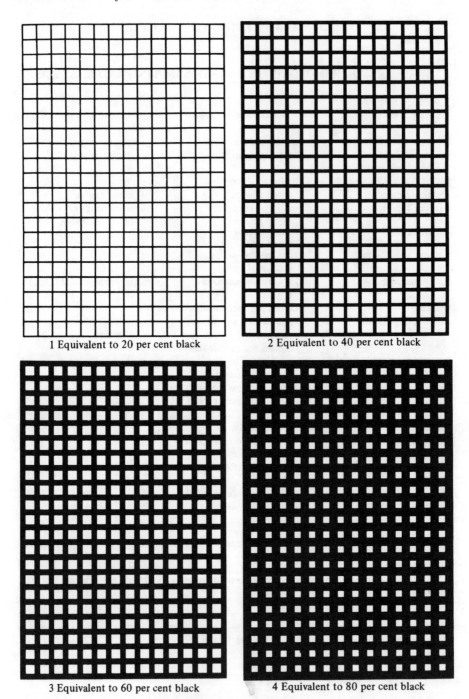

1 Equivalent to 20 per cent black   2 Equivalent to 40 per cent black

3 Equivalent to 60 per cent black   4 Equivalent to 80 per cent black

1.1-33   Ringelmann smoke charts.

Note:

1. Light shining on chart should be the same light that is shining on smoke being examined; for best results, sun should be behind observer.
2. Match smoke as closely as possible with corresponding grid on chart.
3. To compute smoke density, use the following formula:

$$\frac{\text{equivalent units of No. 1 smoke} \times 0.20}{\text{number of observations}} = \text{percentage smoke density}$$

Source: U.S. Bureau of Mines Information Circular 8333, 1967.

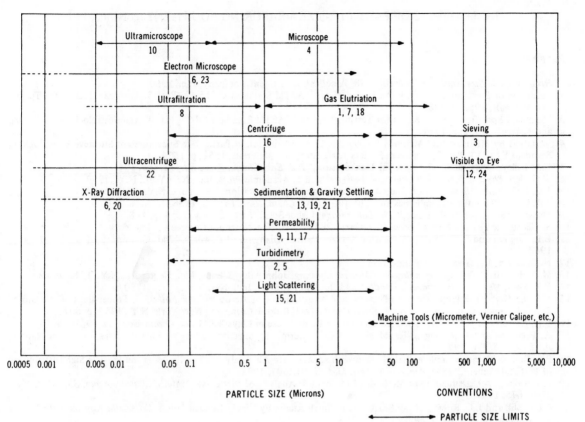

1.1-34   **LIMITS OF PARTICLE-SIZE MEASURING EQUIPMENT** (continued)

REFERENCES

1. *Roller Particle Size Analyzer,* Bulletin 2148, American Instrument Company, Inc., 1950.
2. "Fineness of Portland Cement by the Turbidimeter," ASTM Standard C-115-42, Part 3, American Society for Testing and Materials, 1952, p. 136.
3. "Standard Specifications for Sieves for Testing Purposes," ASTM Standard E11-39, Part 4, American Society for Testing and Materials, 1952, p. 1211.
4. "Analysis by Microscopical Methods for Particle Size Distribution of Particulate Substances of Subsieve Sizes," ASTM Standard E20-51T, Part 3, American Society for Testing and Materials, 1952, p. 1574.
5. E.A. Bailey, "Particle Size by Spectral Transmission," *Ind. Eng. Chem., Anal. Ed., 18:*365, 1946.
6. W.G. Berl, *Physical Method in Chemical Analysis,* Vol. 1, Academic Press, Inc., New York, N.Y., 1950, p. 65.
7. J.M. Dalla Valle, *Micromeritics,* 2nd ed., Pitman Publishing Corporation, New York, N.Y., 1948, p. 88.
8. J.D. Ferry, "Ultrafilter Membranes and Ultrafiltration," *Chem. Rev., 18:*373, 1936.
9. "New Accuracy for Particle Size," *The Laboratory,* Fisher Scientific Company, 1954, p. 148.
10. S. Glasstone, *Textbook of Physical Chemistry,* 2nd ed., D. Van Nostrand Company, New York, N.Y., 1946, p. 1236.
11. E.L. Gooden and C.M. Smith, "Measuring Average Particle Diameter of Powders," *Ind. Eng. Chem., Anal. Ed., 12:*479, 1940.
12. N. Greeman, Jr., Private Communication.
13. M.D. Hassialis, "Testing," *Handbook of Mineral Dressing,* John Wiley & Sons, Inc., New York, N.Y., 1945, p. 19.
14. R.T. Kent, *Mechanical Engineer's Handbook,* John Wiley & Sons, Inc., New York, N.Y., 1950, p. 21.
15. V.K. La Mer, "The Preparation, Collection, and Measurement of Aerosols," *Air Pollution,* Proceedings of the United States Technical Conference on Air Pollution, McGraw-Hill Book Company, New York, N.Y., 1952, p. 607.
16. F.H. Norton and S.J. Speil, "The Measurement of Particle Size in Clays," *J. Amer. Ceram. Soc., 21:*89, 1938.
17. A. Pechukas and F.W. Gage, "Rapid Method for Determining Specific Surface of Fine Particles," *Ind. Eng. Chem., Anal. Ed., 18:*370, 1946.
18. P.S. Roller, "Measurement of Particle Size with an Accurate Air Analyzer: The Fineness and Particle Size Distribution of Portland Cement," *Proc. Amer. Soc. Test. Mat., 32*(II):607, 1932.
19. P.S. Roller, "A Classification of Methods of Mechanical Analysis of Particulate Materials," *Proc. Amer. Soc. Test. Mat., 37*(II):675, 1937.
20. C.G. Shull and L.C. Roess, "X-ray Scattering at Small Angles by Finely Divided Solids, I, General Approximate Theory and Applications," *J. Appl. Phys., 18:*295, 1947.
21. D. Sinclair, "Measurement of Particle Size and Size Distribution," *Handbook on Aerosols,* U.S. Atomic Energy Commission, 1950, p. 97.
22. T. Svedberg, "The Ultracentrifuge and Its Field of Research," *Ind. Eng. Chem., Anal. Ed., 10:*113, 1938.
23. V.K. Zworykin *et al., Electron Optics and the Electron Microscope,* John Wiley & Sons, Inc., New York, N.Y., 1945, p. 123.
24. J.K. Robertson, *Introduction to Physical Optics,* 3rd ed., D. Van Nostrand Company, New York, N.Y., 1941, p. 284.

1.1-35   COLLECTION EFFICIENCY BY PARTICLE SIZE
FOR SELECTED FILTER PAPERS[a]

Flow rate: 0.1 linear meter per minute

Particle diameter, μm	Whatman No. 1	Whatman No. 4	Whatman No. 41	Whatman No. 42	MSA Type S
<0.4	57[b]	23[b]	23[b]	99[b]	48[b]
0.4–0.6	58	24	31	97	47
0.6–0.8	67	25	59	98	77
0.8–1.0	92	77	74	99	92
1.0–2.0	94	63	63	99	93
>2.0	100	100	100	100	100

[a]Efficiency for particle retention is in per cent by count. Particles greater than 0.4 μm collected by high-speed cascade impactor.
[b]DOP test values used for particles below 0.4 μm diameter.

Source: W.J. Smith and N.F. Surprenant, *Amer. Soc. Test. Mater. Proc.*, 53:1122, 1954. Copyright ASTM. Reprinted with permission.

1.1-36   COLLECTION CHARACTERISTICS
OF SAMPLING FILTER MEDIA

Flow rate: 0.1 linear meter per minute

Filter medium	Atmospheric dust count efficiency,[a] %	DOP efficiency, %	Head loss constant,[b] cm Hg(cm^2 min)/1
Whatman 1	50.0	57.0	2.4
Whatman 4	15.0	23.0	0.68
Whatman 32	99.1	99.5	–
Whatman 40	85.1	84.0	–
Whatman 41	26.5	23.0	–
Whatman 41H	24.0	19.0	–
Whatman 42	98.8	99.2	10
Whatman 44	97.0	98.6	–
Whatman 50	92.0	97.0	–
Whatman 540	67.0	65.0	–
S & S 640	13.0	15.0	–
HV 70, 9 mil	96.5	96.5	–
HV 70, 18 mil	99.5	99.3	1.65
MSA Type S	46.0	48.0	–
Millipore Type HA		99.9+	7.6
Millipore Type AA		99.9+	2.6
S & S Ultrafilter	No particles	–	–
Glass fiber paper	found after	99.99+	–
CC-6	6 hours'	99.9+	0.88
AEC-1	running	99.9+	–
AEC glass asbestos		99.9+	–
AEC all-glass		99.9+	–

[a]Dust particles were 1 μm and smaller. Efficiency was determined by particle count, using a sonic impactor. Values are an average of four tests.
[b]Based on data in *Air Pollution*, A.C. Stern, Ed., Vol. 2, Academic Press, New York, N.Y., 1968, p. 25.

Source: W.J. Smith and N.F. Surprenant, *Amer. Soc. Test. Mater. Proc.*, 53:1122, 1954. Copyright ASTM. Reprinted with permission.

1.2   EFFECTS OF AIR POLLUTION

1.2.1   BIOLOGICAL EFFECTS ON HUMANS

1.2-1   SUMMARY OF BIOLOGICAL EFFECTS ON HUMAN SUBJECTS

Pollutant	Effect	Single concentration	Average concentration	Exposure	Other	Reference
Arsenic	Lethal	2 mg/kg			Arsenical dermatitis	1
Asbestos	Epidemiological significance	Asbestosis	>5 mppcf >20 μm long	0–20 years		2
Beryllium	Pulmonary changes caused by Be + F				Berylliosis	
Cadmium	Severe distress	14.5 mg		8 hours		3
Carbon monoxide	Epidemiological significance	30 ppm			Synergistic in PO_2 depression	4
	Discomfort	900 ppm	900 ppm	1 hour		3
	Severe distress	100 ppm	100 ppm	9 hours		3
	Lethal	100 ppm	100 ppm	15 hours		3
		4 000 ppm		<1 hour		
Formaldehyde	Odor	1 ppm	10 ppm	5–18 years		3
	Epidemiological significance					3
	Discomfort	2–3 ppm				3
	Mucosa	4–5 ppm				3
	Severe distress	10–20 ppm				3
Hydrogen fluoride	Odor	50 mg/m³	100 mg/m³	1 minute	Fluorosis	3
	Epidemiological significance	2–5 ppm	100 mg/m³	1 minute	Fluoride (promotes or accelerates lung disease)	3,5
Hydrocarbons	Discomfort		26 mg/m³	Several minutes	HC + O_3 → tumorigen	3

Substance	Effect	Concentration	Concentration	Time	Notes	
Hydrogen sulfide	Odor	0.025 ppm				3
	Central nervous system		0.3 ppm / 0.1 ppm	1 hour		4
	Pulmonary function		250–600 ppm		Prolonged exposure	3
Hydrogen sulfide	Mucosa	50–300 ppm				3
	Pathological		100 ppm	2–15 minutes	Loss of smell	4
Hydrogen sulfide	Lethal	400–700 ppm		½–1 hour		3
Hydrogen sulfide (possibly with mercaptans)					+ influenza → cancer Antagonizes pollutants (strictly speaking, not detrimental to health)	
Inorganic particulates	Pulmonary sclerosis					
Lead, ingested	Central nervous system		1.2 mg/day	10 years	Lead symptoms	4
Lead, inhaled	Histological		6 μg/m^3/day		Long term exposure	4
	Pathological		6 μg/m^3/day (bone)		Long term exposure	4
	Lethal		4.53 mg/100 g (adult) 17.0 mg/100 g (child)			3
Mercury	Acute illness		1.2–8.5 mg/m^3			3
Nitrogen dioxide	Lethal	500 ppm	500 ppm	48 hours		3
	Mild accelerator of lung tumors				NO$_2$ + microorganisms (pneumonia) + HNO$_3$ (bronchiolitis, fibrosa obliterans) + tars (smokers, lung cancer)	1

1.2-1   SUMMARY OF BIOLOGICAL EFFECTS ON HUMAN SUBJECTS (continued)

Pollutant	Effect	Single concentration	Average concentration	Exposure	Other	Reference
Organic particulates (asthmagenic agents)	Asthma					
Ozone	Odor	<0.02–0.05 ppm		Instantaneous	Accelerated aging	6,7
	Pulmonary function		0.60–0.80 ppm	120 minutes	Reduction in steady state pulmonary diffusing capacity (DL_{CO})	8
	Discomfort		0.05–0.10 ppm	13–30 minutes		6,7
	Mucosa		0.30–1.00 ppm	15–60 minutes	O_3 + micro-organisms (lung-tumor accelerator)	6,9
	Severe distress		1.5–2.0 ppm	120 minutes	Impaired lung function: severe fatigue, chest pains; coughing for 2 weeks	10
	Histological		0.2–0.25 ppm	30 minutes	Sphering of red blood cells	11
	Other		0.60–0.80 ppm	120 minutes	Substernal soreness and tracheal irritation 6–12 hours after exposure, disappearing within 24 hours	8

Substance	Effect	Concentration	Concentration	Averaging time	Remarks	Ref.
PAN[a]	Pulmonary function	>0.30 ppm		5 minutes	Significant increase in oxygen uptake during light exercise	12
Sulfur dioxide, sulfur trioxide	Odor	0.5–0.7 ppm		1 second		13
	Taste	0.3–0.1 ppm		A few seconds		13
	Epidemiological significance	0.20 ppm	0.015 ppm	24 hours (annual average)		13
	Pulmonary function	1.6 ppm		10 minutes	SO_2, SO_3 + particulates aggravate lung disease	13
	Discomfort	5 ppm				
	Severe distress	5–10 ppm		10 minutes (sensitive subjects)		13
Total oxidant	Pulmonary function		0.138 (mean of daily maxima NKI)	1 week	b	14
Total sulfates	Odor	600 $\mu g/m^3$	350 $\mu g/m^3$	10 minutes	1 μm diameter particles	15
	Pulmonary function					16
	Discomfort	5 000 $\mu g/m^3$				17
Suspended particulate and settleable matter	Epidemiological significance		500 $\mu g/m^3$	24 hours	In presence of 24-hour August SO_2 concentration of 0.15 ppm	18,19
	Discomfort		≥150 mg/m^3	24 hours	Continuous source, dustfall	20
			1.5 mg/cm^2	30 days		

1.2-1   SUMMARY OF BIOLOGICAL EFFECTS ON HUMAN SUBJECTS (continued)

Pollutant	Effect	Single concentration	Average concentration	Exposure	Other	Reference
Suspended particulate and settleable matter (Cont.)	Other				Particles <5 μm and concentration >10 particles/cm^3 do not have complete elimination from lungs	21

[a] Peroxyacyl nitrate.

[b] Increase in oxygen uptake and pulmonary resistance; decrease in blood oxygen tension levels of emphysematous patients during light exercise when compared to same patients breathing filtered air for 1 week.

Source: M. Feldstein, "Air Pollution," in *Handbook of Analytical Toxicology*, I. Sunshine, Ed., The Chemical Rubber Co., Cleveland, O., 1969; and *Research into Environmental Pollution*, Technical Report Series No. 406, World Health Organization, Geneva, Switz., 1968.

REFERENCES

1. *Air Pollution*, A.C. Stern, Ed., Vol. 1, Academic Press, New York, N.Y., 1962.
2. *Occupational Diseases*, U.S. Public Health Service Publication 1097, p. 51.
3. F.A. Patty, *Industrial Hygiene and Toxicology*, Vol. 2, John Wiley & Sons, New York, N.Y., 1962.
4. *California Standards for Ambient Air Quality and Motor Vehicle Emissions*, Department of Public Health, Bureau of Air Sanitation, 1964.
5. E.J. Largent, *Fluorosis*, Ohio State University Press, 1961.
6. A. Henschler et al., *Archiv für Gewerbepatholgie und Gewerbehygiene, 17*:547, 1960.
7. W.W. Witheridge and C.P. Yaglou, *J. Amer. Soc. Heat. Vent. Eng., 45*:509, 1939.
8. W.A. Young, D.B. Shaw, and D.V. Bates, *J. Appl. Physiol., 19*:765, 1964.
9. S. Wilska, *Acta Chem. Scand., 5*:359, 1951.
10. S.S. Griswold, L.A. Chambers, and H.L. Motley, *Arch. Ind. Health, 15*:108, 1957.
11. R. Brinkman, H.B. Lamberts, and T.S. Veninga, *Lancet, 7325*:133, January 18, 1964.
12. L.E. Smith, *Arch. Environ. Health, 10*:161, February 1965.
13. *Air Quality Criteria for Sulfur Oxides*, U.S. Public Health Service Publication 1619, 1967.
14. J.E. Remmers and O.J. Balchum, Paper 65-43, Presented at the 58th Annual Meeting of the Air Pollution Control Association, June, 1965, Toronto, Canada.
15. K.A. Bushtueva, in *Limits of Allowable Concentrations of Atmospheric Pollutants*, V.A. Rayatonov, Ed., Book 3, U.S. Department of Commerce, Washington, D.C., 1957.
16. M.O. Amdur, L. Silverman, and P. Drinker, *AMA Arch. Ind. Hyg. Occup. Med., 16*:305, 1952.
17. M.O. Amdur et al., *Ann. Occup. Hyg., 3*: 71, 1961.
18. A.F. Martin, *Proc. Roy. Soc. Med.* (London), *57*:969, 1964.
19. J.L. Burn and J. Pemberton, *Int. J. Air Water Pollut.* (London), *7*:5, 1963.
20. *To Control Local Air Pollution from Sources of Particulate or Gaseous Matter Emissions*, Regulation IV, Commonwealth of Pennsylvania, Department of Health, Harrisburg, Pa., Adopted March 15, 1966.
21. C.N. Davis, *Brit. Med. Bull., 19*:49, 1963.

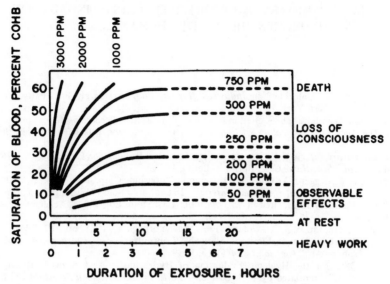

1.2-2   Acute effects of carbon monoxide.[a]

[a]Acute effects of COHb levels in the blood depend on the amount of CO in the atmosphere, duration of exposure, and type of physical activity.

Source: Reprinted with permission from P.C. Wolf, *Environ. Sci. Technol.*, 5(3):213, March 1971. Copyright 1971, American Chemical Society.

1.2-3   SUMMARY OF REPORTED EFFECTS OF INHALATION OF HYDROGEN CHLORIDE BY HUMANS

Concentration, ppm[a]	Exposure time	Effects or comments	Reference
50–100		Work is impossible	1,2
10–50		Work is difficult but possible	1,2
10		Work is undisturbed	1,2
1 300–2 000	Few min	Lethal	3,4
1 000–1 300	30–60 min	Dangerous	3
50–100	60 min	Intolerable	3,5,6,7
35		Irritation of throat after short exposure	7
1 000–2 000		Brief exposures are dangerous	7
10		Irritation	8
5		No organic damage	8
10		Odor threshold value	9
0.067–0.134		Odor threshold value	10,11
0.402		Concentration for threshold reflex effect on optical chronaxie	10,11
0.134		Concentration for threshold reflex effect on eye sensitivity to light	10,11
0.335		Concentration for threshold effect on digito-vascular toxicity	10,11
0.067–0.134		Threshold concentrations of change in the rhythm and depth of respiratory movement	10,11
1–5		Odor threshold value	1

[a]1 ppm = 1 470 $\mu g/m^3$ at 25°C.

Source: *Preliminary Air Pollution Survey of Hydrochloric Acid,* U.S. Department of Health, Education, and Welfare, National Air Pollution Control Administration, Raleigh, N.C., 1969.

1.2-3   SUMMARY OF REPORTED EFFECTS OF INHALATION
OF HYDROGEN CHLORIDE BY HUMANS (continued)

REFERENCES

1. F.F. Heyroth, "Hydrogen Chloride," in *Industrial Hygiene and Toxicology*, F.A. Patty, Ed., 2nd ed., Vol. II, Interscience Books, Inc., New York, N.Y., 1963.
2. L. Matt, *Dissertation*, Wurzburg, 1889, cited in F. Flury and F. Zernik, *Schädliche Gase*, Springer, Berlin, 1931.
3. M.B. Jacobs, *The Analytical Toxicology of Industrial Inorganic Poisons*, Interscience Books, Inc., New York, N.Y., 1967, p. 140.
4. *Handbook of Laboratory Safety*, N.V. Steere, Ed., The Chemical Rubber Co., Cleveland, O., 1967, p. 498.
5. F. Flury and F. Zernik, *Schädliche Gase*, Springer, Berlin, 1931.
6. Y. Henderson and H.W. Haggard, *Noxious Gases*, Reinhold Publishing Corp., New York, N.Y., 1943.
7. N.I. Sax, *Dangerous Properties of Industrial Materials*, 2nd ed., Reinhold Publishing Corp., New York, N. Y., 1963.
8. *VDI 2106*, Part 2, February 1963, cited in *Air Pollution*, A. C. Stern, Ed., Vol. 3, Academic Press, New York, N. Y., 1968, p. 660.
9. *Air Pollution*, A.C. Stern, Ed., Vol. 2, Academic Press, New York, N. Y., 1968, pp. 101, 325.
10. E.V. Elfimova, "Determination of Limit of Allowable Concentration of Hydrochloric Acid Aerosol (Hydrogen Chloride) in Atmospheric Air," *Gig. Sanit.*, 24(1):13, 1959.
11. E.V. Elfimova, "Data for the Hygienic Evaluation of Hydrochloric Acid Aerosol (Hydrochloride Gas) as an Atmospheric Pollutant," Translated by B. S. Levine, *U.S.S.R. Literature on Air Pollution and Related Occupational Diseases*, 9:18, 1962.

1.2-4   LEAD PARAMETERS IN THE HUMAN BODY

	Organ system			
	Total body	Bone	Liver	Kidneys
Intake, g/day	4×10^{-4}			
Average concentration, g/g(wet)	1.1×10^{-6}	$<10^{-6}$	2×10^{-6}	$<1.4 \times 10^{-7}$
Biological half-life, days	1 460	3 650	1 947	531
Fraction from GI tract to blood	0.08			
Fraction in organ of reference of total in body	1.0	0.7	0.1	0.05
Fraction from blood to organ of reference	1.0	0.28	0.08	0.14
Fraction reaching organ by ingestion	0.08	0.02	0.006 4	0.01
Fraction reaching organ by inhalation	0.29	0.08	0.023	0.04

Source: R. Grundy, *J. Air Pollut. Contr. Assn.*, *19*(9):731, 1969. With permission.

1.2-5 COMPARISON OF TOXICOLOGIC ACTIONS OF NO_2 AND O_3

Analogies	Dissimilarities
Acute lung damage (edema and hemorrhage)	NO_2 approx 1/15 as toxic, acutely
Age-related susceptibility	Differences in species susceptibility
	Threshold of acute lung response >50 ppm NO_2
	No threshold of response, O_3
Continuous exposure more damaging than intermittent exposure	No persistent change from intermittent exposures up to level of 25 ppm NO_2
Lung-tumor acceleration	NO_2 less potent tumor accelerator than O_3
Tolerance production	NO_2 slow to develop, short-acting tolerance
	O_3 readily develops, long-lasting tolerance
Exacerbation of pulmonary infection	Increased repiratory frequency, NO_2
Antibody production (denaturation)	Decreased respiratory frequency, O_3
	NO_2 increases O_2 consumption; O_3 decreases it
Denaturation of collagen and elastin	NO_2 reversible; O_3 irreversible (probably)
Antibody formation	Effective prophylactic agents differ

Source: H. E. Stokinger and D. L. Coffin, in *Air Pollution,* A. C. Stern, Ed., Vol. 1, Academic Press, New York, N. Y., 1968, p. 474. With permission.

1.2.1.1   HUMAN PHYSIOLOGY

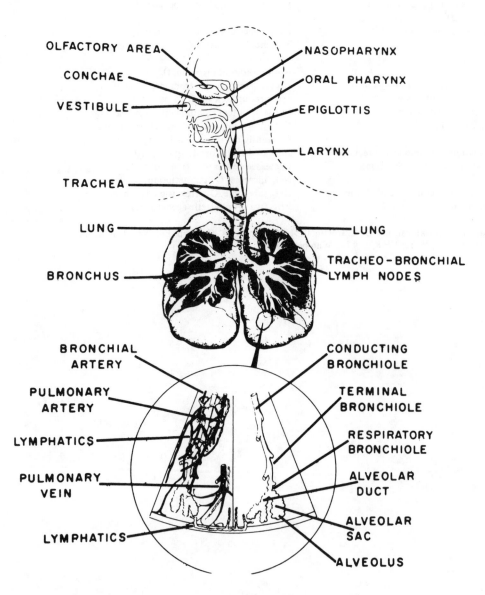

1.2-6   Lung lobule.[a]

[a]Larger particles, about 60 μm in size, tend to deposit in the bronchi and in conducting bronchioles (secondary bronchi). Such particles would not reach the terminal bronchioles. Finer particles would penetrate further into the airways. For example, particles of 2 μm would have virtually no deposition through the trachea, bronchi, and conducting and terminal bronchioles, with deposition occurring in respiratory bronchials and alveolar ducts. Such particles would not quite reach the alveoli.

Source: *Effects of Inhaled Radioactive Particles,* National Academy of Sciences, Publication 848, 1961.

1.2-7   REPRESENTATION OF RESPIRATORY TRACT

Region	Number	Volume, cm³	Relative volume, V	Diameter (2R), cm	Length, cm	Cross-sectional area, cm²	Velocity, cm/sec	Passage time (r'), sec	Fraction passing
Mouth	1	20	0.04	2	7	3	100	0.07	1.00
Pharynx	1	20	0.04	3	3	7	45a	0.07	0.96
Trachea	1	24	0.06	1.6	11	2	150	0.07	0.92
Prim. bronchi	2	10	0.02	1.0	6.5	1.6	190	0.03	0.86
Sec. bronchi	12	4	0.01	0.4	3	1.5	200	0.015	0.84
Ter. bronchi	100	5	0.01	0.2	1.5	3.1	100	0.015	0.83
Quart. bronchi	770	7	0.015	0.15	0.5	14	22	0.02	0.82
Terminal bronchioles	5.4×10^4	45	0.10	0.06	0.3	170	2	0.15	0.81
Resp. bronchi	1.1×10^5	33	0.07	0.05	0.15	300	1.4	0.10	0.72
Alveolar ducts	2.6×10^7	(160)b	(0.63)	(0.02)b	0.02	8 000	—	—	0.65
Alveolar sacs	5.2×10^7	(730)b	—	(0.03)b	(0.03)b	—	—	—	—

aGlottis velocity = 150.
bValues as estimated. These values are corrected to 2.5 liters at end of expiration.

Source: Reprinted with permission from H.D. Landahl, "On the Removal of Airborne Droplets by the Human Respiratory Tract: I: The Lung," *Bull. Math. Biophysics, 12*:43, 1950. Copyright ©1950, Pergamon Press, Elmsford, N.Y.

1.2-8   SUBDIVISIONS OF LUNG VOLUME: MAN

DEFINITIONS

Inspiratory reserve volume	= maximal volume that can be inspired from end-tidal inspiration
Tidal volume	= volume of gas inspired and expired during each respiratory cycle
Expiratory reserve volume	= maximal volume that can be expired from resting expiratory level
Residual volume	= volume of gas in lungs at end of maximal expiration
Inspiratory capacity	= maximal volume that can be inspired from resting expiratory level
Functional residual capacity	= volume of gas in lungs at resting expiratory level
Vital capacity	= maximal volume that can be expired after maximal inspiration
Total lung capacity	= volume of gas in lungs at end of maximal inspiration

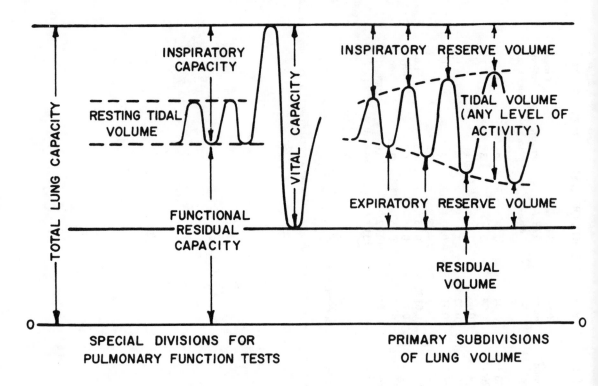

Source: *Respiration and Circulation,* P.L. Altman and D.S. Dittmer, Eds., Federation of American Societies for Experimental Biology, Bethesda, Md., 1971. With permission.

1.2-9   VALUES USEFUL IN PULMONARY PHYSIOLOGY

Values are for a healthy, resting, recumbent young man (1.7 m² surface area) breathing air at sea level, unless other conditions are indicated. Values may change with posture, size, sex, and altitude. There is variation among members of a homogeneous group under standard conditions.

Variable	Abbreviation	Value
Lung Volumes (BTPS)		
Inspiratory capacity	IC	3 600 ml
Expiratory reserve volume	ERV	1 200 ml
Vital capacity	VC	4 800 ml
Residual volume	RV	1 200 ml
Functional residual capacity	FRC	2 400 ml
Thoracic gas volume	TGV	2 400 ml
Total lung capacity	TLC	6 000 ml
(Residual volume/total lung capacity) x 100	(RV/TLC) x 100	20%
Ventilation (BTPS)		
Tidal volume	V_T	500 ml
Respiratory frequency	f	12 breaths/min
Minute volume	$\dot{V}$	6 000 ml/min
Dead space volume	V_D	150 ml
Alveolar ventilation	$\dot{V}_A$	4 200 ml/min
Tests for Distribution of Inspired Gas		
Single-breath test (% increase in N_2 for 500 ml expired alveolar gas)		$< 1.5\% \ N_2$
Pulmonary nitrogen emptying rate (7-min test)		$< 2.5\% \ N_2$
Helium closed circuit (mixing efficiency related to perfect mixing)		76%
Alveolar Gas		
O_2 partial pressure	$P_{A_{O_2}}$	104 mm Hg
CO_2 partial pressure	$P_{A_{CO_2}}$	40 mm Hg
Diffusion and Gas Exchange (STPD)		
O_2 consumption	$\dot{V}_{O_2}$	240 ml/min
CO_2 output	$\dot{V}_{CO_2}$	192 ml/min
Respiratory exchange ratio (CO_2 output:O_2 consumption)	R	0.08
Pulmonary diffusing capacity for O_2	DL_{O_2}	> 15 ml O_2/(min-mm Hg)
Pulmonary diffusing capacity for CO, steady state	DL_{CO}	17 ml CO/(min-mm Hg)
Pulmonary diffusing capacity for CO, single breath	DL_{CO}	25 ml CO/(min-mm Hg)
Fractional CO uptake		53%
Maximal diffusing capacity[a]	D_L	60 ml/(min-mm Hg)

[a]Subject exercising.

1.2-9 VALUES USEFUL IN PULMONARY PHYSIOLOGY (continued)

Variable	Abbreviation	Value
Mechanics of Breathing		
Maximal voluntary ventilation, at BTPS	MVV	170 liters/min
Forced expiratory volume, 1 sec, as % of total vital capacity	$(FEV_{1.0}/VC) \times 100$	83%
Forced expiratory volume, 3 sec, as % of total vital capacity	$(FEV_{3.0}/VC) \times 100$	97%
Maximal expiratory flow rate, for 1 liter at ATPS	MMEF	> 400 liters/min
Maximal inspiratory flow rate, for 1 liter at ATPS		> 300 liters/min
Compliance of lungs	CL	0.2 liter/cm H_2O
Compliance of lungs and thorax	C(L + T)	0.1 liter/cm H_2O
Airway resistance	AWR	1.6 cm H_2O/liter (sec)
Pulmonary resistance		1.9 cm H_2O/liter (sec)
Work of quiet breathing		0.5 kg/m (min)
Maximal work of breathing		10 kg/m (breath)
Maximal inspiratory and expiratory pressures		60–100 mm Hg
Alveolar Ventilation/Pulmonary Capillary Blood Flow		
Alveolar ventilation, in liters per min/blood flow, in liters per min	$\dot{V}_A/\dot{Q}$	0.8
(Physiologic shunt/cardiac output) x 100		< 7%
(Physiologic dead space/tidal volume) x 100	$(VD/VT) \times 100$	< 30%
Pulmonary Circulation		
Pulmonary artery pressure		25/8 mm Hg
Pulmonary capillary blood pressure (wedge)		8 mm Hg
Pulmonary capillary blood flow	$\dot{Q}_c$	5 400 ml/min
Pulmonary capillary blood volume	$\dot{Q}_c$	90 ml
Arterial Blood		
O_2 saturation (% saturation of Hb with O_2)	Sa_{O_2}	97.1%
O_2 tension	Pa_{O_2}	95 mm Hg
CO_2 tension	Pa_{CO_2}	40 mm Hg
Alveolar-arterial PO_2 difference		9 mm Hg
Alveolar-arterial PO_2 difference, 12–14% O_2		10 mm Hg
Alveolar-arterial PO_2 difference, 100% O_2		35 mm Hg
O_2 saturation, 100% O_2		100%[b]
pH		7.4

[b]Plus 1.9 ml O_2 dissolved per 100 ml blood.

Source: *Respiration and Circulation,* P. L. Altman and D. S. Dittmer, Eds., Federation of American Societies for Experimental Biology, Bethesda, Md., 1971. With permission.

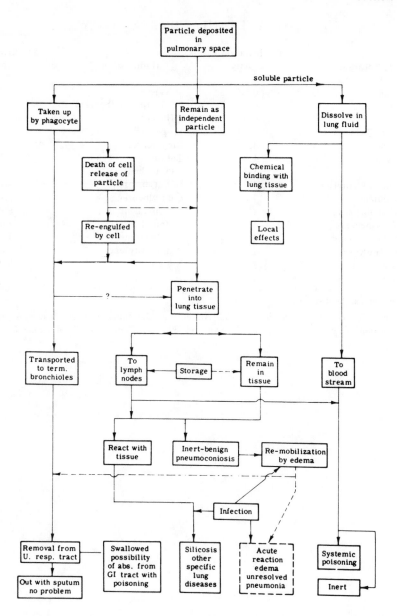

1.2-10   Possible fate of particle after its deposit in pulmonary space.

Source: T.F. Hatch, in *The Air We Breathe,* S.M. Farber and R.H.C. Wilson, Eds., Charles C Thomas, Springfield, Ill., 1961. Courtesy of Charles C Thomas, Publisher.

1.2-11   EYE IRRITATION REACTIVITY[a]

Hydrocarbon	Eye irritation reactivity	Hydrocarbon	Eye irritation reactivity
n-Butane	0	m-Xylene	2.9
n-Hexane	0	1,3,5-Trimethylbenzene	3.1
Isooctane	0.9	1-Hexene	3.5
tert-Butylbenzene	0.9	Propylene	3.9
Benzene	1.0	Ethylbenzene	4.3
Ethylene	1.0	Toluene	5.3
1-Butene	1.3	n-Propylbenzene	5.4
Tetramethylethylene	1.4	Isobutylbenzene	5.7
cis-2-Butene	1.6	n-Butylbenzene	6.4
Isopropylbenzene	1.6	1,3-Butadiene	6.9
sec-Butylbenzene	1.8	α-Methylstyrene	7.4
2-Methyl-2-butene	1.9	Alkylbenzene	8.4
trans-2-Butene	2.3	β-Methylstyrene	8.9
o-Xylene	2.3	Styrene	8.9
p-Xylene	2.5		

[a]Based on a scale of 0–10.

Source: Reprinted with permission from A.P. Altshuller and J.P. Bufalini, "Photochemical Aspects of Air Pollution: A Review," *Environ. Sci. Technol.,* 5(1):39, 1971. Copyright 1971, American Chemical Society.

1.2.2   BIOLOGICAL EFFECTS ON ANIMALS

1.2-12   SUMMARY OF BIOLOGICAL EFFECTS ON ANIMALS

Pollutant	Effect	Subject	Single concentration	Average concentration	Time of average	Other	Reference
Arsenic	Severe distress	Sheep		50-400 ppm in vegetation	Several weeks	Death in some animals	1
Beryllium	Lethal	Bee	0.02–0.1				1
	Pulmonary function	Animals		0.04 mg/m^3 ($BeSO_4$)			2
	Lethal	Animals	10 mg/m^3 (BeF_2) i50–300 mg/kg		15 days		2
Cadmium	Lethal	Rabbit					2
Formaldehyde	Lethal	Rat		800 ppm	30 minutes		2
Hydrogen fluoride	Lethal	Rabbits and guinea pigs		1 000 mg/m^3	30 minutes		2
Mercury	Severe distress	Rabbit		28.6 mg/m^3	4 hours		2
Nitrogen dioxide	Odor	Mice, rabbits, cats	5 ppm		5 seconds		2
	Discomfort	Mice, rabbits, cats	10–20 ppm				2
	Mucosa	Mice, rabbits, cats	10–20 ppm				2
	Severe distress	Mice, rabbits, cats	20–100 ppm				2
	Pathological	Mice	8 ppm	8 ppm	8 weeks		3
	Lethal	Man	500 ppm	500 ppm	48 hours		1
Nitrogen oxide	Lethal	Mice	2 500 ppm		6–7 minutes		2
	Lethal	Mice	320 ppm		60 minutes		4
PAN[a]	Lethal	Mouse		105 ppm	120 minutes		5
Sulfur dioxide	Lethal	Rabbit		50 ppm	30 days/6 hour day		6
	Central nervous system	Animal	0.20 ppm		10 seconds several times		6
Total oxidant	Other	Animals		1 000 ppm	8–16 hours	100% kill	6
	Histological	Mouse		>0.40 ppm (NKI)	2–3 hours		7
	Pathological	Mouse		0.40 (mean of daily maxima NKI)	16 months		8

1.2-12   SUMMARY OF BIOLOGICAL EFFECTS ON ANIMALS (*Continued*)

aPeroxyacyl nitrate.

Source: M. Feldstein, "Air Pollution," in *Handbook of Analytical Toxicology*, I. Sunshine, Ed., The Chemical Rubber Co., Cleveland, O., 1969.

REFERENCES

1. *Air Pollution*, A. C. Stern, Ed., Vol. 1, Academic Press, New York, N.Y., 1962.
2. F. A. Patty, *Industrial Hygiene and Toxicology*, Vol. II, John Wiley & Sons, New York, N.Y., 1962.
3. *Documentation of Threshold Limits*, American Conference of Governmental Industrial Hygienists, Cincinnati, O., 1962.
4. M. A. Jaffe, *Arch. Environ. Health, 16*:241, February 1968.
5. J. T. Middleton, L. O. Emik, and O. C. Taylor, *J. Air Pollut. Contr. Assn., 15*:476, 1965.
6. *Air Quality Criteria for Sulfur Oxides*, U.S. Public Health Service Publication 1619, U.S. Government Printing Office, 1967.
7. R. Bils, *Arch. Environ. Health, 12*:689, 1966.
8. M. B. Gardner, *Arch. Environ. Health, 12*:305, 1966.

1.2.3   BIOLOGICAL EFFECTS ON PLANT LIFE

1.2-13   SUMMARY OF BIOLOGICAL EFFECTS ON VEGETATION[1,2]

Pollutant	Subject	Effect	Single concentration	Average concentration	Time of average	Other	Reference
Arsenic	Plants	Plant leaf symptoms, necrosis		50 μm/droplet K_2CO_3/As_2O_3 solution	Hours		3
Chlorine	Plants	Bleaching between veins, leaf abscission		0.1 ppm	2 hours	Epidermis and mesophyll of mature leaf affected	4
Ethylene	Tomato	Plant leaf symptoms	0.1 ppm		48 hours	Epinasty	
	Orchard Carnation	Plant chlorosis	0.05 ppm 0.1 ppm		6 hours 6 hours	Abnormal sepal Flower fails to open Flower dropping	
		Plant growth altered		See Ref. 5			
Formaldehyde	Petunia	Plant leaf symptoms, necrosis		>0.2 ppm	2 days		6
Hydrogen fluoride	Plants	Dwarfing and lower yield. Leaf abscission. Tip and marginal markings on leaves. Weed color is brown, ivory, to light-tan. Middle-aged leaves, or those fully expanded and mature, were most easily marked. 0.5-5 ppb exposure for several hours					7, 8
Hydrogen sulfide	Plants	Markings occur between veining network on broad-leaved weeds; on narrow-leaved weeds a general powdery appearance of the leaf					8

1.2-13   SUMMARY OF BIOLOGICAL EFFECTS ON VEGETATION (continued)

Pollutant	Subject	Effect	Single concentration	Average concentration	Time of average	Other	Reference
		occurs between the tip and the bend. Color of markings is usually white to tan. Youngest leaves always show the greatest amount of marking. Very often the growing point is killed					
Nitrogen dioxide	Plants	Plant leaf symptoms	3 ppm	2.5	4 hours	Symptoms similar to SO_2. Irregular, white, or brown collapsed lesion on intercostal tissue and near leaf margin	9, 10
Ozone	Plants	Palisade of leaf is affected. Growth suppression and changes in pigmentation, such as flecks, stippling, bleaching, and bleached spotting, occur. There is early abscission, and conifer needles become brown and necrotic. Time, 4 hours; concentration, 0.03 ppm. See References 11-21					
Peroxyacetyl nitrate (PAN)	Plants	Spongy cells of leaves are affected. Glazing, silvering, or bronzing occurs on the lower leaf surface. Time, 6 hours; average concentration, 0.01 ppm. See References 20, 21, 23, and 24					
Sulfur dioxide	Plants	Plant leaf symptoms	>0.25 ppm	0.28 ppm	24 hours	Bleached spots, bleached areas between veins, bleached margin	10, 26
		Plant chlorosis		0.03 ppm	Annual average		10, 26
		Plant growth altered		0.05−0.20 ppm	24 hours; for	Growth suppression, early	10, 26

Pollutant	Measurement	Sensitive plants	Concentration	Exposure time	Exposure period	Effect			Reference
Total oxidant	Plants								See References 11, 14–16, 27
Total sulfates	Plant leaf symptoms	Alfalfa, sugar beets	30-65 ppm	4 hours	growing season	abscission, reduction in yield	Affects mesophyll cells	Droplet size, 5-15 μm	28

REFERENCES

1. M. Feldstein, "Air Pollution," *Handbook of Analytical Toxicology*, I. Sunshine, Ed., The Chemical Rubber Co., Cleveland, O., 1969.
2. I.J. Hindawi, *Air Pollution Injury to Vegetation*, U.S. Department of Health, Education, and Welfare, National Air Pollution Control Administration, Raleigh, N.C., 1970, p. 41.
3. Bay Area Air Pollution Control District, San Francisco, Calif.
4. E. Brennan, I.A. Leone, and R.H. Daines, "Chlorine as a Phytotoxic Air Pollutant," *Int. J. Air Water Pollut., 9*:791, 1965.
5. A.C. Stern, *Air Pollution*, Vol. 1, Academic Press, New York, N.Y., 1962.
6. E.G. Brennan et al., *Science, 143*:818, 1964.
7. M.D. Thomas, *Effects of Air Pollution on Plants*, World Health Organization Monograph Series, No. 46, Columbia University Press, New York, N.Y., 1961.
8. B. Benedict, *Proceedings of the 3rd National Air Pollution Symposium*, April 18-20, 1955, Pasadena, Calif.
9. O.C. Taylor, "Oxidant Air Pollution as Phytotoxicants," presented at the 57th Annual Meeting of Air Pollution Control Association, Houston, Texas, 1967.
10. M.D. Thomas and R.H. Hendricks, "Effects of Air Pollutants on Plants," in *Air Pollution Handbook*, P.L. Magill, F.R. Holden, and C. Ackley, Eds., McGraw-Hill Book Company, New York, N.Y., 1956.
11. H.E. Heggestad, *J. Air Pollut. Contr. Assn., 16*:691, 1966.
12. H.A. Menser, H.E. Heggestad, and O.E. Street, *Phytopathology, 53*:1304, 1963.
13. W.W. Heck, unpublished data.
14. W.W. Heck, J.A. Dunning, and I.J. Hindawi, *Science, 151*:577, 1966.
15. F.D.H. MacDowall, E.I. Mukammai, and A.F.W. Cole, *Canad. J. Plant Sci., 44*:410, 1964.
16. C.R. Berry and L.A. Ripperton, *Phytopathology, 52*:724, 1962.
17. M.C. Ledbetter, P.W. Zimmerman, and A.E. Hitchcock, *Contr. Boyce Thompson Inst., 20*:275, 1959.
18. A.C. Hill et al., *Phytopathology, 51*:356, 1961.
19. D. Sechler and D.R. Davis, *Plant Disease Reporter, 48*:919, 1964.
20. O.C. Taylor, unpublished data.
21. W.M. Dugger et al., *Plant Physiol., 37*:487, 1962.
22. W.W. Heck et al., "Interaction of Environmental Factors on the Sensitivity of Plants to Air Pollution," *J. Air Pollut. Contr. Assn., 15*:511, November 1965.
23. W.M. Dugger, Jr., et al., *Plant Physiol., 38*:468, 1963.
24. W.W. Thompson, W.M. Dugger, Jr., and J. Palmer, *Bot. Gaz., 126*:66, 1965.
25. E.F. Darley et al., "Plant Damage by Pollution Derived from Automobiles," *AMA Arch. Environ. Health, 6*:761, June 1963.
26. *Air Quality Criteria for Sulfur Oxides*, U.S. Public Health Service Publication 1619, U.S. Government Printing Office, Washington, D.C., 1967.
27. J.T. Middleton, J.B. Kendrick, and E.F. Darley, *Proceedings of the 3rd National Air Pollution Symposium*, 1955, Pasadena, Calif.
28. M.D. Thomas, R.H. Hendricks, and G.R. Hill, *Air Pollution*, L.C. McCabe, Ed., McGraw-Hill Book Company, New York, N.Y., 1952.

1.2-14   SENSITIVITY OF SELECTED PLANTS TO OZONE

Crops[a]

Alfalfa
  Medicago sativa L.
Barley
  Hordeum vulgare L.
Bean
  Phaseolus vulgaris L.
Clover, red
  Trifolium pratense L.
Corn, sweet
  Zea mays L.
Grass, bent
  Agrostis palustris Huds.
Grass, brome
  Bromus inermis Leyss.

Grass, crab
  Digitaria sanguinalis L.
Grass, orchard
  Dactylis glomerata L.
Muskmelon
  Cucumis melo L.
Oat
  Avena sativa L.
Onion
  Allium cepa L.
Peanut
  Arachis hypogaea L.
Potato
  Solanum tuberosum L.

Radish
  Raphanus sativus L.
Rye
  Secale cereale L.
Spinach
  Spinacea oleracea L.
Tobacco
  Nicotiana tabacum L.
Tomato
  Lycopersicon esculentum Mill.
Wheat
  Triticum aestivum L.

Trees, Shrubs, and Ornamentals

Alder
  Alnus species
Apple, crab
  Malus baccata Borkh.
Aspen, quaking
  Populus tremuloides Michx.
Box elder
  Acer negundo L.
Bridalwreath
  Spiraea prunifolia Sieb. & Zucc.
Carnation
  Dianthus caryophyllus L.
Catalpa
  Catalpa speciosa Warder

Chrysanthemum
  Chrysanthemum species
Grape
  Vitis vinifera L.
Lilac
  Syringa vulgaris L.
Locust, honey
  Gleditsia triacanthos L.
Maple, silver
  Acer saccharinum L.
Oak, gambel
  Quercus gambelii
Petunia
  Petunia hybrida Vilm.

Pine, eastern white
  Pinus strobus L.
Pine, ponderosa
  Pinus ponderosa Laws.
Privet
  Ligustrum vulgare L.
Snowberry
  Symphoricarpos albus Blake
Sycamore
  Platanus occidentalis L.
Weeping willow
  Salix babylonica L.

[a]Generally, the crops listed are more sensitive than the trees and shrubs.

Source: *Recognition of Air Pollution Injury to Vegetation: A Pictorial Atlas,* J.S. Jacobson and A.C. Hill, Eds., Air Pollution Control Association and National Air Pollution Control Association, Pittsburgh, Pa., 1970.   With permission.

1.2-15   ACCUMULATION OF SELENIUM IN PLANTS

Plant	Se content, ppm (avg)
Pear, leaf	0.45
Apple, leaf	0.53
Radish, leaf	0.28
Lettuce, leaf	0.60
Corn	0.32
Wheat	0.29

Source: Reprinted with permission from W.B. Dye *et al.,* "Fluorometric Determination of Selenium in Plants and Animals with 3,3'-Diaminobenzidine," *Anal. Chem., 35*:1687, 1963. Copyright 1963, American Chemical Society.

1.2-17 LEAD IN SURFACE DIRT AT SAMPLING SITES

City	Site and specific area	Lead, % by weight	
Detroit	Lodge-Ford Freeway Interchange		
	Street near interchange	Thru 40 mesh screen	0.15
		On 40 mesh screen	0.01
	Shoulder of road	Thru 40 mesh screen	0.13
		On 40 mesh screen	0.06
	Grand Circus Park		
	Parking lot	Pulverized	0.21
	Between curb and sidewalk	Pulverized	0.06
New York	10th Ave. and 40th St. at curb		0.18
	(not an air-sampling site)		
Los Angeles	Pico Boulevard		
	Grassy area between curb and		
	sidewalk	Pulverized	0.86
	Street at curb	Pulverized	0.58
	Harbor-Santa Monica Free. Int.		
	Street at curb near interchange	Pulverized	0.30
	Paved area under interchange	Pulverized	0.68
	Unpaved area under interchange	Pulverized	0.01
	Santa Monica		
	Grassy area between curb and		
	sidewalk	Pulverized	0.14
	Street at curb	Pulverized	0.54
	Monrovia		
	Street at curb	Pulverized	0.29
	Flower bed between curb		
	and sidewalk	Pulverized	0.26
New York[1]	Street dirt		0.26
Cincinnati[2]	Top soil (six samples)		0.001 6-0.036
Zurich,	Streets with motor traffic (171 samples)		0.01-0.032
Switzerland[3]	Streets with no motor traffic (7 samples)		0.000 0-0.001 8
Mexico[3]	Dust in rural, non-industrial areas		0.000 06-0.000 6

Source: J. Calucci, *J. Air Pollut. Contr. Assn., 19*(4):260, 1969. With permission.

REFERENCES

1.   *Air Conservation,* The Report of the Air Conservation Commission of the AAAS, 1965, p. 125.
2.   J. Cholak, L. J. Shafer, and T. D. Sterling, "The Lead Content of the Atmosphere," *J. Air Pollut. Contr. Assn., 11*:281, 303, June 1966.
3.   *Report of the Swiss Leaded-Gasoline Commission to the Federal Council on Its Activities during the Period 1947-1960,* Mitteilungen aus dem Gebrete der Levensmitteluntersuchung und Hygiene, Bern, Switz., *52*:135, 1961.

1.3 EMISSION SOURCES

1.3.1 GENERAL EMISSION SOURCES

Sources of air pollution have been identified, and, in many cases, these are well characterized. Data are available on the sources of air pollution and emission factors for these sources, that is, the rates of air-pollutant emissions from various sources as a function of their size, characteristics, etc. These data are presented in the section that follows. The data may be used to estimate total rates of emission from various geographic areas.

1.3-1 SUMMARY OF AIR POLLUTION SOURCES

Category	Examples	Pollutants
Chemical plants	Petroleum refineries, pulp mills, superphosphate fertilizer plants, cement mills	Hydrogen sulfide, oxides of sulfur, fluorides, organic vapors, particles, odors
Crop spraying and dusting	Pest and weed control	Organic phosphates, chlorinated hydrocarbons, arsenic, lead
Crushing, grinding, screening	Road-mix plants	Mineral and organic particulates
Demolition	Urban renewal	
Field burning	Stubble and slash burning	Smoke, fly ash, and soot
Frost-damage control	Smudge pots	
Fuel burning	Home heating and power plants	Oxides of sulfur and of nitrogen, carbon monoxide, smoke, and odors
Fuel fabrication	Gaseous diffusion	Fluoride
Inks	Photogravure and printing	
Metallurgical plants	Smelters, steel mills, aluminum refineries	Metal fumes (lead, arsenic, and zinc), fluorides, and oxides of sulfur
Milling	Grain elevators	
Motor vehicles	Autos, buses, and trucks	Oxides of sulfur and of nitrogen, carbon monoxide, smoke, and odors
Nuclear-device testing	Atmospheric explosions	Radioactive fallout (strontium–90, cesium–137, carbon–14)
Nuclear fission	Nuclear reactors	Argon–41
Ore preparation	Crushing, grinding, and screening	Uranium and beryllium dust
Refuse burning	Community and apartment-house incinerators, open-burning dumps	Oxides of sulfur, smoke, fly ash, organic vapors, and odors
Solvent cleaning	Dry cleaning, degreasing	
Spent-fuel processing	Chemical separation	Iodine–131
Spray painting	Automobile assembly, furniture and appliances finishing	Hydrocarbons and other organic vapors
Waste recovery	Metal scrap yards, auto-body burning, rendering plants	Smoke, soot, organic vapors, and odors

Source: Reprinted by special permission from *Chemical Engineering,* October 14, 1968, Copyright © 1968, by McGraw-Hill, Inc., New York, N.Y. 10020.

1.3-2   SOURCES OF BENZO(a)PYRENE (BaP) EMISSIONS

Estimated U.S. Emissions

Source	BaP emission factor	Annual consumption or production	Estimated annual BaP emission, tons
Heat generation	$\mu g/10^6$ Btu	10^{15} Btu	
Coal			
Residential			
Hand-stoked	1 400 000	0.26	400
Underfeed	44 000	0.20	9.7
Commercial	5 000	0.51	2.8
Industrial	2 700	1.95	5.8
Electric generation	90	6.19	0.6
Oil	200	6.79	1.5
Gas	100	10.57	1.2
Total			421.6
Refuse burning	$\mu g/ton$	10^6 tons	
Incineration			
Municipal	5 300	18	0.1
Commercial	310 000	14	4.8
Open burning			
Municipal refuse	310 000	14	4.8
Grass, leaves	310 000	14	4.8
Auto components	26 000 000	0.20	5.7
Total			20.2
Industries	$\mu g/bbl$	10^6 bbl	
Petroleum catalytic			
Cracking (catalyst regeneration)			
FCC[a]			
No CO boiler[b]	240	790	0.21
With CO boiler	14	790	0.012
HCC[c]			
No CO boiler	218 000	23.3	5.6
With CO boiler	45	43.3	0.002 4
TCC[d] (air lift)			
No CO boiler	90 000	131	13.0
With CO boiler	<45	59	<0.002 9
CC (bucket lift)			
No CO boiler		119	0.004 1
With CO boiler	<31	0	0
Asphalt road mix	50 $\mu g/ton$	187 000 tons	0.000 010
Asphalt air blowing	<10 000 $\mu g/ton$	4 400 tons	<0.000 048
Carbon-black manufacturing	Atmospheric samples indicate that BaP emissions		
Steel and coke manufacturing	from these processes are not extremely high.		
Chemical complex			
Total			18.8
Motor vehicles	$\mu g/gal$	10^{10} gal	
Gasoline			
Automobiles	170	4.61	8.6
Trucks	>460	2.01	>10
Diesel	690	0.257	2.0
Total			>20.6
Total (all sources tested)			481

1.3-2   SOURCES OF BENZO(a) PYRENE (BaP) EMISSIONS (continued)

[a]FCC: fluid catalytic cracker.
[b]CO boiler: carbon monoxide waste heat boiler.
[c]HCC: Houdriflow catalytic cracker.
[d]TCC: Thermofor catalytic cracker.

Source: R.P. Hangebrauck, D.J. von Lehmden, and J.E. Meeker, *Sources of Polynuclear Hydrocarbons in the Atmosphere,* U.S. Public Health Service Publication No. 999-AP-33, 1967.

1.3.2   FUEL PROPERTIES RELATED TO POLLUTION EMISSIONS

Combustion in stationary sources and in transportation is responsible for a large fraction of the total emissions of air pollutants. Evaporation of volatile fuels contributes still more. Hence, the section that follows treats properties of fuels that are related to emission factors in stationary combustion and in transportation.

1.3-3   SELENIUM AND SULFUR CONTENT OF VARIOUS FUELS

Analytical Results of Fuels Analyzed for Selenium

Samples	Se, μg/g	S, mg/g	Se/S ratio, x 10^{-4}
Raw petroleum 1	0.95	20.9	0.45
Raw petroleum 2	0.89	21.3	0.42
Raw petroleum 3	0.50	16.1	0.31
Raw petroleum 4	0.80	18.1	0.44
Raw petroleum 5	0.95	19.6	0.49
	Avg 0.92	**Avg 18.4**	**Avg 0.42**
Heavy petroleum A-1	0.95	12.0	0.79
Heavy petroleum A-2	1.05	8.6	1.22
Heavy petroleum A-3	1.30	12.2	1.09
	Avg 1.10	**Avg 10.9**	**Avg 1.03**
Heavy petroleum B-1	0.50	16.5	0.55
Heavy petroleum B-2	0.70	17.6	0.63
Heavy petroleum B-3	1.05	16.5	0.46
	Avg 0.75	**Avg 16.9**	**Avg 0.55**
Heavy petroleum C-1	1.65	18.9	0.87
Heavy petroleum C-2	0.89	20.2	0.44
Heavy petroleum C-2	0.80	18.9	0.42
	Avg 1.12	**Avg 19.3**	**Avg 0.58**
Coal 1	1.30	2.38	5.50
Coal 2	1.05	2.50	4.20
	Avg 1.18	**Avg 2.44**	**Avg 4.85**
Petroleum 1	0.50	3.45	1.50
Soot 1[a]	4.30	138	0.31
Soot 2[b]	0.50	30	0.17

[a]Mechanically collected.
[b]Electrically collected.

Source: Reprinted with permission from Y. Hashimoto, J. Hwang, and S. Yanagisawa, *Environ. Sci. Technol., 4*(2):157, 1970. Copyright 1970, American Chemical Society.

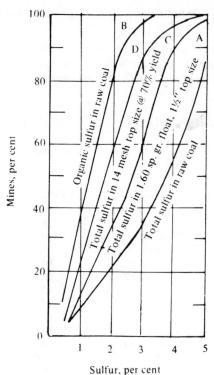

1.3-4   Potential reduction of sulfur content in U.S. coals.

This graph depicts the levels of total and organic sulfur in samples of raw coal collected by the Bureau of Mines from 200 eastern U.S. coal beds. As shown, only 6% of the samples contained less than 1% total sulfur (curve A), and only 35% contained 1% or less organic sulfur (curve B), which sets a theoretical limit for physical separation techniques. About 12% of the samples could be cleaned to 1.0% total sulfur by separating fractions heavier than 1.60 specific gravity, at 1½ in. top size, if yield were of no concern (curve C). The final curve (D) depicts the number of mines that could provide a 1.0% product of 70% yield if an efficient coal-washing job could be done on minus 14 mesh material, a capability not yet available.

Source: Reprinted with permission from "Sulfur Codes Pose Dilemma for Coal," *Environ. Sci. Technol.*, 4(12):1104, 1970. Copyright 1970, American Chemical Society.

1.3.3 GENERAL AND INDUSTRIAL EMISSION FACTORS

1.3-5 ESTIMATION OF PARTICULATE EMISSIONS BY MAJOR INDUSTRIAL AND STATIONARY COMBUSTION SOURCES

Emission factors, i.e., the rates of emission by a particular operation per unit of that operation, enable estimation of total emission in a given area. This is illustrated in the following detailed estimate of particulate emissions in the United States with allowance for particulate controls.

Source (units)	Annual production or consumption, P	Particulate emission factor, lb/unit e_f	Efficiency of control,[b] C_c	Application of control,[a] C_t	Net control,[c] $C_c \times C_t$	Emissions, ton/yr, E
1. Fuel combustion						
A. Coal (tons)						
1. Electric utility						
a. Pulverized	258 400 000	16A = 190[d]	0.92	0.97	0.89	2 710 000
b. Stoker	9 900 000	13A = 146	0.80	0.87	0.70	217 000
c. Cyclone	28 700 000	3A = 35	0.91	0.71	0.64	182 000
Total from electric utility coal						3 109 000
2. Industrial boilers						
a. Pulverized	20 000 000	16A = 170[e]	0.85	0.95	0.81	322 000
b. Stoker	70 000 000	13A = 133	0.85	0.62	0.52	2 234 000
c. Cyclone	10 000 000	3A = 31	0.82	0.91	0.75	39 000
Total from industrial coal						2 595 000
B. Fuel oil (gallons)						
1. Electric utility	7.18×10^9	0.010	0	0	0	36 000
2. Industrial						
a. Residual	7.51×10^9	0.023	0	0	0	87 000
b. Distillate	2.36×10^9	0.015	0	0	0	18 000
Total from fuel oil						141 000
C. Natural gas and LPG (mil scf)						
1. Electric utility	3.14×10^6	15	0	0	0	24 000
2. Industrial	9.27×10^6	18	0	0	0	84 000
Total from natural gas and LPG						108 000
Total from fuel combustion						5 953 000

1.3-5   ESTIMATION OF PARTICULATE EMISSIONS BY MAJOR INDUSTRIAL AND STATIONARY COMBUSTION SOURCES (continued)

Source (units)	Annual production or consumption, P	Particulate emission factor, lb/unit e_f	Efficiency of control,[b] C_c	Application of control,[a] C_t	Net control,[c] $C_c \times C_t$	Emissions, ton/yr, E
2. Crushed stone, sand, and gravel						
A. Crushed stone	681 000 000	17	0.80	0.25	0.20	4 554 000
B. Sand and gravel	918 000 000	0.1	–	–	0	46 000
Total from crushed stone, sand, and gravel						4 600 000
3. Agricultural operations						
A. Grain elevators (tons grain handled)	177 000 000	27	0.70	0.40	0.28	1 700 000
B. Cotton gins (bales)	11 000 000	12	0.80	0.40	0.32	45 000
C. Alfalfa dehydrators (tons dry meal)	1 600 000	50	0.85	0.50	0.42	23 000
Total from agricultural operations						1 768 000
4. Iron and steel						
A. Ore crushing (tons of ore)	82 000 000	2	0	0	0	82 000
B. Materials handling (tons of steel)	131 000 000	10	0.90	0.35	0.32	446 000
C. Pellet plants	–	–	–	–	–	80 000
D. Sinter plants (tons of sinter)	51 000 000	42	0.90	1.0	0.90	107 000
E. Coke manufacture (tons of coal)						
1. Beehive	1 300 000	200	0	0	0	130 000
2. By-product	90 000 000	2	0	0	0	90 000
F. Blast furnace (tons of iron)	88 800 000	130	0.99	1.0	0.99	58 000
G. Steel furnaces (tons of steel)						
1. Open-hearth	65 800 000	17	0.97	0.41	0.40	337 000
2. Basic-oxygen	48 000 000	40	0.99	1.0	0.99	10 000
3. Electric-arc	16 800 000	10	0.99	0.79	0.78	18 000
H. Scarfing	131 000 000	3	0.90	0.75	0.68	63 000
Total from iron and steel						1 421 000
5. Cement (tons)						
A. Wet process	43 600 000					
1. Kilns		167	0.94	0.94	0.88	435 000
2. Grinders, driers, etc.		25	0.94	0.94	0.88	65 000
B. Dry process	31 000 000					
1. Kilns		167	0.94	0.94	0.88	310 000
2. Grinders, driers, etc.		67	0.94	0.94	0.88	124 000
Total from cement						934 000

6. Forest products						
A. Wigwam burners (tons of waste)	27 500 000	10	0	0	0	137 000
B. Sawmills	—	-	—	—	—	No est.
C. Pulp mills (tons of pulp)	37 900 000					
1. Recovery furnace		150	0.92	0.99	0.91	256 000
2. Lime kilns		45	0.95	0.99	0.94	51 000
3. Dissolving tanks		5	0.90	0.33	0.30	66 000
4. Park boilers		—	—	—	—	82 000
D. Particleboard, etc.	—		—	—	—	74 000
Total from forest products						666 000
7. Lime						
A. Crushing, screening (tons of rock)	28 000 000	24	0.80	0.25	0.20	264 000
B. Rotary kilns (tons of lime)	16 200 000	180	0.93	0.87	0.81	294 000
C. Vertical kilns	1 800 000	7	0.97	0.40	0.39	4 000
D. Materials handling	18 000 000	5	0.95	0.80	0.76	11 000
Total from lime						573 000
8. Clay (tons)						
A. Ceramic						
1. Grinding	4 722 000	76	0.80	0.75	0.60	72 000
2. Drying	7 870 000	70	0.80	0.75	0.60	110 000
B. Refractories						
1. Kiln-fired						
a. Calcining	688 000	200	0.80	0.80	0.64	25 000
b. Drying	1 032 000	70	0.80	0.80	0.64	13 000
c. Grinding	3 440 000	76	0.80	0.80	0.64	47 000
2. Castable	550 000	225	0.90	0.85	0.77	14 000
3. Magnesite	120 000	250	0.80	0.70	0.56	7 000
4. Mortars						
a. Grinding	120 000	76	0.80	0.75	0.60	2 000
b. Drying	120 000	70	0.80	0.75	0.60	2 000
5. Mixes	249 000	76	0.80	0.75	0.60	4 000
C. Heavy clay products						
1. Grinding	4 740 000	76	0.80	0.75	0.60	72 000
2. Drying	7 110 000	70	0.80	0.75	0.60	100 000
Total from clay						468 000

1.3-5   ESTIMATION OF PARTICULATE EMISSIONS BY MAJOR INDUSTRIAL AND STATIONARY COMBUSTION SOURCES (continued)

Source (units)	Annual production or consumption, P	Particulate emission factor, lb/unit e_f	Efficiency of control,[b] C_c	Application of control,[a] C_t	Net control,[c] $C_c \times C_t$	Emissions, ton/yr, E
9. Primary nonferrous						
A. Aluminum						
1. Grinding of bauxite (tons of bauxite)	13 000 000	6	—	—	0.80	8 000
2. Calcining of hydroxide (tons of alumina)	5 840 000	200	—	—	0.90	58 000
3. Reduction cells (tons of aluminum)						
a. H. S. Soderberg	800 000	144	0.40	1.0	0.40	35 000
b. V. S. Soderberg	700 000	84	0.64	1.0	0.64	10 000
c. Prebake	1 755 000	63	0.64	1.0	0.64	20 000
4. Materials handling	3 300 000	10	0.90	0.35	0.32	11 000
Total from primary aluminum						142 000
B. Copper						
1. Ore crushing (tons of ore)	170 000 000	2	0	0	0	170 000
2. Roasting (tons of copper)	575 000	168	0.85	1.0	0.85	7 000
3. Reverberatory furnace (tons of copper)	1 437 000	206	0.95	0.85	0.81	28 000
4. Converters (tons of copper)	1 437 000	235	0.95	0.85	0.81	32 000
5. Materials handling (tons of copper)	1 437 000	10	0.90	0.35	0.32	5 000
Total from primary copper						242 000
C. Zinc						
1. Ore crushing (tons of ore)	18 000 000	2	0	0	0	18 000
2. Roasting (tons of zinc)						
a. Fluid-bed	765 000	2 000	0.98	1.0	0.98	15 000
b. Ropp, multi-hearth	153 000	333	0.85	1.0	0.85	4 000
3. Sintering (tons of zinc)	612 000	180	0.95	1.0	0.95	3 000
4. Distillation (tons of zinc)	612 000	—	—	—	—	15 000
5. Materials handling (tons of zinc)	1 020 000	7	0.90	0.36	0.32	2 000
Total from primary zinc						57 000
D. Lead						
1. Ore crushing (tons of ore)	4 500 000	2	0	0	0	4 000
2. Sintering (tons of lead)	467 000	520	0.95	0.90	0.86	17 000
3. Blast furnace	467 000	250	0.85	0.98	0.83	10 000
4. Dross reverberatory furnace (tons of lead)	467 000	20	—	—	0.50	2 000

9. Primary nonferrous (*continued*)

5. Materials handling (tons of lead)	467 000	5	0.90	0.35	0.32	1 000
Total from primary lead						34 000
Total from primary nonferrous						475 000

10. Fertilizer and phosphate rock

A. Phosphate rock (tons of rock)	38 000 000		—	—	—	53 000
B. Fertilizers						
1. Ammonium nitrate (tons of NH_4NO_3)	2 800 000		—	—	—	28 000
2. Urea (tons of urea)	1 000 000		—	—	—	10 000
3. Phosphates						
a. Acid manufacture (tons of P_2O_5)	4 370 000		—	—	—	19 000
b. Granulation (tons of gran. matl.)	18 000 000		—	—	—	190 000
4. Ammonium sulfate (tons of amm. sulfate)	2 700 000		—	—	—	27 000
Total from fertilizers and phosphate rock						327 000

11. Asphalt

A. Paving material (tons)	251 000 000					
1. Driers		32	0.97	0.99	0.96	161 000
2. Secondary sources		8	0.97	0.99	0.96	40 000
B. Roofing material (tons of asphalt)	6 264 000					
1. Blowing		4	—	—	0.50	3 000
2. Saturator		—	—	—	—	14 000
Total from asphalt						218 000

12. Ferroalloys (tons of ferroalloy)

A. Blast furnace	591 000	410	0.99	1.00	0.99	1 000
B. Electric furnace	2 119 000	240	0.80	0.50	0.40	150 000
C. Materials handling	2 710 000	10	0.90	0.35	0.32	9 000
Total from ferroalloys						160 000

13. Iron foundries

A. Furnaces (tons of metal)	18 000 000	16	0.80	0.33	0.27	105 000
B. Materials handling						
1. Coke, limestone, etc. (lb/ton of metal)		5	0.80	0.25	0.20	37 000
2. Sand (tons of sand)	10 500 000	0.3	0	0	0	1 000
Total from iron foundries						143 000

1.3-5 ESTIMATION OF PARTICULATE EMISSIONS BY MAJOR INDUSTRIAL AND STATIONARY COMBUSTION SOURCES (continued)

Source (units)	Annual production or consumption, P	Particulate emission factor, lb/unit e_f	Efficiency of control,[b] C_c	Application of control,[a] C_t	Net control,[c] $C_c \times C_t$	Emissions, ton/yr, E
14. Secondary nonferrous						
A. Copper						
1. Material preparation						
a. Wire burning (tons of insulated wire)	300 000	275	0	0	0	41 000
b. Sweating furnaces (tons of scrap)	64 000	15	0.95	0.20	0.19	–
c. Blast furnaces (tons of scrap)	287 000	50	0.90	0.75	0.68	2 000
2. Smelting and refining (tons of scrap)	1 170 000	70	0.95	0.60	0.57	17 000
Total from secondary copper						60 000
B. Aluminum						
1. Sweating furnaces (tons of scrap)	500 000	32	0.95	0.20	0.19	6 000
2. Refining furnaces (tons of scrap)	1 015 000	4	0.95	0.60	0.57	1 000
3. Chlorine fluxing (tons of Cl used)	136 000	1 000	–	–	0.25	51 000
Total from secondary aluminum						58 000
C. Lead						
1. Pot furnaces (tons of scrap)	53 000	0.8	0.95	0.95	0.90	1 000
2. Blast furnaces	119 000	190	0.95	0.95	0.90	3 000
3. Reverberatory furnaces	554 000	100	0.95	0.95	0.90	4 000
Total from secondary lead						
D. Zinc						
1. Sweating furnaces						
a. Metallic scrap (tons of scrap)	52 000	12	0.95	0.20	0.19	3 000
b. Residual scrap	210 000	30	0.95	0.20	0.19	2 000
2. Distillation furnace (tons of zinc recovered)	233 000	45	0.95	0.60	0.57	5 000
Total from secondary nonferrous						127 000
15. Coal cleaning						
A. Thermal driers (tons dried)	73 000 000	–	–	1.0	–	94 000

	Production	No.	Application of control[a]	Efficiency of control[b]	Net control[c]	Emissions
16. Carbon black						
A. Channel process (lb/ton)	71 000	2 300	0	0	0	82 000
B. Furnace process						
1. Gas	156 000	—	1.00	—	—	5 000
2. Oil	1 180 000	—	1.00	—	—	6 000
Total from carbon black						93 000
17. Petroleum						
A. FCC units (bbl of feed)	1.19×10^9	—	1.0	—	—	45 000
18. Acids						
A. Sulfuric						
1. New acid (tons of 100% H_2SO_4)						
a. Chamber	1 000 000	5	0	—	0	2 000
b. Contact	27 000 000	2	0.90	0.95	0.85	4 000
2. Spent-acid concentrators (tons of spent acid)	11 200 000	30	0.85	0.95	0.80	8 000
B. Phosphoric						
1. Thermal process (tons of P_2O_5)	1 020 000	134	1.0	0.97	0.97	2 000
Total from acids						16 000
TOTAL FROM MAJOR INDUSTRIAL SOURCES						18 081 000

Type Boiler	Elec. Util.[d]	Industrial[e]
Pulverized	11.9%	10.6%
Stoker	11.2%	10.2%
Cyclone	11.8%	10.3%

[a] *Application of control* is defined as that fraction of the total production that has controls.

[b] *Efficiency of control* is defined as the average fractional efficiency of the control equipment, prorated on the basis of production capacity.

[c] *Net control* is defined as the overall level of control; it is the product of the application of control multiplied by the efficiency of control.

[d,e] Average ash content of coal used, determined by phone survey.

Source: A. E. Vandegrift *et al.*, "Particulate Air Pollution in the United States," *J. Air Pollut. Contr. Assn., 21*:321, 1971. With permission.

1.3-6   TYPICAL RATES OF PARTICULATE EMISSION

Emission Factors for Selected Categories of Uncontrolled Sources

Emission source	Emission factor
Natural gas combustion	
Power plants	15 lb/million ft³ of gas burned
Industrial boilers	18 lb/million ft³ of gas burned
Domestic and commercial furnaces	19 lb/million ft³ of gas burned
Distillate oil combustion	
Industrial and commercial furnaces	15 lb/thousand gallons of oil burned
Domestic furnaces	8 lb/thousand gallons of oil burned
Residual oil combustion	
Power plants	10 lb/thousand gallons of oil burned
Industrial and commercial furnaces	23 lb/thousand gallons of oil burned
Coal combustion	
Cyclone furnaces	2X (ash percent) lb/ton of coal burned
Other pulverized coal-fired furnaces	13-17X (ash percent) lb/ton of coal burned
Spreader stokers	13X (ash percent) lb/ton of coal burned
Other stokers	2-5X (ash percent) lb/ton of coal burned
Incineration	
Municipal incinerator (multiple chamber)	17 lb/ton of refuse burned
Commercial incinerator (multiple chamber)	3 lb/ton of refuse burned
Commercial incinerator (single chamber)	10 lb/ton of refuse burned
Flue-fed incinerator	28 lb/ton of refuse burned
Domestic incinerator (gas-fired)	15 lb/ton of refuse burned
Open burning of municipal refuse	16 lb/ton of refuse burned
Motor vehicles	
Gasoline-powered engines	12 lb/thousand gallons of gasoline burned
Diesel-powered engines	110 lb/thousand gallons of diesel fuel burned
Gray iron cupola furnaces	17.4 lb/ton of metal charged
Cement manufacturing	38 lb/barrel of cement produced
Kraft pulp mills	
Smelt tank	20 lb/ton of dried pulp produced
Lime kiln	94 lb/ton of dried pulp produced
Recovery furnaces[a]	150 lb/ton of dried pulp produced
Sulfuric acid manufacturing	0.3-7.5 lb acid mist/ton of acid produced
Steel manufacturing	
Open-hearth furnaces	1.5-20 lb/ton of steel produced
Electric-arc furnaces	15 lb/ton of metal charged
Aircraft, 4-engine jet	7.4 lb/flight
Food and agricultural	
Coffee roasting, direct-fired	7.6 lb/ton of green coffee beans
Cotton ginning and incineration of trash	11.7 lb/bale of cotton
Feed and grain mills	6 lb/ton of product
Secondary metal industry	
Aluminum smelting, chlorination-lancing	1 000 lb/ton of chlorine
Brass and bronze smelting, reverberatory furnace	26.3 lb/ton of metal charged

[a]With primary stack gas scrubber.

Sources: *Air Quality Criteria for Particulate Matter,* AP-49, National Air Pollution Control Administration, January 1969; and *Control Techniques for Particulate Air Pollutants,* AP-51, National Air Pollution Control Administration, January 1969.

1.3-7   ODOR CONCENTRATION MEASURED IN VARIOUS PLANTS

Application	Exhaust flow, scfm	Average odor[a] concentration, odor units/scf	Average emission rate, odor units/min	Remarks
Rubber processing	6 900	50	350 000	Controlled by direct fume incinerator
Coffee roaster	3 600	2 000	7 200 000	Uncontrolled effluent from roasters
Rendering plant	29 000	1 500-25 000	55 000 000 730 000 000	Uncontrolled effluent from dryer
Pulp mill	200 000	<10	<2 000 000	Controlled by recovery furnace
Pulp mill	200 000	17	3 400 000	Controlled by recovery furnace
Pulp mill	200 000	2 000	400 000 000	Recovery furnace intentionally upset
Pulp mill	200 000	2 500-11 000	500 000 000 2 200 000 000	Effluent from cascade evaporator

[a]Based on syringe dilution technique.

Source: D. M. Benforado, W. J. Rotella, and D. L. Horton, "Development of an Odor Panel for Evaluation of Odor Control Equipment," *J. Air Pollut. Contr. Assn., 19*(2):101, 1969.   With permission.

1.4 AIR POLLUTION CONTROL MEASURES

1.4.1 AIR QUALITY CRITERIA

1.4-1 NATIONAL AIR QUALITY STANDARDS

Standards Protective of Human Health

	Level not to exceed	
Pollutant	$\mu g/m^3$	ppm
SO_2	80^a	0.03
	365^b	0.14
Particulate matter	75^c	
	260^b	
Carbon monoxide	10^d	9
	40^e	35
Photochemical oxidants	160^e	0.08
Hydrocarbons	160^f	0.24
Nitrogen oxides	100^a	0.05

[a]Annual arithmetic mean.
[b]Maximum 24-hr concentration not to be exceeded more than once a year.
[c]Annual geometric mean.
[d]Maximum 8-hr concentration not to be exceeded more than once a year.
[e]Maximum 1-hr concentration not to be exceeded more than once a year.
[f]Maximum 3-hr concentration (6-9 a.m.) not to be exceeded more than once a year.

Source: Reprinted with permission from "National Air Quality Standards Finalized," *Environ. Sci. Technol.*, 5(6):503, 1971. Copyright 1971, American Chemical Society.

1.4-2 EMISSION STANDARDS FOR ARSENIC
IN EFFLUENT AIR OR GASES

Location		Standard
Czechoslovakia		0.03 kg/hr
Great Britain	< 5 000 cfm	115 000 $\mu g/m^3$
Great Britain	> 5 000 cfm	46 000 $\mu g/m^3$
New South Wales and Queensland, Australia		23 000 $\mu g/m^3$

Source: Based on A.C. Stern, in *Air Pollution*, A.C. Stern, Ed., Vol. 3, Academic Press, New York, N.Y., 1968, p. 683. With permission.

1.4-3   STANDARDS FOR AIR POLLUTION ALERTS

Alert status	Air concentrations			Duration sustained levels of air concentrations, hours		N. Y.-N. J. metro. area meteorological high air pollution potential forecast for next hours	Action plan
	SO_2, ppm	CO, ppm	Smoke level, COHS				
Air pollution watch	0.5	10+	5.0	1	and/or	36	A
First	0.7+	10+	7.5	4[a]	and	36	1
Second	1.5+	20+	9.0	2	and	12	2
Third	2.0+	30+	10.0	1	and	8	3

Comments: The standards tabulated above are predicated on the presumption than an air pollution alert should be based on the following criteria:

1. The concentration of any two pollutants of three above mentioned with sustained levels measured at selected test sites for periods in excess of the duration indicated except air pollution watch as described below under Plan A.

2. A meteorological forecast reporting that high air pollution potential conditions will persist for the tabulated period of time. This would indicate that the levels of concentration will be present for that period of time.

3. The levels of SO_2, CO, and smoke measured and confirmed for the tabulated periods of time, together with the forecast of weather duration of continued high air pollution potential, provide the basis for the alert status.

Watch and Alert Status and Action:

Plan A—Upon receipt of a high air pollution potential forecast for the next 36 hours, or all three pollutants over the levels shown above for one hour, the Interstate Sanitation Commission will call an air pollution watch and notify cooperating agencies. If both conditions are met, a Public Announcement of the watch will be made by the Commission.

Plans 1 through 3—When air pollution measurements exceed the standards for an alert and the meteorological forecast indicates profound stable air conditions for the period of time tabulated in the above table an alert will be recommended by the Interstate Sanitation Commission to the Commissioners of Health of New York and New Jersey or their designee.

[a]Any one pollutant concentration in excess of eight hours.

Source: T. Glenn, *J. Air Pollut. Contr. Assn., 16*(1):23, 1966. With permission.

1.4.2   *AIR POLLUTION CONTROL EQUIPMENT*

The data tabulated in this section pertain to the principles of operation of air pollution control equipment, effectiveness of control measures, and costs of control equipment. Cost data are included as originally provided in the source material, to indicate the approximate magnitude of costs that may be expected with actual applications. Because costs vary with time and locality, it is advisable that current quotations be sought for precise cost estimates.

1.4-4   AIR POLLUTION CONTROL DEVICES FOR AEROSOLS

Type	Principle of operation	Examples of contaminants controlled	Remarks
Settling chamber	Compartment permitting gravity settling of dusts	Wood, grain, mineral dust	Simple design, low efficiency, used as precleaner
Centrifugal collectors, dry	Single cyclone	Wood, grain, mineral dusts	Simple construction, low efficiency
	Multiclone	Catalyst dust, fly ash	Relatively high efficiency
	Impeller	Foundry dust	High efficiency, small space required
Filters, fibrous, cloth, or viscous	Tubular, bag house	Metallurgical fumes and fine dusts	High efficiency over wide range of particle size
	Screen or frame	Ceramic dust, foundry dust	For small or intermediate size operations
	Tubular bag with reverse jet cleaning	Carbon black, flour dust, grain dust	High filter ratios but bag wear increased
Electrical precipitators	Single stage, high voltage	Metallurgical fumes, catalyst dust	High efficiency under severe conditions, high cost
	Two stage, low voltage	Oil and acid mists	High efficiency, limited use
Wet collectors (scrubbers)	Spray chamber	Rock dust, acid mist	Limited use, high nozzle pressure
	Centrifugal	Rock and sand dusts, some mists	Many variations in design
	Inertial	Grinding dust, foundry dust	Similar to cyclone with liquid tangential sprays
	Venturi scrubber	Chemical fumes, acid mists	High efficiency, high first cost. High velocity carrier stream through venturi

Source: *Contamination Control Handbook,* NASA SP-5076, Technology Utilization Division, National Aeronautics and Space Administration, Washington, D.C., 1969, p. V-15.

1.4-5   SOURCES OF PARTICULATE POLLUTION

Emission Sources and Methods of Control

Industry or process	Source of emissions	Particulate matter	Method of control
Iron and steel mills	Blast furnaces, steel-making furnaces, sintering machines	Iron oxide, dust, smoke	Cyclones, baghouses, electrostatic precipitators, wet collectors
Gray-iron foundries	Cupolas, shake-out systems, core making	Iron oxide, dust smoke, oil, grease, metal fumes	Scrubbers, dry centrifugal collectors
Metallurgical (nonferrous)	Smelters and furnaces	Smoke, metal fumes, oil, grease	Electrostatic precipitators, fabric filters
Petroleum refineries	Catalyst regenerators, sludge incinerators	Catalyst dust, ash from sludge	High-efficiency cyclones, electrostatic precipitators, scrubbing towers, baghouses
Portland cement	Kilns, dryers, material-handling systems	Alkali and process dusts	Fabric filters, electrostatic precipitators, mechanical collectors
Kraft paper mills	Chemical-recovery furnaces, smelt tanks, lime kilns	Chemical dusts	Electrostatic precipitators, venturi scrubbers
Acid manufacture-phosphoric, sulfuric	Thermal processes, phosphate rock acidulating, grinding, and handling systems	Acid mist, dust	Electrostatic precipitators, mesh mist eliminators
Coke manufacturing	Charging and discharging oven cells, quenching, materials handling	Coal and coke dusts, coal tars	Meticulous design, operation, and maintenance
Glass and glass fiber	Raw-materials handling, glass furnaces, fiberglass forming and curing	Sulfuric acid mist, raw materials dusts, alkaline oxides, resin aerosols	Glass fabric filters, afterburners
Coffee processing	Roasters, spray dryers, waste heat boilers, coolers, conveying equipment	Chaff, oil, aerosols, ash from chaff burning, dehydrated coffee dusts	Cyclones, afterburners, fabric filters

Source: *Control Techniques for Particulate Air Pollutants,* AP-51, National Air Pollution Control Administration, January 1969.

1.4-6 EFFICIENCY OF DEVICES
FOR REMOVING BIOLOGICAL PARTICLES,
FROM AIR[a]

Filter	Bacterial removal to be expected, %
Ultra-high-efficiency filters	99.99+
High-efficiency filters	90–99
Medium-efficiency filters	60–90
Roughing filters: fibrous, metallic, oiled, and screen types	10–60
Electrostatic precipitators	60–90
Air washers and scrubbers (low-pressure-drop type)	20–90

[a]Particle sizes of $1-5$ μm.

Source: *Biological Handbook for Engineers,* NASA CR-61237, National Aeronautics and Space Administration, George C. Marshall Space Flight Center, Huntsville, Ala., 1968, p. 4.

1.4-7   AIR POLLUTION CONTROL DEVICES FOR GASES

Type	Principle of operation	Examples of contaminants controlled	Remarks
Absorbers	Packed towers	Malodors from rendering and chemical plants	Solutions—oxidizing agents
	Plate towers	Gases and vapors from refinery and chemical plants	Solution—absorption oil (oil is stripped and recirculated)
	Spray towers and chambers	Hydrogen sulfide from thermal and catalytic cracking plants	Solutions—ethanolamines, thylox, potassium phosphate, sodium phenolate (solutions are regenerated)
		Sulfur dioxides from flue gases, chemical plants	Solution—water solutions of sodium sulfite, ammonium sulfite, and ammonium sulfate (replenished), dimethylamine (regenerated)
		Nitric acid from chemical plants	Solution—alkaline solution
Adsorption	Condensation on surface of a solid	Organic solvent vapors and malodors	Adsorbents—activated charcoal, silica gel, activated alumina
Incineration	Flares	Hydrocarbons from refineries and oil fields	Venturi burners or steam injection for smokeless combustion
	Fume burners	Hydrogen sulfide from refineries and chemical plants	Sulfur dioxide is product of combustion
		Gases and malodors from chemical plants, refineries, food processing	High operating temperatures, $900°-1\,600°\,F$
	Catalytic combustion	Organic vapors, carbon monoxide, oil vapors, ammonia	Catalysts—platinum, nickel operating temperatures, $600°-900°F$
Vapor recovery	Vapor sphere collection	Gasoline, crude oil, other volatiles from storage tanks	Vapors may be compressed and liquefied, or burned as fuel or flared
Floating roofs	Reduces tank breathing losses	Gasoline, crude oil, other volatiles from storage tanks	Closure seals required for good efficiency

Source: *Contamination Control Handbook,* NASA SP-5076, Technology Utilization Division, National Aeronautics and Space Administration, Washington, D.C., 1969.

1.4.2.1 FILTERS

1.4-8   FILTER EFFICIENCY RATINGS

The efficiency rating of air filters is a measure of particulate matter in the air collected by the filter media. This measure is expressed as a percentage factor based on the method employed in testing the efficiency. The three standard tests commonly used follow:

a. **Weight Test (Synthetic Dust).** This test compares the weight of a test dust that passes through the test filter with the weight of the dust introduced into the airstream just ahead of the filter. This test is insensitive to fine particles and is not intended for testing high-efficiency (HEPA) filters.

b. **Discoloration Test (National Bureau of Standards).** This test is known as dust stop, photometric, or blackness test of atmospheric dust. This test compares the degree of discoloration produced on two filter papers, one through which cleaned air is drawn and the other, uncleaned air. This test is insensitive to fine particles and is not intended for testing high-efficiency (HEPA) filters.

c. **DOP Test.** A dioctylphthalate (DOP) smoke is introduced on the upstream side of the filter, and the downstream side is measured with an aerosol photometer to determine the total amount of smoke that passes the filter media. The DOP smoke consists of a homogeneous mixture of DOP vapor and clean air in a volume ratio of 1:4. The concentration is adjusted to approximately 40 grains per thousand feet^3 and the particle size to 0.3 micrometer. A HEPA filter should not permit more than 0.03 per cent of the 0.3-micrometer size smoke to penetrate through the filter media for a 99.97 per cent efficiency rating.

The differing factors in the test methods do not permit direct comparison of efficiency ratings. The following tabulation lists some typical efficiency ratings for each test and demonstrates the contrast in these rating factors.

Weight test, per cent	Discoloration test, per cent	DOP test, per cent
a	a	99.97
a	90–95	80–85
99	80–85	50–60
96	30–35	20–30
76	8–12	2–5

[a]Test not practical for more accurate reading.

Source: *Contamination Control Handbook,* NASA SP-5076, Technology Utilization Division, National Aeronautics and Space Administration, Washington, D.C., 1969, p. V-29.

1.4-9 RECOMMENDED FABRIC AND MAXIMUM FILTERING VELOCITY FOR DUST AND FUME COLLECTION IN REVERSE-JET BAGHOUSES

Material	Fabric	Filtering velocity, ft/min
Aluminum oxide	Cotton sateen	9
Bauxite	Cotton sateen	8
Carbon, banbury mixer	Wool felt	7[a]
Carbon, calcined	Cotton sateen, wool felt	7[a]
Carbon, green	Orlon felt	5
Cement, finished	Cotton sateen	9
Cement, milling	Cotton sateen	7
Cement, raw	Cotton sateen	7
Chrome (ferro), crushing	Cotton sateen	9
Clay, green	Cotton sateen	8
Clay, vitrified silicious	Cotton sateen	10
Enamel, porcelain	Cotton sateen	10
Flour	Cotton sateen	10[a]
Grain	Wool felt, cotton sateen	12
Graphite	Wool felt	5[a]
Gypsum	Cotton sateen, orlon felt	8
Lead oxide fume	Orlon felt, wool felt	6[a]
Lime	Cotton sateen	8
Limestone, crushing	Cotton sateen	9
Metallurgical fumes	Orlon felt, wool felt	6[a]
Mica	Cotton sateen	9
Paint pigments	Cotton sateen	8
Phenolic molding powders	Cotton sateen	8
Polyvinylchloride (PVC)	Wool felt	7[a]
Refractory brick sizing, after firing	Cotton sateen	10
Sand scrubber	Cotton sateen, wool felt	7[a]
Silicon carbide	Cotton sateen	10
Soap and detergent powder	Dacron felt, orlon felt	9[a]
Soybean	Cotton sateen	10
Starch	Cotton sateen	8
Sugar	Cotton sateen, wool felt	8[a]
Talc	Cotton sateen	9
Tantalum fluoride	Orlon felt	6[a]
Tobacco	Cotton sateen	9
Wood flour	Cotton sateen	8
Wood sawing	Cotton sateen	9
Zinc, metallic	Orlon felt, dacron felt	8
Zinc, oxide	Orlon felt	6[a]
Zirconium oxide	Orlon felt	7

[a]Decrease 1 ft/min if concentration is great or particle size small.

Source: American Air Filter Co., Inc., Bulletin No. 279C, Louisville, Ky. With permission.

1.4.2.2   WET COLLECTORS

1.4-10   WET SCRUBBER CAPABILITIES

Type of scrubber	Usual range of particle sizes, μm	Normal draft loss, in. water	Maximum efficiency, %	Typical applications
Open-spray tower	>10.0	¾–2	85	Iron pyrite roasting
Packed tower	>10.0	1–6	85	Absorption of fluorine compounds
Wet centrifugal	> 2.5	2–6	95	Phosphate dust from fertilizer plants
Flooded bed	> 2.5	2–8	95	Primary washer for blast-furnace gas
Orifice	> 2.5	<6	95	Rock and cement dust
Wet dynamic	> 2.0	None	98	Heavy concentration of fine particles
Venturi	> 0.5	6–80	99+	Oxygen converter and ferrous cupola gas, acid mist, lime dust, coal dust, incinerators
Flooded disc	> 0.5	6–70	99+	Same as for venturi

Source: Reprinted by special permission from "Air Pollution Control," *Chemical Engineering* (Deskbook Issue), June 21, 1971, p. 135. Copyright © 1971, by McGraw-Hill, Inc., New York, N.Y. 10020.

1.4.2.3 GAS ADSORPTION DEVICES

1.4-11 TYPES OF ADSORBENTS

Solids possessing adsorptive properties exist in great variety. Some of these solids and their industrial uses for air purification and air pollution control are as follows:

Activated carbon	Solvent recovery, elimination of odors, purification of gases
Alumina	Drying of gases and air
Bauxite	Drying of gases and liquids
Silica gel	Drying and purification of gases

The granular adsorbents are used in packed columns. The air or exhaust gases to be treated are passed through the packed adsorbent bed. Depending on the application and economics, spent adsorbent may be regenerated for reuse, or it may be discarded and replaced with fresh adsorbent.

Source: *Contamination Control Handbook,* NASA SP-5076, Technology Utilization Division, National Aeronautics and Space Administration, Washington, D.C., 1969.

1.4.2.4   ABSORPTION DEVICES

1.4-12   EFFECTIVENESS OF VARIOUS COMPOUNDS IN ABSORBING SULFUR DIOXIDE FROM GASES

TABLE A   90% ABSORBENCY OF 3 000 PPM SO_2 AT 265° F[a,b]

Absorbent	Crystalline phase, X-ray analysis	Purity, wt %	Bulk density, g/cc	SO_2 absorbed, g/100 g absorbent	Preparation
Manganese oxide	$MnO_{1.88}$	90	0.14	33	$MnSO_4 \xrightarrow[NH_3]{(NH_4)_2S_2O_8}$ Ppt washed and dried at 130°C
Cobalt oxide	Co_3O_4	97	0.46	25	$CoSO_4 \xrightarrow{Na_2CO_3}$ Ppt washed, dried at 130°C, and heated in vacuo at 300–340°C for 20 hr
Manganese oxide	$MnO_{1.88}$		0.50	23	$MnSO_4 \xrightarrow{Electrolysis}$ Ppt washed, dried at 130°C, and heated in vacuo at 300–340°C for 20 hr
Manganese oxide	$\gamma\text{-}Mn_2O_3$	96	0.67	19	$MnSO_4 \xrightarrow{Na_2CO_3}$ Ppt washed, dried at 130°C, and heated in vacuo at 300–340°C for 20 hr
Aluminum-sodium oxide	Al_2O_3 Na_2O	73 25	0.54	18	$Al_2(SO_4)_3 \xrightarrow{Na_2CO_3}$ Ppt washed, dried at 130°C, and heated with H_2 at 600–640°C for 20 hr
Hopcalite	CuO MnO_2	11 79	0.93	13	Dried at 130°C

Note: Bismuth oxide ($\alpha\text{-}Bi_2O_3$), molybdenum oxide (MoO_3), lead oxide (PbO), zinc oxide (ZnO), and calcium hydroxide ($Ca(OH)_2$) absorbed less than 1 g of sulfur dioxide for 100 g of charge.

a The 265°F is close to the stack discharge temperature of power plant steam generators.
b Hourly space velocity of gas. 1 050 hr^{-1}; mesh size of absorbent, 8–24.

Cobalt oxide	Co_3O_4	100	0.66	12	$CoSO_4 \xrightarrow[\text{NaOH}]{\text{NaOCl}}$ Ppt washed, dried at 130°C, and heated in vacuo at 300–340°C for 20 hr
Chromium-sodium oxides	Cr_2O_3 Na_2O	70 26	0.91	12	$Cr_2(SO_4)_3 \xrightarrow{Na_2CO_3}$ Ppt washed, dried at 130°C, and heated with H_2 at 600–640°C for 20 hr
Nickel oxide	NiO	91	0.74	9	$NiSO_4 \xrightarrow{Na_2CO_3}$ Ppt washed, dried at 130°C, and heated in vacuo at 300–340°C for 20 hr
Aluminum-potassium oxides	$\gamma\text{-}Al_2O_3$ K_2O	73 21	0.61	6	$Al_2(SO_4)_3 \xrightarrow{K_2CO_3}$ Ppt washed, dried at 130°C, and heated with H_2 at 600–640°C for 20 hr
Nickel oxide	NiO	90	1.49	6	$NiSO_4 \xrightarrow[\text{NaOH}]{\text{NaOCl}}$ Ppt washed, dried at 130°C, and heated in vacuo at 300–340°C for 20 hr
Sodium carbonate	Na_2CO_3	99	0.98	5	Solution of sodium carbonate dried at 130°C, and heated in vacuo at 300–340°C for 20 hr
Sodium stannate		95	0.91	4	Sodium stannate, dried at 130°C, and heated in vacuo at 300–340°C for 20 hr
Iron oxide	$\alpha\text{-}Fe_2O_3$	93	0.98	3	$Fe(NO_3)_3 \xrightarrow{Na_2CO_3}$ Ppt washed, dried at 130°C, and heated at 300–340°C in a stream of nitrogen for 3 hr
Sodium aluminate	$NaAlO_2$	96	0.90	3	Solution of sodium aluminate, dried at 130°C, and heated in vacuo at 300–340°C for 20 hr at 300–340°C for 20 hr
Cadmium oxide	CdO	97	1.13	1	$CdSO_4 \xrightarrow{Na_2CO_3}$ Ppt washed, dried at 130°C, and heated in vacuo at 370–400°C for 20 hr

1.4-12   EFFECTIVENESS OF VARIOUS COMPOUNDS IN ABSORBING SULFUR DIOXIDE FROM GASES (continued)

TABLE A   90% ABSORBENCY OF 3 000 PPM SO$_2$ AT 265°F[a,b]

Absorbent	Crystalline phase, X-ray analysis	Purity, wt %	Bulk density, g/Co	SO$_2$ absorbed, g/100 g absorbent	Preparation
Copper oxide	CuO	99	0.89	1	CuSO$_4$ $\xrightarrow{\text{Na}_2\text{CO}_3}$ Ppt washed, compressed at 4 000 psi, dried at 130°C, and heated in vacuo at 300–340°C for 20 hr
Potassium carbonate		98	0.89	1	Solution of potassium carbonate, dried at 130°C, and heated in vacuo at 300–340°C for 20 hr

Table B   90% ABSORBENCY OF 3 000 PPM SO$_2$ AT 625°F[a,b]

Absorbent	Crystalline phase, X-ray analysis	Purity, wt %	Bulk density, g/Co	SO$_2$ absorbed, g/100 g absorbent	Preparation
Manganese oxide	MnO$_{1.88}$	94	0.13	7.1	MnSO$_4$ $\xrightarrow[\text{NH}_3]{(\text{NH}_4)_2\text{S}_2\text{O}_8}$ Ppt washed, dried at 130°C, and heated in vacuo at 300–340°C for 20 hr
Manganese oxide	γ-Mn$_2$O$_3$	96	0.67	68	MnSO$_4$ $\xrightarrow{\text{Na}_2\text{CO}_3}$ Ppt washed, dried at 130°C, and heated in vacuo at 300–340°C for 20 hr
Hopcalite	MnO$_2$ CuO	79 11	0.92	57	Dried in vacuo at 300–340°C

Note: Aluminum oxide (γ-Al$_2$O$_3$), bismuth oxide (α-Bi$_2$O$_3$), calcium oxide (CaO), magnesium oxide (MgO), molybdenum oxide (MoO$_3$), zinc oxide (ZnO), and potassium carbonate (K$_2$CO$_3$) absorbed less than 1 g of sulfur dioxide for 100 g of charge.

[a] The 625°F is the approximate temperature of flue gases at the inlet of the air preheater.
[b] Hourly space velocity of gas, 1 050 hr^{-1}; mesh size of adsorbent, g–24.

Material	Formula		%			Preparation
Copper oxide	CuO		99	0.89	56	$CuSO_4 \xrightarrow{Na_2CO_3}$ Ppt washed, compressed at 4 000 psi, dried at 130°C, and heated in vacuo at 300–340°C for 20 hr
Manganese oxide	$MnO_{1.88}$			0.50	53	$MnSO_4 \xrightarrow{Electrolysis}$ Ppt washed, dried at 130°C, and heated in vacuo at 300–340°C for 20 hr
Cobalt oxide	Co_3O_4		97	0.46	47	$CoSO_4 \longrightarrow$ Ppt washed, dried at 130°C, and heated in vacuo at 300–340°C for 20 hr
Cobalt oxide	Co_3O_4		100	0.66	44	$CoSO_4 \xrightarrow[\text{NaOH}]{NaOCl}$ Ppt washed, dried at 130°C, and heated in vacuo at 300–340°C for 20 hr
Lead oxide	PbO		99	1.23	18	$Pb(NO_3)_2 \xrightarrow{Na_2CO_3}$ Ppt washed, dried at 130°C, and heated in vacuo at 300–340°C for 20 hr
Aluminum-sodium oxides	Al_2O_3	Na_2O	73 / 25	0.54	17	$Al_2(SO_4)_3 \xrightarrow{Na_2CO_3}$ Ppt washed, dried at 130°C, and heated with H_2 at 600–640°C for 20 hr
Chromium-sodium oxides		Cr_2O_3 / Na_2O	70 / 26	0.91	16	$Cr_2(SO_4)_3 \xrightarrow{Na_2CO_3}$ Ppt washed, dried at 130°C, and heated with H_2 at 600–640°C for 20 hr
Sodium aluminate	$NaAlO_2$		96	0.90	10	Solution of sodium aluminate, dried at 130°C, and heated in vacuo at 300–340°C for 20 hr
Nickel oxide	NiO		91	0.74	9	$NiSO_4 \xrightarrow{Na_2CO_3}$ Ppt washed, dried at 130°C, and heated in vacuo at 300–340°C for 20 hr

1.4-12   EFFECTIVENESS OF VARIOUS COMPOUNDS IN ABSORBING SULFUR DIOXIDE FROM GASES (continued)

TABLE B   90% ABSORBENCY OF 3 000 PPM SO_2 AT 625°[1];a,b

Absorbent	Crystalline phase, X-ray analysis	Purity, wt %	Bulk density, g/cc	SO_2 absorbed, g/100 g absorbent	Preparation
Nickel oxide	NiO	90	1.49	7	$NiSO_4 \xrightarrow{NaOCl / NaOH}$ Ppt washed, dried at 130°C, and heated in vacuo at 300–340°C for 20 hr
Aluminum-potassium oxides	Al_2O_3 K_2O	73 21	0.61	6	$Al_2(SO_4)_3 \xrightarrow{K_2CO_3}$ Ppt washed, dried at 130°C, and heated with H_2 at 600–640°C for 20 hr
Cadmium oxide	CdO	97	1.13	5	$CdSO_4 \xrightarrow{Na_2CO_3}$ Ppt washed, dried at 130°C, and heated in vacuo at 370–400°C for 20 hr
Sodium stannate		95	0.91	5	Sodium stannate, dried at 130°C, and heated in vacuo at 300–340°C for 20 hr
Sodium carbonate	Na_2CO_3	99	0.98	4	Solution of sodium carbonate, dried at 130°C, and heated in vacuo at 300–340°C for 20 hr
Iron oxide	α-Fe_2O_3	93	0.98	3	$Fe(NO_3)_3 \xrightarrow{Na_2CO_3}$ Ppt washed, dried at 130°C, at 300–340°C in a stream of nitrogen for 3 hr
Calcium hydroxide	$Ca(OH)_2$ CaO	81 19	0.36	2	$Ca(NO_3)_2 \xrightarrow{NaOH}$ Ppt washed, dried at 130°C, and heated in vacuo at 300–340°C for 20 hr

Source: D. Bienstock and F. J. Field, "Bench-scale Investigation on Removing Sulfur Dioxide from Flue Gases," *J. Air Pollut. Contr. Assn., 10:*121, 1960. With permission.

1.4.2.5   INCINERATORS, AFTERBURNERS, EXHAUST SYSTEMS

1.4-13   CONTROL OF ODORS BY INCINERATION

Application	Incinerator temperature, °F	Average odor concentration[a] in incinerator, odor units/scf		Exhaust gas flow, scfm	Effect of incineration on odor strength, % reduction
		Inlet	Outlet		
Wire enameling	1 000	1 300	2 100		−61
Oven, portable unit	1 200	2 500	350		86
Field test	1 400	1 300	70		97
Glass fiber	1 009	550	625	14 000	−14
Curing-oven field	1 250	380	53	14 000	86
	1 352	255	25	14 000	90
Abrasive wheel	1 200	800	10		98
Curing-oven laboratory test	1 400	1 600	32		98
Automobile paint	1 350	260	14		95
	1 350	650	10		98
Bake-oven field test	1 450	170	10		94
	1 450	680	18		97
Hard-board curing	1 400	1 000	40		96
Oven laboratory test	1 500	1 400	15		98

[a]Based on syringe dilution technique.

Source: D.M. Benforado, W.J. Rotella, and D.L. Horton, "Development of an Odor Panel for Evaluation of Odor Control Equipment," *J. Air Pollut. Contr. Assn., 19*(2):101, 1969. With permission.

1.4-14   EXHAUST REQUIREMENTS FOR VARIOUS OPERATIONS

Operation	Exhaust arrangement	Remarks
Abrasive blast rooms	Tight enclosures with air inlets (generally in roof)	For 60–100 fpm downdraft or 100 fpm crossdraft in room
Abrasive blast cabinets	Tight enclosure	For 500 fpm through all openings, and a minimum of 20 air changes per minute
Bagging machines	Booth or enclosure	For 100 fpm through all openings for paper bags; 200 fpm for cloth bags
Belt conveyors	Hoods at transfer points enclosed as much as possible	For belt speeds less than 200 fpm, $V = 350$ cfm/ft belt width with at least 150 fpm through openings. For belt speeds greater than 200 fpm, $V = 500$ cfm/ft belt width with at least 200 fpm through remaining openings
Bucket elevator	Tight casing	For 100 cfm/ft^2 of elevator casing cross-section (exhaust near elevator top and also vent at bottom if over 35 ft high)
Foundry screens	Enclosure	Cylindrical–400 fpm through openings; and not less than 100 cfm/ft^2 of cross-section; flat deck–200 fpm through openings, and not less than 25 cfm/ft^2 of screen area
Foundry shakeout	Enclosure	For 200 fpm through all openings, and not less than 200 cfm/ft^2 of grate area with hot castings and 150 cfm/ft^2 with cool castings
Foundry shakeout	Side hood (with side shields when possible)	For 400–500 cfm/ft^2 grate area with hot castings and 350–400 cfm/ft^2 with cool castings
Grinders, disc and portable	Downdraft grilles in bench or floor	For 200–400 fpm through open face, but at least 150 cfm/ft^2 of plan working area
Grinders and crushers	Enclosure	For 200 fpm through openings
Mixer	Enclosure	For 100–200 fpm through openings
Packaging machines	Booth	For 50–100 fpm
	Downdraft	For 75–150 fpm
	Enclosure	For 100–400 fpm
Paint spray	Booth	For 100–200 fpm indraft, depending upon size of work, depth of booth, etc.
Rubber rolls (calendars)	Enclosure	For 75–100 fpm through openings
Welding (arc)	Booth	For 100 fpm through openings

Source: *Air Pollution Engineering Manual,* J. Danielson, Ed., U.S. Department of Health, Education, and Welfare, National Center for Air Pollution Control, Cincinnati, O., 1967, p. 31.

1.4.3   INDUSTRIAL CONTROLS

1.4.3.1   PETROLEUM INDUSTRY

1.4-15   SOURCES AND CONTROL OF AIR CONTAMINANTS FROM CRUDE-OIL PRODUCTION FACILITIES

Phase of operation	Source	Contaminant	Acceptable control
Well drilling, pumping	Gas venting for production rate test	Methane	Smokeless flares, wet-gas-gathering system
	Oil well pumping	Light hydrocarbon vapors	Proper maintenance
	Effluent sumps	Hydrocarbon vapors, H_2S	Replacement with closed vessels connected to vapor recovery
Storage, shipment	Gas-oil separators	Light hydrocarbon vapors	Relief to wet-gas-gathering system
	Storage tanks	Light hydrocarbon vapors, H_2S	Vapor recovery, floating roofs, pressure tanks, white paint
	Dehydrating tanks	Hydrocarbon vapors, H_2S	Closed vessels, connected to vapor recovery
	Tank truck loading	Hydrocarbon vapors	Vapor return, vapor recovery, vapor incineration, bottom loading
	Effluent sumps	Hydrocarbon vapors	Replacement with closed vessels connected to vapor recovery
	Heaters, boilers	H_2S, HC, SO_2, NO_X, particulate matter	Proper operation, use of gas fuel
Compression, absorption, dehydrating, water treating	Compressors, pumps	Hydrocarbon vapors, H_2S	Mechanical seals, packing glands vented to vapor recovery
	Scrubbers, KO pots	Hydrocarbon vapors, H_2S	Relief to flare or vapor recovery
	Absorbers, fractionators, strippers	Hydrocarbon vapors	Relief to flare or vapor recovery
	Tank truck loading	Hydrocarbon vapors, H_2S	Vapor return, vapor recovery, vapor incineration, bottom loading
	Gas odorizing	H_2S mercaptans	Positive pumping, adsorption
	Waste-effluent treating	Hydrocarbon vapors	Enclosed separators, vapor recovery or incineration
	Storage vessels	Hydrocarbon vapors, H_2S	Vapor recovery, vapor balance, floating roofs
	Heaters, boilers	Hydrocarbon, SO_2, NO_X, particulate matter	Proper operation, substitute gas as fuel

Source: *Air Pollution Engineering Manual,* J. Danielson, Ed., U.S. Department of Health, Education, and Welfare, National Center for Air Pollution Control, Cincinnati, O., 1967, p. 562.

1.4-16   CONTROL MEASURES FOR REDUCTION OF AIR CONTAMINANTS FROM PETROLEUM REFINING

Source	Control method
Storage vessels	Vapor recovery systems; floating-roof tanks; pressure tanks; vapor balance; painting tanks white
Catalyst regenerators	Cyclones – precipitator, CO boiler; cyclones – water scrubber; multiple cyclones
Accumulator vents	Vapor recovery; vapor incineration
Blowdown systems	Smokeless flares – gas recovery
Pumps and compressors	Mechanical seals; vapor recovery; sealing glands by oil pressure; maintenance
Vacuum jets	Vapor incineration
Equipment valves	Inspection and maintenance
Pressure relief valves	Vapor recovery; vapor incineration; rupture discs; inspection and maintenance
Effluent-waste disposal	Enclosing separators; covering sewer boxes and using liquid seal; liquid seals on drains
Bulk-loading facilities	Vapor collection with recovery or incineration; submerged or bottom loading
Acid treating	Continuous-type agitators with mechanical mixing; replace with catalytic hydrogenation units; incinerate all vented cases; stop sludge burning
Acid sludge storage and shipping	Caustic scrubbing; incineration; vapor return system; disposal at sea
Spent-caustic handling	Incineration; scrubbing
Doctor treating	Steam strip spent doctor solution to hydrocarbon recovery before air regeneration replace treating unit with other, less objectionable units (Merox)
Sour-water treating	Use sour-water oxidizers and gas incineration; conversion to ammonium sulfate
Mercaptan disposal	Conversion to disulfides; adding to catalytic cracking charge stock; incineration; using material in organic synthesis
Asphalt blowing	Incineration; water scrubbing (nonrecirculating type)
Shutdowns, turnarounds	Depressure and purge to vapor recovery

Source: *Air Pollution Engineering Manual,* J. Danielson, Ed., U.S. Department of Health, Education, and Welfare, National Center for Air Pollution Control, Cincinnati, O., 1967, p. 565.

Section 2
Water Sources and Quality

2.1 QUANTITY OF WATER

2.1-1 ESTIMATED WORLD WATER SUPPLY

Water item	Volume, thousands		Percent of total water
	mi^3	km^3	
Water in land areas			
Fresh water lakes	30	125	0.009
Saline lakes and inland seas	25	104	0.008
Rivers (average instantaneous volume)	0.3	1.25	0.0001
Soil moisture and vadose water	16	67	0.005
Ground water to depth of 4,000 m	2,000	8,350	0.61
Icecaps and glaciers	7,000	29,200	2.14
Total, in land area (rounded)	9,100	37,800	2.8
Atmosphere	3.1	13	0.001
World ocean	317,000	1,320,000	97.3
Total, all items (rounded)	326,000	1,360,000	100
Annual evaporation[a]			
From world ocean	85	350	0.026
From land areas	17	70	0.005
Total	102	420	0.031
Annual precipitation			
On world ocean	78	320	0.024
On land areas	24	100	0.007
Total	102	420	0.031
Annual runoff to oceans from rivers and icecaps	9	38	0.003
Ground water outflow to oceans[b]	0.4	1.6	0.0001
Total	9.4	39.6	0.0031

[a] Evaporation (420,000 km^3) is a measure of total water participating annually in the hydrological cycle.
[b] Arbitrarily set equal to about 5 percent of surface runoff.

Source: U.S. Geological Survey, 1967.

2.1-2 DISTRIBUTION OF WATER IN THE CONTERMINOUS U.S.

	Area, mi^2	Volume, mi^3	Annual circulation, million acre ft/yr	Detention, period, yr
Frozen water				
Glaciers	200	16	1.3	40
Ground ice		(seasonal only)		
Liquid water				
Fresh water lakes[a]	61,000	4,500	150	100
Salt lakes	2,600	14	4.6	10
Average in stream channels	–	12	1,500	0.03
Ground water				
Shallow	3,000,000	15,000	250	200
Deep	3,000,000	15,000	5	10,000
Soil moisture (3 ft root zone)	3,000,000	150	2,500	0.2
Gaseous water				
Atmosphere	3,000,000	45	5,000	0.03

[a] U.S. part of Great Lakes only.

Source: U.S. Geological Survey.

2.1-3   U.S. FRESH WATER RESERVES[1]

Source of water	Amount,[a] 10^6 acre ft	Reserve life,[b] yr
Atmosphere	176	0.6
Soil moisture	635	2.2
Rivers	45	0.1
Reservoirs	365	1.3
Lakes	13,000	46
Ground water	47,500	168
Total		215

[1] Adapted from MacKichan[2,3] and Ackerman and Lof.[4]
[a] In terms of current recoverable reserve discovered.
[b] To support entire national water withdrawal.

REFERENCES

1. L. Koenig, *J. Am. Water Works Assn., 55*(1):59, 1963.
2. K.A. MacKichan, *J. Am. Water Works Assn., 49*(4):369, 1957.
3. K.A. MacKichan, *J. Am. Water Works Assn., 53*(10):1211, 1961.
4. E.A. Ackerman and G.O. Löf, *Technology in American Water Development*, published for Resources for the Future, Inc. by the Johns Hopkins Press, 1959, p. 12.

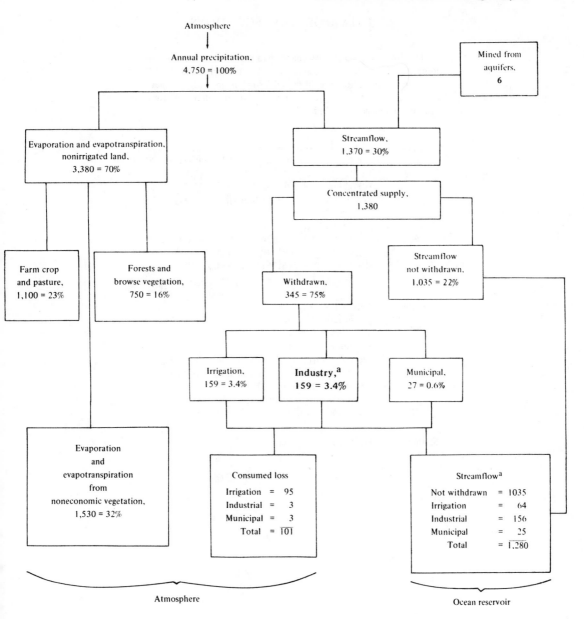

2.1-4   Average distribution of precipitation in the Continental U.S.

Legend:
All figures in units of million acre ft
  one acre ft = 43,560 ft^3
  one acre ft = 326,700 gal
Total precipitation = 1,552 x 10^{12} gal/year
                or 4,200 x 10^9 gal/day

[a] The same water may be reused at points spaced along a single stream.

Source: A. Wolman, *Water Resources,* National Academy of Sciences, National Research Council, Committee on Natural Resources, 1962.

2.1-5   EVAPORATION EQUATIONS

The following equations enable pan and lake evaporation to be computed from climatic data at first order weather stations. Daily values of evaporation are obtained using mean daily temperature and vapor pressure data together with data on solar radiation and wind movement as specified.

For pan evaporation, the expression is

$$E_p = \left\{ \exp\left[(T_a - 212)(0.1024 - 0.01066 \ln R)\right] - 0.001 + 0.025 (e_s - e_a)^{0.88} (0.37 + 0.0041 U_p) \right\} \times \left\{ 0.025 + (T_a + 398.36)^{-2} \ 4.7988 \times 10^{10} \exp\left[-7482.6/(T_a + 398.36)\right] \right\}^{-1}$$

For lake evaporation, the expression is

$$E_L = \left\{ \exp\left[(T_a - 212)(0.1024 - 0.01066 \ln R)\right] - 0.001 + 0.0105 (e_s - e_a)^{0.88} (0.37 + 0.0041 U_p) \right\} \times \left\{ 0.015 + (T_a + 398.36)^{-2} \ 6.8554 \times 10^{10} \exp\left[-7482.6/(T_a + 398.36)\right] \right\}^{-1}$$

The terms in these expressions are

E_p = pan evaporation, in.
E_L = lake evaporation, in.
T_a = air temperature, °F
e_a = vapor pressure, in. mercury at temperature T_a
e_s = vapor pressure, in. mercury at temperature T_d
T_d = dew point temperature, °F
R = solar radiation, langleys/day
U_p = wind movement, mi/day

Source: U.S. Weather Bureau, 1962.

2.2 CONSTITUENTS IN WATER

2.2-1 MAJOR CHEMICAL CONSTITUENTS IN WATER

Sources, Concentrations, and Effects upon Usability

Constituent	Major sources	Concentration in natural water	Effect upon usability of water	Concentration in public water supplies of 100 cities, mg/l				Drinking water should contain less than concentration shown if more suitable supplies are or can be made available[a]
				Untreated water		Treated water		
				Max	Min	Max	Min	
Silica, SiO_2	Feldspars, ferromagnesium and clay minerals, amorphous silica cachert, opal	Ranges generally from 1.0–30 mg/l, although as much as 100 mg/l is fairly common; as much as 4,000 mg/l is found in brines	In the presence of calcium and magnesium, silica forms a scale in boilers and on steam turbines that retards heat and fluid flow; the scale is difficult to remove. Silica may be added to soft water to inhibit corrosion of iron pipes	72	0.0	72	0.0	

Iron, Fe	**Natural sources**
	Igneous rocks
	  Amphiboles, ferromagnesian micas, ferrous sulfide (FeS), ferric sulfide or iron pyrite (FeS_2), magnetite (Fe_3O_4)
	Sandstone rocks
	  Oxides, carbonates, and sulfides of iron clay minerals
	Manmade sources
	Well casing, piping, pump parts, storage tanks, and other objects of cast iron and steel which may be in contact with the water
	Industrial wastes
		Generally less than 0.50 mg/l in fully aerated water. Ground water having a pH less than 8.0 may contain 10 mg/l, rarely as much as 50 mg/l may occur. Acid water from thermal springs, mine wastes, and industrial wastes may contain more than 6,000 mg/l	More than 0.1 mg/l precipitates after exposure to air; causes turbidity, stains plumbing fixtures, laundry and cooking utensils, and imparts objectionable tastes and colors to foods and drinks. More than 0.2 mg/l is objectionable for most industrial uses	1.90	0.00	1.30	0.00	0.3

2.2-1　MAJOR CHEMICAL CONSTITUENTS IN WATER (continued)

Constituent	Major sources	Concentration in natural water	Effect upon usability of water	Concentration in public water supplies of 100 cities, mg/l				Drinking water should contain less than concentration shown if more suitable supplies are or can be made available[a]
				Untreated water		Treated water		
				Max	Min	Max	Min	
Manganese, Mn	Manganese in natural water probably comes most often from soils and sediments. Metamorphic and sedimentary rocks and mica biotite and amphibole hornblende minerals contain large amounts of manganese	Generally 0.20 mg/l or less. Ground water and acid mine water may contain more than 10 mg/l. Water at the bottom of a stratified reservoir may contain more than 150 mg/l	More than 0.2 mg/l precipitates upon oxidation; causes undesirable tastes, deposits on foods during cooking, stains plumbing fixtures and laundry, and fosters growths in reservoirs, filters, and distribution systems. Most industrial users object to water containing more than 0.2 mg/l	0.60	0.00	2.5	0.00	0.05
Calcium, Ca	Amphiboles, feldspars, gypsum, pyroxenes, aragonite, calcite, dolomite, clay minerals	As much as 600 mg/l in some western streams, brines may contain as much as 75,000 mg/l	Calcium and magnesium combine with bicarbonate, carbonate, sulfate, and silica to form heat-retarding, pipe-clogging scale in boilers and in other heat-exchange equipment. Calcium and magnesium combine with ions of fatty acid in soaps to form soapsuds; the more calcium and magnesium, the more soap required to form suds. A high concentration of magnesium has a laxative effect, especially on new users of the supply	145	0.0	145	0.0	
Magnesium, Mg	Amphiboles, olivine, pyroxenes, dolomite, magnesite, clay minerals	As much as several hundred mg/l in some western streams; ocean water contains more than 1,000 mg/l and brines may contain as much as 57,000 mg/l		120	0.0	120	0.0	
Sodium, Na	Feldspars (albite); clay minerals; evaporites, such as halite (NaCl) and mirabilite (Na₂SO₄·10H₂O); industrial wastes	As much as 1,000 mg/l in some western streams; about 10,000 mg/l in sea water; about 25,000 mg/l in brines	More than 50 mg/l sodium and potassium in the presence of suspended matter causes foaming, which accelerates scale formation and corrosion in boilers. Sodium and potas-	177	1.1	198	1.1	

Constituent	Source	Significance					
Potassium, K	Feldspars (orthoclase and microcline), feldspathoids, some micas, clay minerals	Generally less than about 10 mg/l; as much as 100 mg/l in hot springs; as much as 25,000 mg/l in brines		30	0.2	30	0.0
Carbonate, CO₃	Limestone, dolomite	Commonly 0 mg/l in surface water; commonly less than 10 mg/l in ground water. Water high in sodium may contain as much as 50 mg/l of carbonate		21	0	26	0
Bicarbonate, HCO₃		Commonly less than 500 mg/l; may exceed 1,000 mg/l in water highly charged with carbon dioxide		380	5	380	0
		Upon heating, bicarbonate is changed into steam, carbon dioxide, and carbonate. The carbonate combines with alkaline earths—principally calcium and magnesium—to form a crustlike scale of calcium carbonate that retards flow of heat through pipe walls and restricts flow of fluids in pipes. Water containing large amounts of bicarbonate and alkalinity is undesirable in many industries					
Sulfate, SO₄	Oxidation of sulfide ores; gypsum; anhydrite; industrial wastes	Commonly less than 1,000 mg/l except in streams and wells influenced by acid mine drainage. As much as 200,000 mg/l in some brines	250	572	0.0	572	0.0
		Sulfate combines with calcium to form an adherent, heat-retarding scale. More than 250 mg/l is objectionable in water in some industries. Water containing about 500 mg/l of sulfate tastes bitter; water containing about 1,000 mg/l may be cathartic					
Chloride, Cl	Chief source is sedimentary rock (evaporites); minor sources are igneous rocks. Ocean tides force salty water upstream in tidal estuaries	Commonly less than 10 mg/l in humid regions; tidal streams contain increasing amounts of chloride (as much as 19,000 mg/l) as the bay or ocean is approached. About 19,300 mg/l in sea water; and as much as 200,000 mg/l in brines	250	540	0.5	540	0.0
		Chloride in excess of 100 mg/l imparts a salty taste. Concentrations greatly in excess of 100 mg/l may cause physiological damage. Food processing industries usually require less than 250 mg/l. Some industries—textile processing, paper manufacturing, and synthetic rubber manufacturing—desire less than 100 mg/l					

2.2-1   MAJOR CHEMICAL CONSTITUENTS IN WATER (continued)

Constituent	Major sources	Concentration in natural water	Effect upon usability of water	Concentration in public water supplies of 100 cities, mg/l				Drinking water should contain less than concentration shown if more suitable supplies are or can be made available[a]
				Untreated water		Treated water		
				Max	Min	Max	Min	
Fluoride, F	Amphiboles (hornblende), apatite, fluorite, mica	Concentrations generally do not exceed 10 mg/l in ground water or 1.0 mg/l in surface water. Concentrations may be as much as 1,600 mg/l in brines	Fluoride concentration between 0.6 and 1.7 mg/l in drinking water has a beneficial effect on the structure and resistance to decay of children's teeth. Fluoride in excess of 1.5 mg/l in some areas causes "mottled enamel" in children's teeth. Fluoride in excess of 6.0 mg/l causes pronounced mottling and disfiguration of teeth	7.0	0.0	7.0	0.0	The recommended control limits depend upon annual averages of maximum daily air temperature and range from 0.6–0.8 mg/l at 79.3°–90.5°F and 0.9–1.7 mg/l at 50.0°–53.7°F
Nitrate, NO_3	Atmosphere; legumes, plant debris, animal excrement, nitrogenous fertilizer in soil and sewage	In surface water not subjected to pollution, concentration of nitrate may be as much as 5.0 mg/l but is commonly less than 1.0 mg/l. In ground water the concentration of nitrate may be as much as 1,000 mg/l	Water containing large amounts of nitrate (more than 100 mg/l) is bitter tasting and may cause physiological distress. Water from shallow wells containing more than 45 mg/l has been reported to cause methemoglobinemia in infants. Small amounts of nitrate help reduce cracking of high-pressure boiler steel	23	0.0	23	0.0	45; In areas in which the nitrate content of water is known to be in excess of the listed concentration, the public should be warned of the potential dangers of using the water for infant feeding

| Dissolved solids | The mineral constituents dissolved in water constitute the dissolved solids | Surface water commonly contains less than 3,000 mg/l; streams draining salt beds in arid regions may contain in excess of 15,000 mg/l. Ground water commonly contains less than 5,000 mg/l; some brines contain as much as 300,000 mg/l | More than 500 mg/l is undesirable for drinking and many industrial uses. Less than 300 mg/l is desirable for dyeing of textiles and the manufacture of plastics, pulp paper, rayon. Dissolved solids cause foaming in steam boilers; the maximum permissible content decreases with increases in operating pressure | 1,580 | 10 | 1,580 | 22 | 500 |

[a]U.S. Public Health Service, 1962.

Source: C.N. Durfor and E. Becker, *Geological Survey Water Supply Paper*, U.S. Department of the Interior, 1964. Table prepared by B.P. Robinson.

2.2-2   MINOR CONSTITUENTS IN WATER

Constituent	Some common sources	Concentration in natural water	Concentration in treated water of public supplies of 100 largest cities, mg/l		Drinking water should contain less than concentration shown if more suitable supplies are or can be made available,[a] mg/l	Concentration in drinking water more than that shown constitutes grounds for rejection of the supply,[a] mg/l
			Max	Min		
Aluminum	Feldspars, micas, and clay minerals; also present as the result of alum treatment in water supplies	Generally less than 0.5 mg/l. Sea water, 7,500 mg/l in acid lakes	1,500	3.3		
Barium	Feldspars, mica, biotite, barite, witherite	Generally less 0.5 mg/l. Sea water, about 0.05 mg/l; brine, as much as 5,000 mg/l	380	1.7		1.0
Boron	Amphiboles, biotite, colemanite, and tourmaline; detergents; industrial wastes	Generally less than a few tenths mg/l. Sea water, about 4.6 mg/l; mineral springs, may exceed 500 mg/l; brine, may exceed 9,000 mg/l	590	2.5		
Chromium	Chromite; industrial pollution	Generally less than 0.1 mg/l. Industrial pollution, as much as 40 mg/l	35	Not detectable		0.05 (hexavalent)
Copper	Native copper, chalcocite, bornite, chalcopyrite, cuprite, many other minerals containing copper. Copper and brass pipes; algae-controlling copper salts; industrial wastes	Generally less than a few tenths mg/l. Sea water, 0.001–0.01 mg/l; industrial pollution (mine water) as much as several thousand mg/l	250	Less than 0.61	1.0	
Lead	Galena, aragonite, feldspars. Dissolution of lead pipes; industrial and mining wastes	Generally less than 0.5 mg/l in streams and in ground water. Sea water, 0.004 mg/l; industrial pollution, several hundred mg/l	62	Not detectable		
Lithium	Mica, amphiboles, pyroxenes	Generally less than 1 mg/l. Sea water, 0.1 mg/l; brine, as much as 80 mg/l	170	Not detectable		
Nickel	Pentlandite, niccolite, chloanthite, garnierite, millerite, olivine, hypersthene	Generally less than 0.1 mg/l	34	Not detectable		
Phosphorus	Apatite; organic wastes; fertilizers; detergents	Generally less than a few mg/l. Sea water, 0.003–0.3 mg/l; industrial pollution, 10 mg/l	370	Not detectable		

	Source	Occurrence in natural water	2.5 pCi/l	Less than 0.1 pCi/l	3 pCi of ^{226}Ra
Radium	Most igneous rocks, sandstones, shales	Uncontaminated water generally contains less than 10 pCi/l	2.5 pCi/l	Less than 0.1 pCi/l	3 pCi of ^{226}Ra
Rubidium	Greisen, lepidolite (concealed in many potassium minerals), potassium feldspar, potassium mica	Generally less than 0.1 mg/l	67	Not detectable	
Strontium	Feldspars, apatite, pyroxenes, amphiboles, celestite, strontianite, aragonite, fossils	Generally less than 5.0 mg/l in freshwater streams; as much as 30 mg/l in ground water. Sea water, about 13 mg/l; brine, as much as 3,500 mg/l	1,200	2.2	
Titanium	Ilmenite, rutile, sphene, phlogopites, amphiboles, biotite, pyroxene	Generally less than 0.1 mg/l	49	Not detectable	
Uranium	Uraninite, carnotite; most igneous rocks, sandstones, and shales	Uncontaminated water generally contains less than 44 mg/l	250	Less than 0.1	
Zinc	Sphalerite; mine waste waters; industrial wastes	Generally in trace amounts. Sea water, 0.005 mg/l; industrial pollution, as much as several thousand mg/l	610	Not detectable	

[a]U.S. Public Health Service, 1962.

REFERENCES

1. C.N. Durfor and E. Becker, *Geological Survey Water Supply Paper*, U.S. Department of the Interior, 1964.
2. K. Rankama and T. G. Sahama, *Geochemistry*, Chicago University Press, 1950.

2.2-3   CONCENTRATION OF TRACE ELEMENTS IN U.S. RIVERS (1958—1962)

Element	Colorado River[a] Range µg/l	Colorado River[a] Frequency of detection percent	Columbia River[b] Range µg/l	Columbia River[b] Frequency of detection percent	Mississippi River[c] Range µg/l	Mississippi River[c] Frequency of detection percent	Missouri River[d] Range µg/l	Missouri River[d] Frequency of detection percent	Ohio River[e] Range µg/l	Ohio River[e] Frequency of detection percent
Ba	20–200	79	4–60	94	20–100	100	10–200	96	20–200	94
Cr	10–30	12	1–10	87	3–20	23	8–10	10	4–10	20
Cu	2–50	61	1–50	100	1–100	96	1–50	85	1–100	91
Fe	10–300	82	2–90	100	6–100	95	4–200	92	5–100	100
Mn	–	0	3–5	13	9–50	11	–	6	6–30	17
Mo	8–40	82	1–9	90	3–10	68	7–20	65	3–10	37
Ni	10–30	9	2–10	48	3–10	23	6–9	8	3–30	31
Pb	–	6	4–50	29	10–50	11	–	3	10–30	11
V	–	9	–	3	–	0	–	2	–	0
Ag	–	0	–	0	–	12	0.2–20	57	0.6–1	60

a Cd (30) and S (100) occurred once each; Be, Sn, Bi, Zn, and Co were never found in positive amounts.
b Other positive values were; Cd (3), Sn (2), and Zn (900, 2300); Be, Sb, Bi, Co. and Co were never found in positive amounts.
c Other positive values were; Sn (3, 8, 10) and Cd (7); Be, Sb, Bi, Co. and Zn were never found in positive amounts.
d Co (8) occurred once; Cd, Be, Sn, Sb, Bi, and Zn were never found in positive amounts were; Be, Bi, Cd, Co, Sb, Sn, and Zn.

Source: Reprinted, with publisher's permission, from *J. Am. Water Works Assn.*, 57(2):156, © 1965, by the American Water Works Association, Inc.

2.3 SOURCES OF WATER

2.3.1 RIVERS

2.3-1 SUMMARY, RAW WATER CHARACTERISTICS

Public Water Supplies

Constituent	River	Lake	Impoundment	Ground water
SiO_2, mg/l	0.00-21	1.1-5.8	1.0-15	4.3-72
Fe, mg/l	0.00-1.9	0.00-0.16	0.0-0.84	0.00-1.2
Ca, mg/l	2.8-110	2.0-44	0.0-76	3.2-121
Mg, mg/l	0.9-28	0.5-11	0.3-29	0.3-120
Na, mg/l	1.9-131	1.6-24	1.1-99	6.1-129
K, mg/l	0.3-7.8	0.7-4.5	0.2-7.9	0.4-30
HCO_3, mg/l	15-316	9-180	4-256	55-364
CO_3, mg/l	0-1	0	0-21	0-5
SO_4, mg/l	0.0-290	0.8-23	0.6-282	0.8-572
Cl, mg/l	0.5-196	2.6-32	1.0-107	2.0-92
F, mg/l	0.0-0.6	0.0-0.3	0.0-1.0	0.0-7.0
NO_3, mg/l	0.0-8.3	0.0-2.2	0.0-4.7	0.0-17
Dissolved solids (180°C), mg/l	31-657	29-233	10-761	68-1,220
Hardness as $CaCO_3$, mg/l	12-368	7-153	2-336	10-738
Noncarbonate hardness as $CaCO_3$, mg/l	0-206	0-54	0-207	0-446
Specific conductance, μmho 25°C	42-1,270	33-387	8-865	108-1,660
pH	6.2-8.4	6.6-8.3	5.8-9.1	6.7-8.7
Color, units	0-115	1-15	0-50	0-76
Beta activity, pCi/l	16-78	12-66		1.8-130
Radium, pCi/l				<0.1-1.9
Uranium, μg/l				<0.1-250
Ag, μg/l	ND[a]-0.48	<0.18-<0.34	ND-0.74	ND-1.5
Al, μg/l	0-2,400	38-90	17-690	2.9-83
B, μg/l	7.0-130	7.8-80	3.6-250	0-500
Ba, μg/l	3.1-340	31-120	6.2-150	4.6-67
Be, μg/l	ND	ND	ND-0.75	ND
Co, μg/l	ND-2.3	ND	ND-0.99	ND
Cr, μg/l	ND-7.8	0.34-3.6	ND-3.8	ND-1.1
Cu, μg/l	1.1-84	6.0-57	0.78-73	<0.8-15
Fe, μg/l	9.0-3,800	14-100	29-850	1.1-6,600
Li, μg/l	0.08-91	0.43-1.7	0-100	ND-96
Mn, μg/l	2.6-380	0-100	0-450	ND-340
Mo, μg/l	ND-7.8	ND-1.5	ND-5.8	ND-68
Ni, μg/l	ND-15	3.2-21	ND-15	ND-<15
P, μg/l	ND	ND	ND	ND
Pb, μg/l	ND-60	6.4-26	ND-94	ND-38
Rb, μg/l	1.6-16	ND-3.5	ND-5.5	ND-3.7
Sn, μg/l	ND	ND	ND-1.1	ND
Sr, μg/l	2.6-610	80-110	2.4-540	19-6,300
Ti, μg/l	0.5-35	<2-<7.6	0.6-11	ND-<5
V, μg/l	ND-5.1	ND	ND-<5.8	ND-74
Zn, μg/l	ND-210	ND-<340	ND-99	ND-<470
Ga, μg/l	ND-0.0X[b]			<0.X
Zr, μg/l	0.X-X		0.X	ND
Yb, μg/l	0.0X-X	ND	0.X	ND
Y, μg/l	0.X-X	ND	0.X	ND

2.3-1   SUMMARY, RAW WATER CHARACTERISTICS (continued)

[a] ND, looked for but not found.
[b] X, semiquantitative determination in digit order shown.

Source: Developed from data in C.N. Durfor and E. Becker, *Geological Survey Water Supply Paper*, U.S. Department of the Interior, 1964.

2.3-2   SUMMARY OF DISSOLVED TRACE ELEMENTS

1,577 Surface Waters, October 1, 1962–September 30, 1967

Element	No. positive occurrences	Frequency of detection, %	Observed values, μg/l		
			Min	Max	Mean
Zinc	1,207	76.5	2	1,183	64
Cadmium	40	2.5	1	120	9.5
Arsenic	87	5.5	5	336	64
Boron	1,546	98.0	1	5,000	101
Phosphorus	747	47.4	2	5,040	120
Iron	1,192	75.6	1	4,600	52
Molybdenum	516	32.7	2	1,500	68
Manganese	810	51.4	0.3	3,230	58
Aluminum[a]	456	31.2	1	2,760	74
Beryllium	85	5.4	0.01	1.22	0.19
Copper	1,173	74.4	1	280	15
Silver	104	6.6	0.1	38	2.6
Nickel	256	16.2	1	130	19
Cobalt	44	2.8	1	48	17
Lead	305	19.3	2	140	23
Chromium	386	24.5	1	112	9.7
Vanadium	54	3.4	2	300	40
Barium	1,568	99.4	2	340	43
Strontium	1,571	99.6	3	5,000	217

[a] 1,464 aluminum analyses.

Source: J.F. Kopp in *Trace Substances in Environmental Health*, Vol. 3, D.D. Hemphill, Ed., University of Missouri Press, Columbia, 1969. With permission.

2.3-3   TRACE ELEMENT COMPOSITION OF SUSPENDED MATERIAL
IN RIVERS

River and State	Suspended load, mg/l	ppm					
		Cr	Ag	Mo	Ni	Co	Mn
Brazos, Texas	954	100	0.4	11	30	20	690
Colorado, Texas	150	82	0.6	10	40	17	780
Red, La.	436	37	0.3	5	6	7	320
Mississippi, Ark.	185	150	0.7	18	100	33	2,300
Tombigbee, Ala.	25	220	1.0	22	200	31	5,900
Alabama, Ala.	54	150	4.0	19	100	34	3,700
Chattahoochie, Ga.	71	190	7.0	20	100	35	2,400
Flint, Ga.	12	210	1.0	28	100	39	5,100
Savannah, S.C.	30	460	2.0	35	250	36	4,400
Wateree, S.C.	37	200	1.5	24	100	34	7,000
Pee Dee, S.C.	188	150	0.4	15	100	23	1,300
Cape Fear, N.C.	61	130	0.7	16	70	21	1,700
Neuse, N.C.	36	380	4.0	22	70	30	3,000
Roanoke, N.C.	33	240	4.9	21	100	45	7,900
James, Va.	41	290	7.0	29	300	60	15,000
Rappahannock, Va.	28	140	1.0	31	80	46	2,200
Potomac, Va.	34	170	1.5	23	400	94	7,700
Susquehanna, Pa.	54	290	15.0	32	>1,000	>500	12,000
Rhone, France Avignon, June 1966	296	150	0.7	14	60	29	820
Rio Maipo, Chile Puente Alto, S. of Santiago, September 1966	41	68	1.0	44	40	76	2,400

Source: K.K. Turekian and M.R. Scott, *Environ. Sci. Technol., 1*(11):940, 1967. With permission.

2.3-4   ORGANICS IN THREE RIVERS

Location	Organics,[a] ppb		
	1958	1959	1960
Mississippi River			
Minneapolis, Minn.	256	209	195
Dubuque, Iowa	268	195	168
Burlington, Iowa	141	125	127
E. St. Louis, Ill.	173	212	206
Cape Girardeau, Mo.	216	181	170
West Memphis, Ark.	85	131	126
New Orleans, La.	74	86	82
Average	177	163	153
Missouri River			
Williston, N.D.	148	59	80
Bismarck, N.D.	134	156	159
Yankton, S.D.	115	139	147
Omaha, Neb.	123	107	177
St. Joseph, Mo.	156	100	139
Kansas City, Kan.	134	89	89
St. Louis, Mo.	96	117	102
Average	129	110	103
Ohio River			
East Liverpool, Ohio	102	122	241
Huntington, W.V.	160	239	225
Cincinnati, Ohio	404	205	355
Evansville, Ind.	77	76	114
Cairo, Ill.	127	95	182
Average	174	147	223

[a] As alcohol extractable.

Source: Reprinted, with publisher's permission, from *J. Am. Water Works Assn.,* *55*(3):369, © 1963 by the American Water Works Association, Inc.

2.3-5   MAJOR CONSTITUENTS OF RIVER WATERS USED FOR WATER SUPPLY

Table A

Location	Source of supply	Percent of supply	SiO$_2$, mg/l	Fe, mg/l	Ca, mg/l	Mg, mg/l	Na, mg/l	K, mg/l	HCO$_3$, mg/l	CO$_3$, mg/l	SO$_4$, mg/l	Cl, mg/l
Birmingham, Ala.	Cahaba River and Lake Purdy	90	5.7	0.03	26	4.9	10	1.7	87	0	31	2.0
Phoenix, Ariz.	Verde River	12	14	0.02	50	14	131	4.8	176	0	51	196
Los Angeles, Cal.	Colorado River[a]	20	8.7	–	84	28	92	4	140	1	285	83
Sacramento, Cal.	Sacramento River	85	21	0.01	10	6.6	10	0.8	69	0	7	8.0
Washington, D.C.	Potomac River	100	5.7	0.00	35	7.9	7.4	1.9	107	0	35	10
Tampa, Fla.	Hillsborough River	97	7.8	0.03	55	5.6	8.0	1.5	150	0	28	16
Atlanta, Ga.	Chattahoochee River	100	8.0	0.07	2.8	1.2	2.3	1.0	15	0	0.0	5.0
Savannah, Ga.	Abercorn Creek	100	10	0.27	4.0	0.9	3.5	1.0	18	0	5.2	4.5
Kansas City, Kans.	Missouri River	100	16	0.01	49	13	26	4.8	169	0	69	14
Topeka, Kans.	Kansas River	100	17	0.03	110	23	72	6.0	316	0	128	95
Louisville, Ky.	Ohio River	100	0.2	0.05	41	9.5	16	2.6	90	0	74	20
New Orleans, La.	Mississippi River	100	5.3	0.01	48	11	17	2.8	148	0	43	26
Detroit, Mich.	Detroit River	100	2.1	0.17	28	7.0	4.1	0.9	92	0	18	8.0
Minneapolis, Minn.	Mississippi River	100	11	0.02	48	16	7.3	1.5	222	0	16	1.6
Kansas City, Mo.	Missouri River	100	7.0	0.10	55	16	35	6.6	196	0	101	12
St. Louis, Mo.	Missouri River	34	6.2	0.20	36	8.8	17	5.1	127	0	47	10
	Mississippi River	66	5.5	0.40	37	8.9	17	5.1	128	0	48	10
Omaha, Neb.	Missouri River	100	16	0.02	74	23	65	6.4	238	0	210	13
Paterson, NJ	Passaic River[b]	47	13	0.47	19	5.8	–	–	63	–	30	15
Philadelphia, Pa.	Delaware River[c]			0.35–1.5					29–50	0–0	21–36	4.0–11
	Schuylkill River[c]			0.34–1.9					50–87	0–0	59–100	8.0–23
Chattanooga, Tenn.	Tennessee River	100	3.8	0.05	23	5.0	8.2	1.1	70	0	13	16

2.3-5   MAJOR CONSTITUENTS OF RIVER WATERS USED FOR WATER SUPPLY (continued)

Location	Source of supply	Percent of supply	SiO2, mg/l	Fe, mg/l	Ca, mg/l	Mg, mg/l	Na, mg/l	K, mg/l	HCO3, mg/l	CO3, mg/l	SO4, mg/l	Cl, mg/l
Nashville, Tenn.	Cumberland River	100	2.7	0.24	22	3.1	2.9	0.9	64	0	17	2.0
Austin, Texas	Colorado River	100	9.9[e]		40	20	39	3.8	187	0	34	55
			10[f]		44	19	(39, braced)		177	0	34	60
El Paso, Texas	Rio Grande River	14	16	0.46	92	20	165	7.8	236	0	290	127
Houston, Texas	San Jacinto River	24	6.6	0.09	9.5	1.5	11	2.0	29	0	2.4	21
Salt Lake City, Utah	Big Cottonwood Creek	33	5.6	0.00	38	12	4.1	1.0	122	0	41	7.0
Richmond, Va.	James River	100	9.5	0.01	12	2.5	3.1	1.0	41	0	7.6	4.0
Seattle, Wash.	Cedar River	100	9.5	0.05	7.0	0.3	1.9	0.3	26	0	2.4	0.5
Cincinnati, Ohio	Ohio River[d]	100	0.00–0.04	–	25–51	5.5–12	–	–	35–58	0	62–132	15–59
Range			0.00–21	0.00–1.9	2.8–110	0.9–28	1.9–131	0.3–7.8	15–316	0–1	0.0–290	0.5–196

[a] Average analysis, July 1, 1960–June 30, 1961.
[b] Analyzed by the Passaic Valley Water Commission.
[c] Maximum value of constituents in monthly average of analyses by the city of Philadelphia Water Department during 1960.
[d] Maximum constituents in monthly analyses by the city of Cincinnati during 1961.
[e] Composite of daily samples, January 1–31, 1962.
[f] Composite of daily samples, June 1–30, 1962.

2.3-5　MAJOR CONSTITUENTS OF RIVER WATERS USED FOR WATER SUPPLY (continued)

Table B

Location	Source of supply	F, mg/l	NO_3, mg/l	Dissolved solids residue at 180°C, mg/l	Hardness as $CaCO_3$, mg/l	Noncarbonate hardness as $CaCO_3$, mg/l	Specific conductance, μmho at 25°C	pH	Color, units	Beta activity, pCi/l
Birmingham, Ala.	Cahaba River and Lake Purdy	0.1	0.3	195.	85.	14.	220.	7.0	0	–
Phoenix, Ariz.	Verde River	0.4	2.7	560	184	40.	1000	7.8	3	–
Los Angeles, Cal.	Colorado River[a]	0.4	1.4	657	323	206	1040	8.4	–	
Sacramento, Cal.	Sacramento River	0.1	0.9	110	52	0	149	7.4	3	
Washington, D.C.	Potomac River	0.1	1.1	172	120	33	279	7.0	5	29[g]
Tampa, Fla.	Hillsborough River	0.2	1.0	214	160	37	339	8.1	25	
Atlanta, Ga.	Chattahoochee River	0.1	0.1	31	12	0	42	6.6	10	13[g]
Savannah, Ga.	Abercorn Creek	0.2	0.7	53	14	0	45	6.6	115	–
Kansas City, Kans.	Missouri River	0.2	8.3	312	174	35	464	7.2	9	70[g]
Topeka, Kans.	Kansas River	0.2	4.9	638	368	109	994	7.5	4	46[g]
Louisville, Ky.	Ohio River	0.3	3.2	221	141	67	370	7.3	.5	38[g]
New Orleans, La.	Mississippi River	0.2	0.2	254	163	42	414	6.8	10	
Detroit, Mich.	Detroit River	0.0	0.5	129	99	24	213	7.6	2	16[g]
Minneapolis, Minn.	Mississippi River	0.2	0.4	228	185	3	266	7.5	17	66[g]
Kansas City, Mo.	Missouri River	0.5	3.8	350	203	42	522	7.8	15	70[g]
St. Louis, Mo.	Missouri River	0.4	3.4	220	126	22	323	7.8	5	48[g]
	Mississippi River	0.4	3.2	222	129	24	324	7.7	5	78[g]
Omaha, Neb.	Missouri River	0.6	0.4	523	279	84	797	7.3	7	78[g]
Paterson, NJ	Passaic River[b]		0.6	152	69			7.1	40	
Philadelphia, Pa.	Delaware River[c]		0.3–0.9	129–345	42–71			7.2–7.3	8–25	24[g]
	Schuylkill River[c]		1.3–1.9	180–320	100–156			7.4–7.6	8–40	30[g]

a　Average analysis, July 1, 1960–June 30, 1961.
b　Analyzed by the Passaic Valley Water Commission.
c　Maximum value of constituents in monthly average of analyses by the city of Philadelphia Water Department during 1960.
g　Maximum beta activity, raw water, July 1, 1961–June 30, 1962.

2.3-5   MAJOR CONSTITUENTS OF RIVER WATERS USED FOR WATER SUPPLY (continued)

Location	Source of supply	F, mg/l	NO_3, mg/l	Dissolved solids residue at 180°C, mg/l	Hardness as $CaCO_3$, mg/l	Noncarbonate hardness as $CaCO_3$, mg/l	Specific conductance, μmho at 25°C	pH	Color, units	Beta activity, pCi/l
Chattanooga, Tenn.	Tennessee River	0.2	2.9	125	78	20	195	8.0	5	58[g]
Nashville, Tenn.	Cumberland River	0.2	2.8	88	68	15	142	7.6	5	
Austin, Texas	Colorado River	0.3	1.2	294	182	30	526	7.6		
		0.3	0.8	306	188	43	526	7.6		
El Paso, Texas	Rio Grande River	0.6	0.2	831	312	118	1270	7.8	10	34[g]
Houston, Texas	San Jacinto River	0.3	0.0	117	30	6	127	6.2	80	
Salt Lake City, Utah	Big Cottonwood Creek	0.2	0.2	170	144	44	291	7.8	5	
Richmond, Va.	James River	0.0	0.8	70	40	6	96	7.3	5	
Seattle, Wash.	Cedar River	0.0	0.2	40	19	0	47	7.4	5	−
Cincinnati, Ohio	Ohio River[d]	0.1–0.4	–	151–377	91–178	–	–	7.2–8.3	–	20[c]
Range		0.0–0.6	0.0–8.3	31–657	12–368	0–206	42–1270	6.2–8.4	0–115	16–78

[a] Average analysis, July 1, 1960–June 30, 1961.
[b] Analyzed by the Passaic Valley Water Commission.
[c] Maximum value of constituents in monthly average of analyses by the city of Philadelphia Water Department during 1960.
[d] Maximum value of constituents in monthly average of analyses by the city of Philadelphia Water Department during 1960.
[e] Maximum constituents in monthly analyses by the city of Cincinnati during 1961.
[f] **Composite of daily samples, June 1–30, 1962.**
[g] Maximum beta activity, raw water, July 1, 1961–June 30, 1962.

Source: C.N. Durfor and E. Becker, *Geological Survey Water Supply Paper*, U.S. Department of the Interior, 1962.

2.3.2 LAKES AND IMPOUNDMENTS

2.3-6 MAJOR CONSTITUENTS OF LAKE WATERS USED FOR WATER SUPPLY

Location	Source of supply	Percent of supply	SiO_2, mg/l	Fe, mg/l	Ca, mg/l	Mg, mg/l	Na, mg/l	K, mg/l	HCO_3, mg/l	CO_3, mg/l	SO_4, mg/l	Cl, mg/l	F, mg/l
Birmingham, Ala.	Inland Lake	10	3.7	0.16	2.0	0.5	2.1	0.9	9.	0	0.8	2.9	0.0
Chicago, Ill.	Lake Michigan	100	1.1	0.03	33	11	3.9	0.7	132	0	20	6.5	0.1
Shreveport, La.	Cross Lake	100	3.7	0.02	10	4.2	24.	2.0	33	0	12	40	0.1
Grand Rapids, Mich.	Lake Michigan	100	1.8	0.01	34	11	4.1	0.9	136	0	21	7.0	0.0
St. Paul, Minn.	Series of lakes	100	2.5	0.02	44	10	6.9	2.8	180	0	14	4.0	0.2
Buffalo, N.Y.	Lake Erie	100	2.0	0.01	38	8.6	9.5	1.4	116	0	23	23	0.1
Rochester, N.Y.	Lake Ontario	21	3.0	0.07[a]	36	9.7	–	–	93	–	17	27	0.3
	Hemlock Lake	79	5.0	0.00	21	5.0	5.2	1.2	62	0	23	7.1	0.0
	Canadice Lake	–	2.9	0.00	12	2.9	3.0	1.0	31	0	19	2.6	0.1
Syracuse, N.Y.	Skaneateles Lake	100	1.7	0.00	34	5.8	1.6	0.9	110	0	17	2.6	0.0
Ft. Worth, Texas	Lake Worth	99	5.8	0.05	44	8.4	20	4.5	153	0	20	32	0.3
Milwaukee, Wisc.	Lake Michigan	100	2.1	0.04	35	4.1	1.0	134	0	19	6.5	0.1
Range			1.1–5.8	0.00–0.16	2.0–44	0.5–11	1.6–24	0.7–4.5	9–180	0	0.8–23	2.6–40	0.0–0.3

Location	Source of supply	NO_3, mg/l	Dissolved solids (residue at 180°C) mg/l	Hardness as $CaCO_3$, mg/l	Noncarbonate hardness as $CaCO_3$ mg/l	Specific conductance μmho at 25°C	pH	Color, units	Beta activity pCi/l
Birmingham, Ala.	Inland Lake	0.4	29	7	0	33	6.8	10	
Chicago, Ill.	Lake Michigan	0.4	153	128	20	259	7.8	1	–
Shreveport, La.	Cross Lake	0.3	142	42	15	231	6.6	10	–
Grand Rapids, Mich.	Lake Michigan	0.5	153	130	18	265	7.8	4	66[b]
St. Paul, Minn.	Series of lakes	0.2	199	153	5	312	7.5	15	22[b]
Buffalo, N.Y.	Lake Erie	0.2	177	131	36	306	8.0	1	
Rochester, N.Y.	Lake Ontario	0.1	233	130	54	–	8.3	–	
	Hemlock Lake	0.7	107	73	22	178	6.9	2	
	Canadice Lake	0.6	62	42	17	104	6.8	2	
Syracuse, N.Y.	Skaneateles Lake	2.2	132	109	19	227	7.1	1	
Ft. Worth, Texas	Lake Worth	0.0	228	144	19	387	7.6		12[b]
Milwaukee, Wisc.	Lake Michigan	0.4	159	129	18	258	8.0	1	
Range		0.0–2.2	29–233	7–153	0–54	33–387	6.6–8.3	1–15	12–66

[a] In solution when collected.

[b] Maximum beta activity, raw water, July 1, 1961 – June 30, 1962.

Source C.N. Durfor and E. Becker, *Geological Survey Water Supply Paper*, U.S. Department of the Interior, 1962.

2.3-7   MINOR CONSTITUENTS OF SURFACE WATERS USED FOR WATER SUPPLY

Lakes

Location	Source of supply	Ag, µg/l	Al, µg/l	B, µg/l	Ba, µg/l	Be, µg/l	Co, µg/l	Cr, µg/l	Cu, µg/l	Fe, µg/l	Li, µg/l	Mn, µg/l	Mo, µg/l
Birmingham, Ala.	Inland Lake	<0.2	72	7.8	80	ND[a]	ND	1.6	24	14	0.75	<2	<0.8
Chicago, Ill.	Lake Michigan	<0.18	46	80	78	ND	ND	0.34	6.0	100	1.3	100	ND
Shreveport, La.	Cross Lake	<0.25	77	15	74	ND	ND	0.52	52	59	0.71	<2.5	<1.7
Grand Rapids, Mich.	Lake Michigan	<0.34	55	51	120	ND	ND	1.3	44	25	1.7	41	1.5
St. Paul, Minn.	Series of lakes	<0.26	41	10	31	ND	ND	2.8	22	28	0.92	3.1	0.77
Buffalo, N.Y.	Lake Erie											10	
Rochester, N.Y.	Lake Ontario		90									0	
	Hemlock Lake		—									0	
	Canadice Lake		—									0	
Syracuse, N.Y.	Skaneateles Lake												
Ft. Worth, Texas	Lake Worth												
Milwaukee, Wisc.	Lake Michigan	<0.24	38	29	36	ND	ND	3.6	57	26	0.43	2.4	1.1
Range		<0.18–<0.34	38–90	7.8–80	31–120	ND	ND	0.34–3.6	6.0–57	14–100	0.43–1.7	0–100	ND–1.5

Location	Source of supply	Ni, µg/l	P, µg/l	Pb, µg/l	Rb, µg/l	Sn, µg/l	Sr, µg/l	Ti, µg/l	V, µg/l	Zn, µg/l	Ga, µg/l	Zr, µg/l	Yb, µg/l	Y, µg/l	Date of sample collection
Birmingham, Ala.	Inland Lake	3.2	ND	14	ND	ND	80	<2	ND	ND					8/29/61
Chicago, Ill.	Lake Michigan	21	ND	11	3.5	ND	85	2.3	ND	ND	ND		ND	ND	8/25/61
Shreveport, La.	Cross Lake	6.2	ND	7.1	ND	ND	84	<2.5	ND	ND					7/27/61
Grand Rapids, Mich.	Lake Michigan	4.1	ND	8.9	<3.4	ND	82	<3.4	ND	<340					8/30/61
St. Paul, Minn.	Series of lakes	5.1	ND	6.4	ND	ND	110	<2.6	ND	<260					7/31/61
Buffalo, N.Y.	Lake Erie														8/22/61
Rochester, N.Y.	Lake Ontario														6/19/61
	Hemlock Lake														6/ 9/61
	Canadice Lake														6/ 9/61
Syracuse, N.Y.	Skaneateles Lake														6/ 7/61
Ft. Worth, Texas	Lake Worth														2/27/62
Milwaukee, Wisc.	Lake Michigan	4.0	ND	26	ND	ND	81	<2.4	ND	ND			ND	ND	8/ 2/61
Range		3.2–21	ND	6.4–26	ND–3.5	ND	80–110	<2–2.3	ND	ND–<340					

[a] ND—looked for but not found.

Source: C.N. Durfor and E. Becker, *Geological Survey Water Supply Paper*, U.S. Department of the Interior, 1962.

2.3-8   MAJOR CONSTITUENTS OF IMPOUNDMENT WATERS USED FOR WATER SUPPLY

Location	Supply	Percent of supply	SiO₂, mg/l	Fe, mg/l	Ca, mg/l	Mg, mg/l	Na, mg/l	K, mg/l	HCO₃, mg/l	CO₃, mg/l	SO₄, mg/l
Mobile, Ala.	Big Creek (Imp)	100	3.2	0.03	0.0	0.6	2.1	0.3	4	0	0.6
San Diego, Cal.	Lake Hodges Reservoir (Imp)[a]	10	11	0.17	87	29	99	5.6	155	0	282
	Lower Otay Reservoir (Imp)[a]	4	12	0.04	27	21	90	6.3	148	21	43
San Francisco, Cal.	Hetch Hetchy Reservoir[b]	72	2.4	0.02	0.7	0.3	1.1	0.2	4.2	0	0.7
Denver, Colo.	Fraser River and Williams Fork	47	8.2	0.06	6.8	1.9	2.0	0.2	24	0	9.0
	South Platte River and Bear Creek	38	6.8	0.08	27	8.0	24	1.8	86	0	38
	South Platte River	15	7.8	0.21	31	12	34	2.0	104	0	47
Bridgeport, Conn.	Easton Reservoir	21	6.5	0.09	7.0	1.1	3.6	1.2	11	0	13
	Hemlocks Reservoir	51	4.0	0.06[d]	6.4	1.6	3.6	1.0	16	0	13
	Trap Falls Reservoir	26	7.7	0.12	7.3	1.6	3.8	1.0	12	0	14
Ft. Wayne, Ind.	St. Joseph River (Imp)	100	8.0	0.56	76	20	9.7	2.6	256	0	67
Baltimore, Md.	Loch Raven Reservoir	55	5.5	0.01	12	3.2	3.7	1.8	43	0	8.4
	Liberty Reservoir	45	5.9	0.00	8.5	2.6	3.6	1.5	25	0	9.0
Boston, Mass.	Quabbin Reservoir[c]	100	1.0	0.04	2.9	0.6	1.7	0.7	5	0	6.9
Flint, Mich.	Flint River (Imp)	100	4.6	0.24	71	24	15	2.4	254	0	60
Newark, N.J.	Wanaque River (Imp)	45	1.9	0.00	8.8	3.4	3.6	0.8	23	0	17
Albany, N.Y.	Hannacrois Creek (Imp)	92	9.3	0.11	11	2.8	1.8	1.0	34	0	13
Charlotte, N.C.	Catawba River (Imp)	100	10	0.00	3.2	1.2	3.8	1.2	20	0	3.8
Greensboro, N.C.	Lake Brandt (Imp)	100	9.4	0.02	4.4	1.7	2.9	2.4	20	0	7.2
Oklahoma City, Okla.	Lake Hefner (Imp)	100	3.2	0.02	58	26	90	7.6	182	0	152
Tulsa, Okla.	Spavinaw Creek (Imp)	100	-	-	-	-	-	-	-	-	-
Providence, R.I.	Scituate Reservoir (Imp)	92	5.0	0.05	2.6	0.6	2.9	0.6	5	0	6.0
Corpus Christi, Texas	Nueces River (Imp)	100	15	0.10	42	8.3	62	7.9	122	0	45
Dallas, Texas	Garza-Little Elm Reservoir (Imp)	67	2.1	0.06	53	6.2	39	4.4	139	0	44
	Grapevine and Garza-Little Elm	26	2.2	0.04	55	7.1	33	4.4	157	0	52
Salt Lake City, Utah	Deer Creek Reservoir (Imp)	24	6.7	0.00	36	11	4.3	1.0	122	0	32
Norfolk, Va.	Lake Wright (Imp)	62	2.3	0.24	12	4.9	14	3.0	29	0	29
	Lake Prince (Imp)	38	5.8	0.62	11	1.6	4.6	2.0	29	0	10
	Lake Burnt Mills		4.0	0.84	3.6	1.4	4.4	1.7	13	0	5.6
Range			1.0–15	0.0–0.84	0.0–76	0.3–29	1.1–99	0.2–7.9	4–256	0–21	0.6–282

[a] Analyzed by the City of San Diego.
[b] Analyzed by the City of San Francisco.
[c] Analyzed by the Metropolitan District Commission.
[d] In solution when collected.

2.3-8   MAJOR CONSTITUENTS OF IMPOUNDMENT WATERS USED FOR WATER SUPPLY   (continued)

Location	Supply	Cl, mg/l	F, mg/l	NO$_3$, mg/l	Dissolved solids (residue at 180°C), mg/l	Hardness as CaCO$_3$, mg/l	Noncarbonate hardness as CaCO$_3$, mg/l	Specific conductance, µmho at 25°C	pH	Color, units
Mobile, Ala.	Big Creek (Imp)	2.3	0.0	0.2	34	2	0	18	5.8	40
San Diego, Cal.	Lake Hodges Reservoir (Imp)[a]	97	0.4	1.0	761	334	207	–	7.9	–
	Lower Otay Reservoir (Imp)[a]	109	0.5	0.4	461	155	0	–	9.1	–
San Francisco, Cal.	Hetch Hetchy Reservoir[b]	1.0	0.0	0.0	10	3	0	8	7.4	1
Denver, Colo.	Fraser River and Williams Fork	1.0	0.3	0.5	40	25	5	58	7.6	0
	South Platte River and Bear Creek	33	1.0	0.5	172	100	30	332	7.5	0
	South Platte River	44	0.9	0.2	232	127	41	448	7.0	0
Bridgeport, Conn.	Easton Reservoir	5.9	0.1	0.8	51	22	13	73	6.6	9
	Hemlocks Reservoir	5.5	0.1	0.4	47	27	14	78	6.9	6
	Trap Falls Reservoir	6.4	0.1	1.0	57	25	15	79	6.6	12
Ft. Wayne, Ind.	St. Joseph River (Imp)	10	0.4	2.6	334	272	62	535	7.6	22
Baltimore, Md.	Loch Raven Reservoir	5.5	0.0	3.3	68	43	8	111	7.0	3
	Liberty Reservoir	5.5	0.1	4.7	59	32	11	91	6.4	5
Boston, Mass.	Quabbin Reservoir[c]	2.1	0.1	0.0	25	10	–	–	6.3	5
Flint, Mich.	Flint River (Imp)	26	0.1	1.1	348	276	68	578	7.4	17
Newark, N.J.	Wanaque River (Imp)	5.6	0.1	0.6	55	36	17	96	6.5	5
Albany, N.Y.	Hannacrois Creek (Imp)	2.1	0.0	0.0	56	39	11	92	7.1	6
Charlotte, N.C.	Catawba River (Imp)	1.5	0.1	0.3	37	13	0	46	7.2	3
Greensboro, N.C.	Lake Brandt (Imp)	3.5	0.1	1.5	49	18	2	59	7.0	10
Oklahoma City, Okla.	Lake Hefner (Imp)	107	0.6	0.5	518	250	101	865	7.8	2
Tulsa, Okla.	Spavinaw Creek (Imp)	–						–		
Providence, R.I.	Scituate Reservoir (Imp)	4.5	0.0	0.0	28	9	5	40	6.1	7
Corpus Christi, Texas	Nueces River (Imp)	96	0.4	0.0	34͜	139	39	597	7.6	0
Dallas, Texas	Garza-Little Elm Reservoir (Imp)	64	0.4	0.5	291	158	44	515	7.5	–
	Grapevine and Garza-Little Elm	42	0.4	1.0	282	166	38	491	7.4	–
Salt Lake City, Utah	Deer Creek Reservoir (Imp)	7.0	0.3	0.3	159	136	36	276	7.8	5
Norfolk, Va.	Lake Wright (Imp)	22	0.1	1.1	123	52	28	174	6.7	35
	Lake Prince (Imp)	9.0	0.1	1.5	73	34	10	102	6.9	32
	Lake Burnt Mills	8.5	0.1	1.0	47	16	6	56	6.7	50
Range		1.0–107	0.0–1.0	0.0–4.7	10–761	2–334	0–207	8–865	5.8–9.1	0–50

[a] Analyzed by the City of San Diego.
[b] Analyzed by the City of San Francisco.
[c] Analyzed by the Metropolitan District Commission.

Note: Imp – Impoundment; Res – Reservoir.

Source: C.N. Durfor and E. Becker, *Geological Survey Water Supply Paper,* U.S. Department of the Interior, 1962.

2.3-9   MINOR CONSTITUENTS OF IMPOUNDMENT WATERS USED FOR WATER SUPPLY

Location	Source of supply	Ag, mg/l	Al, mg/l	B, mg/l	Ba, mg/l	Be, mg/l	Co, mg/l	Cr, mg/l	Cu, mg/l	Fe, mg/l	Li, mg/l
Mobile, Ala.	Big Creek (Imp)[a]										
San Diego, Cal.	Lake Hodges Reservoir (Imp)[a]										
	Lower Otay Reservoir (Imp)[a]										
San Francisco, Cal.	Hetch Hetchy Reservoir[b]										
Denver, Colo.	Fraser River and Williams Fork	ND[c]	23	3.8	6.2	ND	ND	<0.07	16	39	0.24
	South Platte River and Bear Creek	<0.26	190	21	58	ND	<2.6	1.0	31	29	4.2
	South Platte River	ND	86	100	72	ND	ND	ND	1.7	58	5.8
Bridgeport, Conn.	Easton Reservoir		–								
	Hemlocks Reservoir		100								
	Trap Falls Reservoir										
Ft. Wayne, Ind.	St. Joseph River (Imp)	<0.54	180	75	130	ND	ND	2.0	70	110	1.6
Baltimore, Md.	Loch Raven Reservoir	0.74	99	16	39	0.75	0.99	2.6	26	280	0.16
	Liberty Reservoir										
Boston, Mass.	Quabbin Reservoir[e]		150								
Flint, Mich.	Flint River (Imp)	<0.53	690	79	150	ND	ND	0.90	40	850	3.0
Newark, N.J.	Wanaque River (Imp)		17						17		
Albany, N.Y.	Hannacrois Creek (Imp)	0.12[f]	59	3.6	10.	ND	ND	0.72	73	39	0.07
Charlotte, N.C.	Catawba River (Imp)	<0.04[f]		250	8.9	ND	<0.37	0.10	0.78	52	0.06
Greensboro, N.C.	Lake Brandt (Imp)		100								0
Oklahoma City, Okla.	Lake Hefner (Imp)	<0.76	130	170	140	ND	ND	1.9	14	39	20
Tulsa, Okla.	Spavinaw Creek (Imp)	<0.19	330	63	63	ND	<1.9	3.8	8.4	63	0.23
Providence, R.I.	Scituate Reservoir (Imp)										
Corpus Christi, Texas	Nueces River (Imp)										
	Garza-Little Elm Reservoir (Imp)										
Dallas, Texas	Grapevine and Garza-Little Elm										
Salt Lake City, Utah	Dear Creek Reservoir (Imp)	<0.42	80	120	130	ND	ND	0.93	12	80	13
Norfolk, Va.	Lake Wright (Imp)		300								100
	Lake Prince (Imp)		200								0
	Lake Burnt Mills		300								0
Range		ND–0.74	17–690	3.6–250	6.2–150	ND–0.75	ND–0.99	ND–3.8	0.78–73	29–850	0–100

2.3-9   MINOR CONSTITUENTS OF IMPOUNDMENT WATERS USED FOR WATER SUPPLY (continued)

Location	Source of supply	Mn, µg/l	Mo, µg/l	Ni, µg/l	P, µg/l	Pb, µg/l	Rb, µg/l	Sr, µg/l	Ti, µg/l	V, µg/l	Zn, µg/l	Date of sample collection
Mobile, Ala.	Big Creek (Imp)											
San Diego, Cal.	Lake Hodges Reservoir (Imp)[a]	230										4/ 5/62
	Lower Otay Reservoir (Imp)[a]	30										—
San Francisco, Cal.	Hetch Hetchy Reservoir[b]											7/ 8/61
Denver, Colo.	Fraser River and Williams Fork	5.5	<0.20	0.7	ND	ND	1.1	4.1	1.0	ND	ND	8/28/61
	South Platte River and Bear Creek	15	3.4	3.9	ND	<2.6	ND	290.	<2.6	ND	ND	8/28/61
	South Platte River	7.2	1.2	ND	ND	<3.6	ND	240	<3.6	ND	ND	8/29/61
Bridgeport, Conn.	Easton Reservoir	10										6/ 1/62
	Hemlocks Reservoir	10[d]										6/ 1/62
	Trap Falls Reservoir	30[d]										6/ 1/62
Ft. Wayne, Ind.	St. Joseph River (Imp)	22	19	9.6	ND	20	ND	540	5.9	ND	ND	8/22/61
Baltimore, Md.	Loch Raven Reservoir	52	0.83	5.8	ND	43	3.3	32	2.7	<3.0	99	8/22/61
	Liberty Reservoir											8/22/61
Boston, Mass.	Quabbin Reservoir[e]											1/15/62
Flint, Mich.	Flint River (Imp)	170	5.8	15	ND	15	ND	260	11	ND	ND	8/31/61
Newark, N.J.	Wanaque River (Imp)	0										7/31/62
Albany, N.Y.	Hannacrois Creek (Imp)	450	ND	2.6	ND	11	1.2	21	0.6	ND	ND	8/17/61
Charlotte, N.C.	Catawba River (Imp)	11	ND	0.7	ND	4.4	1.2	2.4	3.5	<1.1	<37	8/15/61
Greensboro, N.C.	Lake Brandt (Imp)	40										1/16/62
Oklahoma City, Okla.	Lake Hefner (Imp)	7.6	4.5	7.6	ND	13	ND	530	<7.6	ND	ND	7/28/61
Tulsa, Okla.	Spavinaw Creek (Imp)	23	1.7	9.2	ND	94	ND	46	3.1	<5.8	<190	7/27/61
Providence, R.I.	Scituate Reservoir (Imp)											4/18/62
Corpus Christi, Texas	Nueces River (Imp)	10										1/31/62
Dallas, Texas	Garza-Little Elm Reservoir (Imp)											3/22/62
	Grapevine and Garza-Little Elm											3/22/62
Salt Lake City, Utah	Deer Creek Reservoir (Imp)	420	2.4	18	ND	9.7	5.5	220	<4.2	ND	ND	7/12/61
Norfolk, Va.	Lake Wright (Imp)	20										1/17/62
	Lake Prince (Imp)	0										
	Lake Burnt Mills	0										
Range		0–450	ND–5.8	ND–15	ND	ND–94	ND–5.5	2.4–540	0.6–11	ND–<5.8	ND–99	

[a] Analyzed by City of San Diego.
[b] Analyzed by City of San Francisco.
[c] ND — looked for but not found.
[d] In solution when collected.
[e] Analyzed by the Metropolitan District Commission.
[f] Spectrographic concentrations based on nonacidified residue on evaporation.

Note: Imp — Impoundment; Res — Reservoir.

Source: C.N. Durfor and E. Becker, *Geological Survey Water Supply Paper*, U.S. Department of the Interior, 1962.

2.3-10 DISTRIBUTION OF PESTICIDES IN A LAKE

		Average residues				
Attributes	No. samples	DDT and related compounds	γ BHC	Toxaphene	Dieldrin	Heptachlor
Water (ppb)[a]	82	0.62 (0–22.0)	0.01 (0–0.15)	0.02 (0–0.32)	trace	trace
Particles (ppm)[b]	33	14.74 (1.8–78.0)	0	0	0	0
Bottom Sediment (ppm)[b]	39	4.44 (0.01–94.0)	trace	0.03 (0–0.30)	trace	trace

[a] $\mu g/l.$
[b] mg/kg.

REFERENCE

1. J.O. Keith and E.G. Hunt, *Trans. 31st North Amer. Wildlife Res. Conf.,* 1966, p. 150.

2.3-11 EFFECT OF VARIOUS SOIL TYPES ON BENTHONIC REGION WATER QUALITY

	Type of soil on reservoir floor							
	Sedimented silt, 2–15% organics		Mineral soil, 8–14% organics		Organic soil (muck), 20–50% organics		Organic soil (peat), 50–90% organics	
	Station no.							
Characteristic	3	8	5	6	1	10	7[a]	9[a]
Mean water depth, ft	88	72	50	30	23	16	17	17
Temperature, °C	13.7	13.5	15.0	16.7	15.3	16.9	16.6	16.8
pH	7.0	7.0	7.1	7.0	7.1	7.1	7.0	7.2
DO, ppm	8.4	8.1	8.5	6.5	8.7	8.1	7.3	8.7
DO saturation, percent	84	80	85	69	90	86	78	93
Total alkalinity as $CaCo_3$, ppm	18	18	18	18	18	20	18	18
Color units	7	7	6	7	8	12	12	8
Turbidity units	1	1	1	1	1	2	1	1
Tannin and lignin content, ppm	0.22	0.26	0.25	0.27	0.24	0.31	0.36	0.30
Orthophosphate content, ppm	0.01	0.02	0.02	0.01	0.01	0.02	0.01	0.01
Total phosphate content, ppm	0.10	0.10	0.12	0.11	0.11	0.12	0.11	0.12
Iron content, ppm	0.05	0.07	0.06	0.11	0.09	0.21	0.20	0.08
Specific conductance at 25°C, $\mu mho/cm$	44	44	45	47	44	49	45	44
Ammonia nitrogen content as N, ppm	0.10	0.04	0.11	0.02	0.06	0.05	0.08	0.09
Organic nitrogen content as N, ppm	0.11	0.10	0.12	0.16	0.13	0.12	0.13	0.16
Nitrate nitrogen content as N, ppm	0.05	0.05	0.05	0.05	0.05	0.04	0.04	0.05
Total nitrogen content as N, ppm	0.26	0.19	0.28	0.23	0.24	0.21	0.25	0.30
Potassium content, ppm	0.2	0.3	0.3	0.4	0.3	0.3	0.4	0.3
Chlorophyll A content, ppb	0.32	0.55	0.93	1.05	0.77	0.98	0.83	0.55

[a] Stations 7 and 9 are in the swamp.

Source: Reprinted, with publisher's permission, from *J. Am. Water Works Assn.,* 57(12):1528, © 1965, by the American Water Works Association, Inc.

2.3-12   SUSPENDED SOLIDS IN NATURAL WATERS

Order of size, μ	General debris		Organic matter	
	Inorganic	Organic	Plant	Animal
$10^6 - 10^5$ Size 5	–	Twigs, leaves, uprooted plants, etc	Large detached clumps of filamentous algae or fungi	Fish, amphibians, etc
$10^5 - 10^4$ Size 4	–	Twigs, leaves, uprooted plants, etc plus animal fragments	Large detached clumps of filamentous algae or fungi	Fish, amphibians, worms, etc
$10^4 - 10^3$ Size 3	–	Twigs, leaves, uprooted plants, etc	Larger filamentous algae (*Characeae, Rhodophyceae*) fragments	Insects, mollusks, also ova, and larvae of larger animals (mollusks, crustaceans, fish, etc) some worms, bryozoa
$10^3 - 10^2$ Size 2	Sand particles	Animal and vegetable fragments	Large diatoms and desmids, etc	Crustaceans, hydroids, many sarcodina and infusoria
$10^2 - 10^1$ Size 1	Fine sand and silt	Animal and vegetable fragments, e.g., sponge spicules	Most algae	Smaller crustaceans, and worms, ova, motile reproductive units, etc, many protozoans
$10^1 - 10^0$ Size 0	Fine silt or clay particles	Animal and vegetable fragments, e.g., sponge spicules	Small algal forms (e.g., *Chlorella*), swarmers, spores, most bacteria, etc	Motile reproductive units or cells

Source: L.D. Bowen, J.W. Lovell, Jr., and D.K.B. Thistlewayte, 3rd Int. Conf. Water Pollut. Res., Water Pollution Control Federation, Alexandria, Va., 1966. With permission.

2.3.3   GROUND WATER

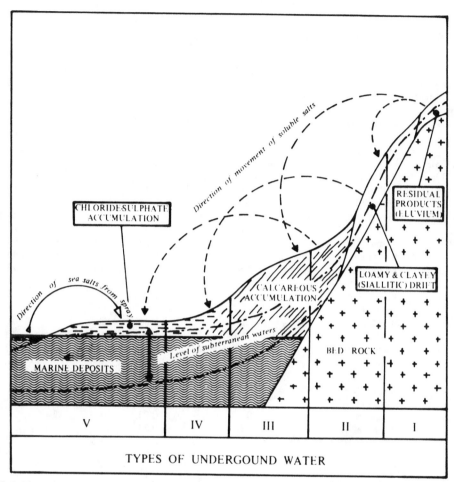

2.3-13   Diagram showing the types of *subterranean waters* in relation to the products of weathering.

Source: I.I. Chebatorev, *Water Water Eng., 54*(10):146, 1950. With permission.

2.3-14   GENETIC TYPES OF SUBTERRANEAN WATERS IN RELATION TO THEIR SALINITY

Diagram symbols	Types of waters[a]	Classes	Ratio of carbonate to chloride	Practical utility
I	Bicarbonate (alkaline) water	Reversed	>3.0	Excellent in quality
II	Bicarbonate-chloride (alkaline-saline) water	Semireversed	3.0−1.5	Good in quality
III	Chloride-bicarbonate (saline-alkaline) water	Mixed	1.5−0.2	Transitional type; usually of good quality, but often by protracted pumping deteriorates in quality
IV	Chloride-sulfate (saline) water	Brackish	<0.2	Doubtful in quality; sometimes potable
V	Chloride (saline) water	Brine	<0.2	Usually poor in quality

[a]   See 2.3-13.

Source: I.I. Chebetarev, *Water Water Eng., 54*(10):146, 1950. With permission.

2.3-15   MAJOR CONSTITUENTS OF GROUND WATERS
USED FOR WATER SUPPLY

Wells and Infiltration Galleries

Location	Source of supply	Percent of supply	SiO_2, mg/l	Fe, mg/l	Ca, mg/l	Mg, mg/l	Na, mg/l	K, mg/l
Montgomery, Ala.	18 wells–Court St.	39	12	0.08	3.5	0.3	51	1.3
	31 wells–Day St.	61	17	0.05	16	1.0	51	1.6
Fresno, Cal.	Composite of wells	100	68	0.00	15	11	17	4.4
Long Beach, Cal.	Well water[a]	–	20	0.04	13	1.0	71	1.0
Jacksonville, Fla.	Well water	100	26	0.15	60	22	14	1.6
Miami, Fla.	Hialeah well field	40	9.5	0.33	88	5.0	22	1.6
	Orr well field	60	4.3	0.00	82	1.8	7.8	0.4
St. Petersburg, Fla.	Cosme well field	100	14.	0.01	70	3.3	6.1	0.7
Savannah, Ga.	Well No. 2	100	53	0.02	28	8.3	9.0	1.7
Honolulu, Haw.	Kaimuki P.S.	13	39	0.74	5.9	6.6	64	2.9
	Beretania P.S.	23	41	0.03	8.7	9.2	35	3.2
	Kalihi underground station	21	42	0.03	3.2	15	34	3.2
Des Moines, Iowa	Infiltration galleries	75	19	0.08	94	33	13	2.8
Topeka, Kans.	Well 2	1	18	0.38	112	18	66	6.4
	Well 3	1	17	0.03	121	21	70	7.0
Wichita, Kans.	Wells in Equus beds	100	22	0.07	66	10	60	3.0
Lincoln, Neb.	Ashland wells– composite	96	34	0.00	58	9.4	25	8.4
	Lincoln wells[b]	4	36	0.04	54	16	25	4.5
Dayton, Ohio	Well water[c]	100	–	0.23	89	33	11	
Memphis, Tenn.	Allen well field	28	8.3	0.72	12	6.1	7.5	0.7
	Sheahan well field	28	8.7	0.75	7.5	3.0	12	0.7
	McCord well field	16	6.8	1.2	7.6	5.1	6.3	0.7
	Parkway well field	28	8.5	0.74	8.4	3.8	17	0.9
Amarillo, Texas	Palo Duro well field		56	0.02	44	39	19	3.8
	McDonald well No. 2		59	0.00	41	37	17	4.9
	Bush well No. 4		72	0.02	39	43	29	6.2
	Westex well No. 3		71	0.01	44	28	27	5.6
	Well 6, section 49		64	0.00	32	34	35	5.6
Lubbock, Texas	City well No. 62, NW well field	7	55	0.00	54	49	129	16
	Well No. 102, shallow water well field	10	43	0.00	98	120	102	30
San Antonio, Texas	Bexar Metropolitan Water District wells		12	0.02	64	17	8.7	1.0
Ranges			4.3–72	0.00–1.2	3.2–121	0.3–120	6.1–129	0.4–30

[a] Average analysis by the City of Long Beach, July 1, 1960–June 30, 1961.
[b] Analyzed by the City of Lincoln.
[c] Average analysis by the City of Dayton of monthly composite, 1960.

2.3-15   MAJOR CONSTITUENTS OF GROUND WATERS
USED FOR WATER SUPPLY (continued)

Wells and Infiltration Galleries

HCO$_3$, mg/l	CO$_3$, mg/l	SO$_4$, mg/l	Cl, mg/l	F, mg/l	NO$_3$, mg/l	Dissolved solids (residue at 180°C), mg/l	Hardness as CaCO$_3$, mg/l
118	0	9.2	8.8	0.4	1.9	154	10
149	0	9.6	13	0.4	2.5	203	44
118	0	6.0	5.8	0.1	13	221	82
182	5	13	20	0.4	–	229	37
187	0	87	17	0.7	0.0	373	240
272	0	16	30	0.3	0.1	333	240
248	0	15	12	0.1	0.6	263	212
232	0	0.8	10	0.1	0.7	229	188
133	0	7.2	6.0	0.5	0.0	184	104
81	0	14	76	0.0	0.7	251	42
78	0	8.6	45	0.1	0.0	184	60
62	0	9.0	63	0.0	1.0	215	70
340	0	89	11	0.3	12	464	371
312	0	131	82	0.4	0.5	611	354
338	0	145	87	0.3	0.3	663	390
252	0	68	41	0.4	0.7	404	206
199	0	68	8.5	0.4	0.0	317	183
248	0	59	15	0.3	5.0	333	162
326	0	83	19	–	–	439	356
78	0	3.8	3.0	0.4	1.2	87	55
61	0	3.0	4.2	0.1	0.9	80	31
55	0	5.2	3.0	0.3	0.8	68	40
82	0	3.4	2.0	0.3	0.9	97	36
316	0	37	5.8	4.0	2.8	368	270
305	0	30	3.8	3.3	3.2	352	254
296	0	78	7.8	2.8	4.3	428	274
292	0	31	6.2	3.3	2.0	362	225
313	0	24	6.8	3.5	4.9	364	220
364	0	189	92	3.9	17	794	336
357	0	572	64	7.0	9.4	1,220	738
236	0	25	17	0.4	3.8	278	230
55-364	0–5	0.8–572	2.0–92	0.0–70	0.0–17	68–1,200	10–738

Source: C.N. Durfor and E. Becker, *Geological Survey Water Supply Paper*, U.S. Department of the Interior, 1965.

2.3-16 MINOR CONSTITUENTS OF GROUND WATERS USED FOR WATER SUPPLY

Wells and Infiltration Galleries

Location	Noncarbonate hardness as CaCO$_3$, mg/l	Specific conductance µmho at 25°C	pH	Color, units	Beta activity, pCi/l	Radium (Ra), pCi/l	Uranium (U), µg/l	Silver (Ag), µg/l	Aluminum (Al), µg/l	Boron (B), µg/l	Barium (Ba), µg/l	Chromium (Cr), µg/l	Copper (Cu), µg/l
Montgomery, Ala.	0	243	7.2	0	–	–	–	ND[a]	83	110	67	ND	4.3
Fresno, Cal.	0	305	7.3	0									
Long Beach, Cal.	0	249	7.5	3	12	<0.1	0.5	ND	6.2	17	34	0.60	1.0
Jacksonville, Fla.	0	–	8.7	76	–	–	–	ND	11	34	15	0.71	0.85
Miami, Fla.	87	504	7.9	5	–	–	–	<0.50	12	71	21	ND	2.2
St. Petersburg, Fla.	17	530	7.8	45	–	–	–	ND	6.9	8.5	5.7	ND	1.4
Savannah, Ga.	9	444	7.7	5	–	–	–	ND	13	13	14	<0.36	<0.8
Honolulu, Haw.	0	382	7.6	10	–	–	–	<0.26	2.9	10	13	0.31	1.8
	0	236	8.0	5	1.8	<0.1	<0.1	<0.34	6.7	24	5.4	1.1	1.3
	0	388	6.7	–	2.6	<0.1	<0.1	<0.26	7.9	28	4.6	1.1	3.1
	0	289	7.0	–	4.6	<0.1	<0.1	<0.29	5.5	35	13	1.1	8.4
Des Moines, Iowa	19	312	6.6	3	–					80			
Topeka, Kans.	92	706	7.8	3						110			
	98	958	7.3	4						160			
Wichita, Kans.	113	1,030	7.4	1						20			
Lincoln, Neb.	0	651	7.9	2						50			
	20	475	7.4	–						0			
Dayton, Ohio	0	–	–	–									
Memphis, Tenn.	89	137	7.6	5									
	0	114	6.8	5									
	0	108	6.9	5									
	0	138	6.7	5									
Amarillo, Texas	12	565	7.4	0									
	4	536	7.1	0									
	32	625	7.1	0									
	0	529	7.1	0									
	0	547	7.2	0									
Lubbock, Texas	38	1,200	7.1	0									
	446	1,660	7.0	0									
San Antonio, Texas	36	461	6.8	0	130	1.9	250	1.5	32	500	30	<1.5	15
Range	0–446	108–1,660	6.7–8.7	0–76	1.8–130	<0.1–1.9	<0.1–250	ND–1.5	2.9–83	0–500	4.6–67	ND–1.1	<0.8–15

2.3-16 MINOR CONSTITUENTS OF GROUND WATERS USED FOR WATER SUPPLY (continued)

Wells and Infiltration Galleries

Location	Iron (Fe), μg/l	Lithium (Li), μg/l	Magnesium (Mn), μg/l	Molybdenum (Mo), μg/l	Nickel (Ni), μg/l	Lead (Pb), μg/l	Rubidium (Rb), μg/l	Strontium (Sr), μg/l	Titanium (Ti), μg/l	Vanadium (V), μg/l	Zinc (Zn), μg/l	Date of sample collection
Montgomery, Ala.	45	0.12	48	ND	ND	ND	3.3	93	<2.4	ND	ND	8/29/61
Fresno, Cal.	4.8	0.62	ND	<0.9	ND	5.7	2.8	19	ND	12	ND	7/19/61
Long Beach, Cal.												7/ 1/60 –
Jacksonville, Fla.	1.1	ND	ND	ND	<4.7	ND	ND	400	ND	ND	<470	6/30/61
Miami, Fla.	230	0.50	35	2.8	ND	31	ND	490	<5.0	ND	ND	8/29/61
St. Petersburg, Fla.	450	<0.41	<4.1	ND	<4.1	4.1	ND	280	ND	ND	ND	8/23/61
Savannah, Ga.	79	0.7	<3.6	ND	<3.6	ND	ND	97	ND	ND	ND	8/21/61
Honolulu, Haw.	15	1.4	<2.6	ND	ND	<2.6	3.7	39	ND	ND	<262	8/30/61
	15	0.74	ND	<0.77	<2.6	0.6	<3.4	50	<1.0	11	ND	8/29/62
	15	0.56	ND	ND	<2.9	4.9	3.3	67	<0.8	7.9	ND	8/29/62
	21	<0.29	3.7			5.8	3.7	60	1.0	12	ND	8/29/62
Des Moines, Iowa			60									1/22/62
Topeka, Kans.			180									3/ 9/62
			150									3/ 9/62
Wichita, Kans.			340									2/13/62
Lincoln, Neb.			0									1/23/62
Dayton, Ohio			10									7/10/61
			20									1960
Memphis, Tenn.			10									9/15/61
			10									9/15/61
			0									9/15/61
			10									9/15/61
Amarillo, Texas			0									2/12/62
			10									2/12/62
			0									2/12/62
												2/12/62
												2/12/62
Lubbock, Texas	6,600		10									2/14/62
San Antonio, Texas		96	75	68		38	ND	6,300	ND	74	ND	2/14/62
												2/ 8/62
Range	1.1–6,600	ND–96	ND–340	ND–68	ND–<15	ND–38	ND–3.7	19–6,300	ND–<5	ND–74	ND–<470	

Note: No data was found for beryllium, cobalt, phosphorus, tin, gallium, zirconium, ytterbium, and yttrium.

[a] ND, looked for but not found.

Source: C.N. Durfor and E. Becker, *Geological Water Supply Paper,* U.S. Department of the Interior, 1965.

2.3-17 APPROXIMATE PERMEABILITIES OF VARIOUS MATERIALS

Typical of material and grain size range, in.	Typical effective size D_{90}		Estimated coefficient of permeability, gal/day ft^2		
	mm	in.	Avg	General range	Max
Clay and silt <0.0029		0.001	10	0 to 25	50
Sand					
Fine (0.029–0.010)	0.10	0.004	175[a]	50 to 350	800
Medium (0.010–0.024)	0.25	0.010	600[a]	100 to 1,200	3,000
Coarse (0.024–0.079)	0.50 to 1.00	0.020 to 0.040	1,500[a]	300 to 2,000	5,000
Gravel					
Fine (0.079–0.375)	2	0.080	2,500[a]	1,000 to 5,000	10,000
Medium (0.257–0.500)	5	0.200	3,500[a]	3,000 to 10,000	20,000
Medium sand and coarse sand mixture (average)	1,000 (U.C. 2.5)				
Coarse sand and fine gravel mixture (average)	1,500 (U.C. 2.5)				

[a] Material is clean, well rounded, and generally uniformly graded (U.C. of 1.5 assumed).

Source: Reprinted, with publisher's permission, from *J. Am. Water Works Assn.*, *59*(10):1292, © 1967 by the American Water Works Association, Inc.

REFERENCE

1. C.F. Tolman, *Ground Water*, McGraw-Hill, New York, 1937, p. 387.

2.3.4 ESTUARINE AND MARINE WATERS

2.3-18 CHEMICALS IN SEA WATER

Constituent	Concentration,[1] mg/kg	Concentration,[2] mg/l	Residence time, years[2],[a]
Chlorine	18,980	19,000	
Sodium	10,560	10,600	260,000,000
Sulfate	2,560		
Magnesium	1,272	1,350	45,000,000
Sulfur	885		
Calcium	400	400	8,000,000
Potassium	380	380	11,000,000
Bicarbonate	142		
Bromine	65	65	
Carbon		28	
Oxygen		8	
Strontium	13	8	19,000,000
Boron	4.6	4.6	
Silicon	0.04−8.6[b]	3.0	8,000
Fluorine	1.4	1.3	
Nitrogen	0.03−0.9	0.8	
Argon		0.6	
Lithium	0.1	0.17	20,000,000
Rubidium	0.2	0.13	270,000
Phosphorus	0.001−0.10	0.07	
Iodine	0.05	0.06	
Barium	0.05	0.03	84,000
Indium		0.02	
Aluminum	0.16−1.9	0.01	100
Iron	0.002−0.02	0.01	140
Zinc	0.005−0.014	0.01	180,000
Molybdenum		0.01	500,000
Selenium	0.004	0.004	
Copper	0.001−0.09	0.003	50,000
Arsenic	0.003−0.024	0.003	
Tin	0.003	0.003	100,000
Lead	0.004−0.005	0.003	2,000
Uranium	0.00015−0.0016	0.003	500,000
Vanadium		0.002	10,000
Manganese	0.001−0.01	0.002	1,400
Titanium	Trace	0.001	160
Thorium	≤0.0005	0.0007	350
Cobalt	0.0001	0.0005	18,000
Nickel	0.0001−0.0005	0.0005	18,000
Gallium		0.0005	1,400
Cesium	~0.002	0.0005	40,000
Antimony		0.0005	350,000
Cerium		0.0004	80
Yttrium		0.0003	7,500
Neon		0.0003	
Krypton		0.0003	
Lanthanum		0.0003	440
Silver		0.0003	2,100,000
Bismuth		0.0002	45,000
Cadmium		0.0001	500,000
Tungsten		0.0001	1,000
Germanium		0.0001	7,000
Xenon		0.0001	

2.3-18   CHEMICALS IN SEA WATER (continued)

Constituent	Concentration,[1] mg/kg	Concentration,[2] mg/l	Residence time, years[2],[a]
Chromium		0.00005	350
Beryllium		0.00005	150
Scandium		0.00004	5,600
Mercury	0.0003	0.00003	42,000
Niobium		0.00001	300
Thallium		0.00001	
Neodymium		0.000009	270
Helium		0.000005	
Gold		0.000004	560,000
Praeseodymium		0.000002	320
Gadolinium		0.000002	260
Dysprosium		0.000002	460
Erbium		0.000002	690
Ytterbium		0.000002	530
Samarium		0.000002	180
Holmium		0.0000007	530
Europium		0.0000004	300
Radium[3]	8×10^{-11}		

[a] Calculated residence time = ratio of amount of element present in the sea to either the gross contribution by rivers or the estimated sedimentation rate.
[b] Silicate.

REFERENCES

1. From U.S. Geological Survey.
2. B.S. Newell, *Australian Fisheries*, Reprint, June 1969.
3. H.U. Sverdup *et al.*, *The Oceans*, Prentice Hall, Inc., 1942.

2.3.5 WASTE WATERS

2.3-19 WATER QUALITY OF RECYCLED WATERS

Reverse Osmosis

Component	Fort Stockton, Texas[a]			Midland, Texas[d]		Kermit, Texas[e]			River Valley Golf Course San Diego, Calif[f]
	Feed, ppm[b]	Product, ppm	Brine, ppm	Feed, ppm	Product, ppm	Feed, ppm	Product, ppm	Brine, ppm	Feed, ppm
Total dissolved solids, calculated	1,690[c]	29[c]	6,600[c]	4,270[c]	30[c]	6,565	348	12,557	4,732
Ca	171	1	660	435	1	276	0	695	395
Mg	63		246	268		156	5	328	265
Na	278	9	1,100	530	9	1,410	124	3,199	925
K									25
Fe	0.10	<0.02	<0.02	0.06	<0.02				8.2
Mn	<0.05	<0.05	<0.05	<0.05	<0.05				3.8
CO_3	0	0	0						
HCO_3	243	2	880	227	1	561	23	1,293	450
SO_2	468	4	1,840	1,920	3	1,396	11	3,337	550
Cl	452	11	1,780	840	11	1,760	185	3,785	2,300
F	0.9	0.1	8.0	5.0	<0.1				0.50
NO_3	13.5	2.0	45	40	5				0.84
Total alkalinity as $CaCO_3$	199	2	720	186	1				370
Total hardness as $CaCO_3$	690	2	2,670	2,190	3	1,333	20	3,085	2,070
pH	7.9	5.9	7.5	7.7	5.9	8.9	8.9	8.7	7.2
OH									0
Boron									0.40
Silicon, SiO_2									30

[a] Samples taken February 11, 1966. Feed water source: municipal water supply, Fort Stockton, Texas. Chemical analyses performed by Texas State Department of Health Laboratories.

[b] Taken from feed one-test one analysis.

[c] Includes all of the bicarbonates.

[d] Samples taken January 16, 1966. Feed water source: Cold Park well No. 9, reservoir, Midland, Texas. Chemical analysis performed by Texas State Department of Health Laboratories.

[e] Feed source was Shell Oil Co. deep well at Kermit, Texas. The analysis was performed by Martin Water Laboratories, Monahaus, Texas, February 19, 1966.

[f] Sample source: Unacidified feed. Collected in November 4, 1966. Conductivity, 7750 μmho/cm at 25°C.

Source: Reprinted, with publisher's permission from *J. Am. Water Works Assn.*, 59(12):1527, © 1967, by the American Water Works Association, Inc.

2.4 NEEDS

2.4-1 ESTIMATED WATER USE IN THE U.S.

Daily Average in Billions of Gallons

							Self supplied uses					
	Total water use		Irrigation[a]		Public water utilities		Rural domestic[b]		Industrial and miscellaneous[c]		Steam electric utilities	
Year	Total	Ground	Total	Ground	Total	Ground	Total	Ground	Total	Ground	Total	Ground
1900	40.19	7.28	20.19	2.22	3.00	1.05	2.00	1.60	10.00	2.40	5.00	0.01
1910	66.44	11.68	39.04	5.27	4.70	1.49	2.20	1.76	14.00	3.15	6.50	0.01
1920	91.54	15.78	55.94	8.17	6.00	1.79	2.40	1.94	18.00	3.87	9.20	0.01
1930	110.50	18.18	60.20	9.09	8.00	2.30	2.90	2.40	21.00	4.37	18.40	0.02
1940	136.43	22.56	71.03	11.22	10.10	2.82	3.10	2.64	29.00	5.86	23.20	0.02
1950	202.70	35.19	100.00	19.80	14.10	3.78	4.60	4.09	38.10	7.47	45.90	0.05
1960	322.90	58.17	135.00	35.24	22.00	5.68	6.00	5.58	61.20	11.57	98.70	0.10
1965	269.62	48.57	110.85	30.04	23.74	5.96	4.08	3.86	46.41	8.63	84.54	0.08
1970	327.30	54.27	119.18	33.13	27.03	6.65	4.34	4.13	55.95	10.24	120.80	0.12
1971	338.84	55.27	120.85	33.64	27.69	6.77	4.39	4.18	57.86	10.55	128.05	0.13
1975	384.95	59.29	127.51	35.70	30.31	7.27	4.59	4.37	65.49	11.79	157.05	0.16
1980	442.63	63.98	135.85	38.17	33.60	7.73	4.85	4.61	75.03	13.28	193.30	0.19

[a] Total take, including delivery losses but not including reservoir evaporation.
[b] Rural farm and nonfarm household and garden use, and water for farm stock and dairies.
[c] For 1900–1960, includes manufacturing and mineral industries, rural commercial industries, air conditioning, resorts, hotels, motels, military, and other state and federal agencies, and other miscellaneous uses; thereafter, includes manufacturing, mining and mineral processing, ordnance, and construction.

Source: U.S. Bureau of Domestic Commerce. 1900–1960 based principally on committee prints, *Water Resources Activities in the United States,* for the Senate Committee on National Water Resources, U.S. Senate, 1960; thereafter, based on the first *National Assessment of the Water Resources Council,* 1968.

2.4-2 ESTIMATED WATER CONSUMPTION AT DIFFERENT TYPES OF ESTABLISHMENTS

Type of establishment	gal/cap day	Type of establishment	gal/cap day
Dwelling units, residential		Institutions	
Private dwellings on individual wells or metered supply	50–75	Average type	75–125
Apartment houses on individual wells	75–100	Hospitals	150–250
Private dwellings on public water supply, unmetered	100–200	Schools	
Apartment houses on public water supply, unmetered	100–200	Day, with no cafeteria, gymnasium, or showers	15[1]
Subdivision dwelling on individual well, or metered supply, per bedroom	150	Day, with cafeteria or lunch room, no gymnasiums or showers	10–20[1]
Subdivision dwelling on public water supply, unmetered, per bedroom	200	Day, with cafeteria and showers	15–25[1]
		Boarding	75–100[1]
Dwelling units, treatment		Theatres	
Hotels	50–100	Indoor, per seat, two showings per day	3
Boarding houses	50	Outdoor, including food stand, per car	
Lodging houses and tourist homes	40	(3 1/3 persons)	3–5
Motels, without kitchens, per unit	100–150		
		Automobile service stations	
Camps		Per vehicle served	10
Pioneer type	25	Per set of pumps	500
Children's, central toilet and bath	40–50		
Day, no meals	15	Stores	
Luxury, private bath	100–150[1]	First 25 ft frontage	450
Labor	35–50	Each additional 25 ft frontage	400
Trailer with private toilet and bath, per unit (2½ persons)[a]	125–150	Country clubs	
		Resident type	100
Restaurants (including toilet)		Transient type, serving meals	17–25
Average	7–10	Offices	10–15
Kitchen wastes only	2½–3	Factories, sanitary wastes,	
Short order	4	per shift	15–35
Short order, paper service	1–2	Self-service laundry, per machine	250–500
Bars and cocktail lounges	2	Bowling alleys, per alley	200
Average type, per seat	35	Swimming pools and	
Average type, 24 hr, per seat	50	beaches, toilet and shower	10–15
Tavern, per seat	20	Picnic parks, with flush toilets	5–10
Service area, per counter seat (toll road)	350	Fairgrounds (based on daily attendance)	1
Service area, per table seat (toll road)	150	Assembly halls, per seat	2
		Airport, per passenger	3–5

[a] Add 125 gallons per trailer space for lawn sprinkling, car washing, leakage, etc.

Note: Water under pressure, flush toilets, and wash basins are assumed provided unless otherwise indicated. These figures are offered as a guide; they should not be used blindly. Add for any continuous flows and industrial usages. Figures are flows per capita per day, unless otherwise stated.

Source: J.A. Salvato, *Public Works, 91*(5), 1960. With permission.

REFERENCE

1. *Report of Public Health Service Committee on Plumbing Standards,* U.S. Department of Health, Education, and Welfare, Public Health Service, 1963.

2.4.1 DOMESTIC

2.4-3 VARIATIONS IN PER CAPITA WATER DEMAND

Ratio, period demand to yearly avg

Year	Yearly avg, gal/cap day	Max month	Max week	Max day	Max hr
1900	60–85	1.15–1.30		1.50–2.00	
1927	50–400	1.25	1.35	1.50	
1940	110			1.50	2.10–2.25
1947	100–120	1.25	1.50	1.75	2.62–3.06
1948	110			1.50–2.00	2.00–3.00
1949	110	1.40		1.50–1.80	2.50–3.50
1953	135	1.28	1.48	1.80	2.70–3.25

Source: Reprinted, with publisher's permission, from *J. Am. Water Works Assn.*, *53*(4):459, © 1961, by the American Water Works Association, Inc.,

2.4-4 PEAK WATER DEMANDS IN RESIDENTIAL AREAS

Lot size, ft²	Type of home				
	Group house 2,000–2,400	Detached dwelling 5,000–7,000	Detached dwelling 9,000–12,000	Detached dwelling 15,000–25,000	Detached dwelling 40,000 and over
No. of observations	63	51	67	70	61
Max day, avg rate, gpd/service	327	228	556	1,170	2,320
Peak hour, gpd/service	625	605	1,050	2,630	5,400
Ratio, peak hour to max day	1.91:1	2.67:1	1.89:1	2.25:1	2.33;1
Ratio, peak hour to avg day	3.84:1	3.30:1	4.63:1	7.90:1	10.30:1
Max day, gpcd	93	64	163	278	564
Avg day, gpcd	48	51	67	80	117
Dwelling units/acre[a]	9.0	4.2	3.0	1.6	0.8
Persons/acre	30.6	16.8	10.50	7.46	3.33
Max hr, gpm/acre	3.8	1.8	2.2	2.9	2.7
Max day, gpm/acre	2.9	0.7	1.2	1.3	1.3

[a] Adjusted by taking into account such areas as streets, public buildings, parks, institutions, and drainage courses.

REFERENCE

1. J.B. Wolfe, *J. Am. Water Works Assn.*, *53*:1251, 1961.

2.4-5   DOMESTIC WATER USE DATA

Item	Daily family use, gal[1]	Daily per capita use, gal[1]	Percent[2]	Percent[2]
Drinking and kitchen	8	2.0	2.3	11
Dishwasher	15	3.75	4.3	
Toilet	96	24.0	27.8	41
Bathing	80	20.0	23.2	37
Laundering	34	8.5	9.8	4
Auto washing	10	2.5	2.8	1
Lawn watering	100	25.0	28.9	3
Garbage disposal unit	3	0.75	0.8	—
Cleaning	—	—	—	3
Total	346	86.50	99.9	100
All uses except toilet and lawn watering	150	37.5		

REFERENCES

1. G.W. Reid, *Southwest Water Works J.*, *46*:18, 1965.
2. Akron Bureau of Water Supply, Akron City Water Works, Akron, Ohio.

2.4-6   APPLIANCE WATER USE

Equipment	gpcd, max
Automatic washer	58.3
Wringer washer	50.2
Disposer, food	81
Dishwasher	91.3
Air conditioner (water cooling)	2.160−2.880[a]
Wading pools	58.5

[a] Per ton of refrigerator capacity.

REFERENCE

1. D.F. Dunn and T.E. Larson, *J. Am. Water Works Assn.*, *55*:441, 1963.

2.4-7 WATER REQUIREMENTS FOR DOMESTIC SERVICE,
PUBLIC BUILDINGS, SCHOOLS, AND CAMPS

Domestic fixtures	
Fill lavatory	2 gal
Fill bath tub	30 gal
Shower bath	30–60 gal
Flush toilet	6 gal
Dishwashing machine	3 gal per load
Automatic laundry machine	30–50 gal per load
Lawn sprinkler	120 gal/hr
1/2 in. hose and nozzle	240–300 gal/hr
5/8 in. hose and nozzle	270–330 gal/hr
3/4 in. hose and nozzle	300–360 gal/hr
Private homes	
For each member of family including kitchen, laundry, and bath	40 gal/day
Public buildings	gal/hr per fixture
Hotel	50
Apartment houses	20
Hospitals	25
Office buildings	40
Mercantile buildings	35
1½ in. fire hose and nozzle	2,400
Day school	50
Schools and camps	gal/cap day
Schools	15–17
Camps	
with hot and cold running water, kitchen, laundry, shower, bath, and flush toilets	45
with flush toilets	25
without running water or flush toilets	5

Source: *The Water Encyclopedia,* D.K. Todd, Ed., Water Information Center, Port Washington, N.Y., 1970. With permission.

2.4.2   COMMERCIAL AND INDUSTRIAL

2.4-8   COMMERCIAL WATER SUPPLY REQUIREMENTS

Establishment	Average demand, gpd	Rate of max hourly demand		Hour of peak occurrence	Ratio of max hourly commercial demand to R-40 demand[a]
		gal/hr	gal/day		
Gasoline service station	636	522	12,500	7–9 p.m.[b]	2.3:1
Washmobile[c]	9,370	3,140	75,500	11–12 a.m.	14:1
Laundromat[d]	3,670	522	12,500	12–1 p.m.	2.3:1
Commercial laundry[e]	3,570	825	19,800	10–12 a.m.	3.7:1
Motel (110 units)	5,500	762	18,250	8–9 a.m.	3.4:1
Apartment (22 units)	4,300	486	11,650	5–6 p.m.	2.2:1

[a] Lot type R-40 peak hourly demand for single service is 5,400 gpd.
[b] Erratic, with lower peaks occurring 1–2 p.m. and 5–7 p.m.
[c] Cars washed at rate of 24 per hour. All facilities–steam cleaning of tires, automatic rinsing–available.
[d] Ten washers. Open 8 a.m.–8 p.m.
[e] Facilities equivalent to ten washers. Open 8 a.m.–8 p.m.

Source: Reprinted, with publisher's permission, from *J. Am. Water Works Assn., 53*(10):1251, © 1961, by the American Water Works Association, Inc.

2.4-9   AIRPORT WATER SUPPLY REQUIREMENTS

Airport	Year	Av for year[a]
Dulles International, Va.	1965	36.5
Kennedy International, N.Y.	1966	39.0
Tulsa International, Okla.	1965	41.2
	1966	34.2
O'Hare International, Chicago, Ill.	1965	23.0
	1966	22.5
Kansas City Municipal, Mo.	1965	30.0
	1966	26.4
Love Field, Dallas, Tex.	1964	54.5
	1965	40.2
	1966	32.2

[a]   Data are in gal per passenger day. Gal × 3.785 = liters.

REFERENCE

1.   E.H. Bryant *et al., J. Water Pollut. Control Fed., 41*:772, 1969.

2.4-10 REQUIRED FIRE FLOW AND STANDARD HYDRANT SPACING

High Value Districts

Population, 1,000's	Flow required, gpm	Duration, hr	Avg area per hydrant, 1,000 ft²	Population, 1,000's	Flow required, gpm	Duration, hr	Avg area per hydrant, 1,000 ft²
1	1,000	4	120	22	4,500	10	85
1.5	1,250	5		27	5,000	10	
2	1,500	6		33	5,500	10	
3	1,750	7		40	6,000	10	80
4	2,000	8	110	55	7,000	10	70
5	2,250	9		75	8,000	10	60
8	2,500	10		95	9,000	10	55
10	3,000	10	100	120	10,000	10	48
13	3,500	10		150	11,000	10	43
17	4,000	10	90	200[a]	12,000	10	40

[a] Over 200,000 population, 12,000 gpm, with 2,000–8,000 gpm additional for a second fire, for a 10 hr duration.

Source: Reprinted, with publisher's permission, from *J. Am. Water Works Assn., 53*(4):458, © 1961, by the American Water Works Association, Inc.

2.4-11 INDUSTRIAL REQUIREMENTS FOR WATER

Product	Unit produced	Water required, gal/unit	Ref.
Airplane	to test	50,000–125,000	1
Alcohol	gal	100	1
		120[a]	6
Aluminum	lb	3.15[b]	6
		160	1,7
Aviation gas	gal	7–10	1
Brewing			
Beer	1 bbl	470	1
		750	7
		298–2,500	8
Whiskey	gal	80	1
Buildings			
Office	person day	27–45	1
Hospital	bed day	135–350	1,6
Hotels	guest room day	300–525	1
Laundries			
Commercial	lb work	4.3–5.7	1
Institutional	lb work	3.	1
Restaurants	meal	0.5–4.0	1
Butadiene	lb	160	1
		10–300[c]	6
Canning			
Apricots	100 cases No. 2 cans	8,000	1
Asparagus	100 cases No. 2 cans	7,000	1
Beans			
Green	100 cases No. 2 cans	3,500	1
Lima	100 cases No. 2 cans	25,000	1
Pork and beans	100 cases No. 2 cans	3,500	
Beets	100 cases No. 2 cans	2,500	1
Corn	100 cases No. 2 cans	2,500	1
Grapefruit			
Juice	100 cases No. 2 cans	500	1
Sections	100 cases No. 2 cans	5,600	1
Peaches, pears	100 cases No. 2 cans	6,500	1
Peas	100 cases No. 2 cans	2,500	1
Pumpkin (squash)	100 cases No. 2 cans	2,500	1
Sauerkraut	100 cases No. 2 cans	300	1
Spinach	100 cases No. 2 cans	16,000	1
Succotash	100 cases No 2 cans	12,500	1
Tomatoes			
Products	100 cases No. 2 cans	7,000	1
Whole	100 cases No. 2 cans	750	1
Cement	ton	750	1,6
Coke	100 tons	360,000	1
Distilling, grain			
Combined wastes	1,000 bu grain mashed		
Thin slop	1,000 bu grain mashed		
Tailings	1,000 bu grain mashed		
Evaporator condensate	1,000 bu grain mashed	600,000	1
Distilling, molasses	1,000 gal 100 proof	8,400	1
Distilling, cooling water	1,000 gal 100 proof	120,000	1
Electric power	kw	80	1
Electric power, steam		80–170[c]	6
generated	kWh	52–170	8,9
Explosives	lb	100+	1
		100	6

2.4-11   INDUSTRIAL REQUIREMENTS FOR WATER (continued)

Product	Unit produced	Water required, gal/unit	Ref.
Gasoline	gal	7–10	1
		8.5	2
		20	6
		8.5–25	8
Iron ore (brown ore)	ton	1,000	1,6
Meat			
Packinghouse	100 hogs killed	550	1
		1,100	2
		5.5–11	8
Slaughterhouse	100 hogs killed	550	1
Stockyards	1 acre	160	1
Milk			
Receiving station	1,000 lbs raw milk	180	1
Bottling works	1,000 lbs raw milk	250	1
		4,500	6
Cheese factory	1,000 lbs raw milk	200	1
		2,000	6
Creamery	1,000 lbs raw milk	110	1
		110–250	2
Condensery	1,000 lbs raw milk	150	1
Dry milk factory	100 lbs raw milk	150	1
General dairy	100 lbs raw milk	340	1
Butter	ton	5,000	6
Oil, edible	gal	22	1
Oil field	100 bbl crude	18,000	1
		15.1–1,500	9
Oil refining	100 bbl	77,000	1,2,6
Paper			
Paper mill	1 ton	39,000	1,6
Paperboard	1 ton	100,000	7
		15,000	2
		15,000–90,000[c]	6
		7,700–80,000	9
Pasteboard	1 ton	14,000	1
Strawboard	1 ton	26,000	1,6
Deinking	1 ton	83,000	1
		5,000–85,000	8
Paper pulp			
Groundwood	1 ton dry	5,000	1
		4,000–50,000[c]	6
Kraft	1 ton dry	93,000	6
Soda	1 ton dry	85,000	1,6
Sulfate	1 ton dry	64,000	1
		70,000	6
Sulfite	1 ton dry	60,000	1
		70,000–133,000[c]	6
Poultry	1 bird	1 (per day)	1
Rail freight	ton mi	0.1	1
Records	1 disc	2.4	1
Smokeless powder	ton	50,000	1
Soap factories	ton	500	1,6
Steam power	ton of coal	60,000–120,000	1
Sugar refineries	lb	0.5	1
Tanning			
Vegetable	100 lb raw hide	800	1
Chrome	100 lb raw hide	800	1

2.4-11 INDUSTRIAL REQUIREMENTS FOR WATER (continued)

Product	Unit produced	Water required, gal/unit	Ref.
Textile			
Cotton	1,000 lb processed	28,000–40,000	8
Sizing	1,000 lb processed	820	1
Desizing	1,000 lb processed	1,750	1
Keiring	1,000 lb processed	1,240	1
Bleaching	1,000 lb processed	300	1
		30,000–80,000	6
Scouring	1,000 lb processed	3,400	1
Mercerizing	1,000 lb processed	30,000	
Dyeing			
Basic	1,000 lb processed	18,000	1
		4,000–8,000	6
Direct	1,000 lb processed	6,400	1
Vat	1,000 lb processed	19,000	1
Sulfur	1,000 lb processed	5,400	1
Developed	1,000 lb processed	14,400	1
Naphthol	1,000 lb processed	4,800	1
Analine black	1,000 lb processed	15,600	1
Print works	1,000 lb processed	4,500	1
Finishing	1,000 lb processed	6	1
Knit goods	lb bleached	8	1
Rayon manufacture	1,000 lb produced	160	1
Rayon hosiery	1,000 lb produced	9,000	1
Woolens	1,000 lb finished	70,000	1,6,8
		40,000–510,000[j]	9
Wool scouring	1,000 lb produced	1,260	2
Steel	ton	65,000	2
		44,000	3
		1,100	4
		15,000–110,000[c,g]	6
		67,000[h]	7
		65,000[i]	8
		6,000–110,000[g]	9
Manufactured ice	ton	244	2
		240–9,000	8
Soft drinks	case	2.5	2
Corn products	bushel	40 (direct)	4
		100–200 (indirect)	4
		140–240[d,e]	6
Ammonium sulfate	ton	200,000	6
Calcium carbide	ton	30,000	6
Carbon dioxide (from flue gas)	ton	20,000	6
Cottonseed oil	gal	20	6
Hydrogen	ton	660,000	6
Oxygen liquid	ft^3	2	6
Soda ash	ton	18,000	6
Sodium chlorate	ton	60,000	6
Sulfuric acid	ton	650–4,875[c]	6
Bread	ton	500–1,000[d]	6
Corn syrup	bu	30–40[d]	6

2.4-11   INDUSTRIAL REQUIREMENTS FOR WATER (continued)

Product	Unit produced	Water required, gal/unit	Ref.
Gelatin (edible)	ton	13,200–20,000[d]	6
Meat			
Packing	ton[f]	4,130	6
Packinghouse operation	hog unit	550	6
Sugar			
Beet sugar	ton	2,160	6
		24,000–34,000	8
Cane sugar	ton	1,000	6
		1,000–28,000	8
		4,000–110,000	9
Petroleum refined products	bbl	150–15,000[c]	6
Synthetic fuel			
Coal hydrogenates	bbl	7,296	6
Coal derivatives	bbl	11,150	6
Natural gas derivatives	bbl	3,736	6
Shale derivatives	bbl	873	6
Rayon	ton	220,000	7
		250,000–400,000	8,9
Caper ammonium yarn	ton	90,000–160,000[d]	6
Viscose yarn	ton	200,000	6
Weave, dye and finish	yd	15	6
Coal and coke			
By-product coke	ton	1,500–3,600[a]	6
		3,600	7
Washing	ton	200	6
Rock wool	ton	5,000	6
Sulfur mining	ton	3,000	6
Synthetic rubber	ton	660,000	7
Buna S	ton	631,400	6
GR-S	ton	28,000–670,000[c]	6
Taconite pellets	ton	12,000	7
Ingot iron	ton	18,000	7

[a] Alcohol industrial (100 proof).
[b] Alumina (Bayer process).
[c] Range from no reuse to maximum recycling.
[d] Range covers various products or processes involved.
[e] Wet milling.
[f] Live animal.
[g] Steel (rolled).
[h] Steel from ingots.
[i] Steel finished.
[j] Per M yards.

REFERENCES

1. H.E. Jordan, *J. Am. Water Works Assn., 38*(1):65, 1946.
2. S.T. Powell and H.E. Bacon, *J. Am. Water Works Assn., 42*(8):777, 1950.
3. C.C. Elder, *J. Am. Water Works Assn., 43*(2):124, 1951.
4. W.J. Lauterbach, *J. Am. Water Works Assn., 44*(11):1033, 1952.
5. Task Group Report, *J. Am. Water Works Assn., 45*(12):1249, 1953.
6. Am. Water Works Assn. Distribution Manual, *J. Am. Water Works Assn., 53*(4):459, 1961.
7. E.H. Ruble, *J. Am. Water Works Assn., 57*(7):831, 1965.
8. R.D. Hoak, *Sew. Ind. Wastes, 25*(12):1438, 1953.
9. R. Hodges *et al., J. Water Pollut. Control Fed., 38*(10):1601, 1966.

2.4-12 MINIMUM NUMBER OF PLUMBING FIXTURES

Type of fixture

Type of building occupancy	Water closets	Urinals	Lavatories	Bathtubs or showers	Drinking fountains	Other fixtures
Assembly – places of worship	No. persons / No. fixtures: 150 women — 1; 300 men — 1	No. persons / No. fixtures: 300 men[a] — 1	1		1	1 slop sink
Assembly – other than places of worship (auditoriums, theaters, convention halls)	No. persons / No. fixtures: 1–100 — 1; 101–200 — 2; 201–400 — 3. Over 400, add 1 fixture for each additional 500 men and 1 for each 300 women	No. persons / No. fixtures: 1–200 — 1; 201–400 — 2; 401–600 — 3. Over 600, add 1 fixture for each 300 men[a]	No. persons / No. fixtures: 1–200 — 1; 201–400 — 2; 401–750 — 3. Over 750, add 1 fixture for each 500 persons		1 for each 300 persons	
Dormitories—school or labor, also institutional	Men: 1 for each 10 persons. Women: 1 for each 8 persons.	1 for each 25 men. Over 150, add 1 fixture for each 50 men[a]	1 for each 12 persons. (Separate dental lavatories should be provided in community toilet rooms. A ratio of 1 dental lavatory to each 50 persons is recommended.)	1 for each 8 persons. For women's dormitories, additional bathtubs should be installed at the ratio of 1 for each 30 women. Over 150 persons add 1 fixture for each 20 persons	1 for each 75 persons	Laundry trays, 1 for each 50 persons. Slop sinks, 1 for each 100 persons
Dwellings – one and two family	1 for each dwelling unit		1 for each dwelling unit	1 for each dwelling unit		Kitchen sink, 1 for each dwelling unit
Dwellings – multiple or apartment	1 for each dwelling unit or apartment		1 for each dwelling unit or apartment	1 for each dwelling unit or apartment		Kitchen sink, 1 for each dwelling unit or apartment. For apartments or multiple dwelling units in excess of 10 apartments or units, 1 double laundry tray for each 10 units or 1 automatic laundry washing machine for each 20 units

a Where urinals are provided for the women, the same number shall be provided as for men.

2.4-12   MINIMUM NUMBER OF PLUMBING FIXTURES (continued)

Type of fixture

Type of building occupancy	Water closets	Urinals	Lavatories	Bathtubs or showers	Drinking fountains	Other fixtures
Industrial – factories, warehouses, foundries, and similar establishments	No. each sex / No. fixtures 1–10 : 1 11–25 : 2 26–50 : 3 51–75 : 4 76–100 : 5 1 fixture for each additional 30 employees.	Where more than 10 men are employed: No. men / No. urinals 11–30 : 1 31–80 : 2 81–160 : 3 161–240 : 4	No. persons / No. fixtures 1–100 : 1–10 Over 100 : 1–15	1 shower for each 15 persons exposed to excessive heat or to occupational hazard from poisonous, infectious, or irritating material	1 for each 75 persons.	
Institutional – other than hospitals or penal institutions (on each occupied story)	1 for each 25 men 1 for each 20 women	1 for each 50 men[a]	1 for each 10 persons	1 for each 10 persons	1 for each 50 persons	
Hospitals Individual room wards Waiting rooms Employees	1 1 for each 8 patients		1 1 for each 10 patients 1	1 1 for each 20 patients	1 for each 100 patients	1 slop sink per floor
Penal institutions Employees Prisoners	Same as public 1 in each cell 1 in each exercise room Same as public	Same as public 1 in each exercise room Same as public	Same as public 1 in each cell 1 in each exercise area Same as public	1 on each cell block floor	Same as public 1 on each cell block floor 1 in each exercise area Same as public	1 slop sink per floor
Public buildings, offices, business, mercantile, storage, and institutional employees	No. each sex / No. fixtures 1–15 : 1 16–35 : 2 36–55 : 3 56–80 : 4 81–110 : 5 111–150 : 6 1 fixture for each additional 40 employees	Urinals may be provided in men's[a] toilet rooms in lieu of water closets but for not more than 1/3 of the required number of water closets	No. employees / No. fixtures 1–15 : 1 16–35 : 2 36–60 : 3 61–90 : 4 91–125 : 5 1 fixture for each additional 45 persons		1 for each 75 persons	1 slop sink per floor

2.4-12   MINIMUM NUMBER OF PLUMBING FIXTURES (continued)

Type of fixture

Type of building occupancy	Water closets		Urinals	Lavatories	Bathtubs or showers	Drinking fountains	Other fixtures
	Boys	Girls					
Schools							
Elementary	1/40	1/35	1/30 boys	1/30 pupils	In gym or pool shower rooms, 1/5 pupils of a class	1/100 pupils but at least 1 per floor	1 slop sink per floor
Secondary	1/75	1/45	1/30 boys	1/30 pupils			
Working men, temporary facilities	1/30 working men		1/30 working men	1/30 working men		1 fixture or equivalent for each 100 working men	

a Where urinals are provided for the women, the same number shall be provided as for men.

Source: *Report of Public Health Service Committee on Plumbing Standards,* U.S. Department of Health, Education, and Welfare, Public Health Service, 1963.

2.4-13   COMBINED HOT AND COLD WATER DEMAND[a]

Fixture	No. fixture units Private use	No. fixture units Public use	Supply control	Ref.
Automatic clothes washer (2 in. standpipe)	1.5−3	Depends on size		3,8
Bar sink	1	2		1
Bathroom group[b]	8		Flushometer valve closet	3,5,6,7
Bathroom group[b]	6		Flush valve closet	3,5,6,7
Bathtub (with or without shower over)	2	4	Faucet	1,2,3,5,6,7
Bed pan washer		10		5
Bidet	3	4		5
Combination fixture	3			5,6,7
Combination sink and tray with food disposal unit	4			3
Combination sink and tray with one 1½ in. trap	2			3
Combination sink and tray with separate 1½ in. trap	3			3
Dental lavatory	1	2		3,5
Dental unit or cuspidor		1		1,3,5
Dishwashing machine	1−1.5	4−6	0.5 in. outlet	2,8
Drinking fountain (each head)	½	1		1,3,5
Electric water cooler	1			5
Floor drains with 2 in. waste	3			3
Garden hose	10			4
Hose bibb or sill cock (standard type)	2−4[c]	4−8[c]		8
House trailer (each)	6	6		1
Kitchen sink, domestic	2	4	Faucet	2,3
Kitchen sink, domestic, with food waste grinder	2			3
Laundry tray or clothes washer (each pair of faucets), 1 or 2 tubs	2	4		1,3,5
Laundry tray, 1 to 3 tubs	3	3	Faucet	2,6,7
Lavatory	1	2	Faucet	1,2,5,6,7
Lavatory, barber, beauty parlor		3		5
Lavatory, surgeons		3		5
Lawn sprinklers (standard type, each head)	1	1		1
Shower (each head)	1½−2	3−4	Mixing valve	2,3,5,6,7,8
Sink, flushing rim		10	Flushometer	1,5
Sink, flushing rim		6	Flush valve	3
Sink, pantry	1			4
Sink, pot or scullery		5		5
Sink, service (P-trap)	2			3
Sink, service (trap standard)	3			3
Sink, slop		1		4
Sink, washup, circular spray		4		1
Sink, washup, each set of faucets		2		1,3,5
Urinal, pedestal or similar type		10	Flushometer	1,5,6,7
Urinal, pedestal, siphon jet blowout		6		3
Urinal, stall, washout		3−5		1,3,5,6,7
Urinal, trough (each 6 ft section)		2		3
Urinal, wall, lip		3−5		1,3,5,6,7
Urinal, wall or stall		3	Flush tank valve	1,5
Water closet	3	5	Flush tank	1,5,6,7
Water closet	4−6	10	Flushometer	1,5,6,7,8

2.4-13   COMBINED HOT AND COLD WATER DEMAND[a] (continued)

| Fixture | No. fixture units | | Supply control | Ref. |
	Private use	Public use		
Water supply outlets for items not listed				
3/8 in.	1	2		1
½ in.	2	4		1
3/4 in.	3	6		1
1 in.	6	10		1
Unlisted fixture drain or trap size				
1¼ in.	1	1		3
1½ in.	2	2		3
2 in.	3	3		3
2½ in.	4	4		3
3 in.	5	5		3
4 in.	6	6		3

[a] One fixture unit is equivalent to a flow of 7.5 gpm.
[b] A bathroom group consists of a water closet, lavatory, and bathtub or shower stall.
[c] Function of size: 3/4 in. 2 fixture units (private), 4 fixture units (public); 1 in. 4 fixture units (private), 8 fixture units (public).

REFERENCES
1. N.S. Moodhe, *J. Am. Water Works Assn.*, *59*(1):43, 1967.
2. H.V. Aldrich, *J. Am. Water Works Assn.*, *53*(3):267, 1961.
3. *Report of Public Health Service Committee on Plumbing Standards*, U.S. Department of Health, Education, and Welfare, Public Health Service, Washington, D.C., 1963.
4. H.E. Babbitt, *Plumbing*, 2nd ed., McGraw-Hill Book Company, Inc., New York, 1950.
5. *The Uniform Plumbing Code for Housing*, Technical Paper 6, Housing and Home Finance Agency, 1948.
6. *Plumbing Manual*, Report BMS66, National Bureau of Standards, 1940.
7. *National Plumbing Code*, ASA A40.8, 1955.
8. A.M. Cunningham, *J. Am. Water Works Assn.*, *64*(1):3, 1972.

2.4.3   IRRIGATION

2.4-14   PROJECTIONS OF IRRIGATED LAND IN THE U.S.

Region	1960	1965	1980	2000	2020
North Atlantic	240	310	380	550	700
South Atlantic-Gulf	850	1,500	1,800	2,750	3,750
Great Lakes	100	140	230	350	470
Ohio	35	55	90	180	260
Tennessee	15	20	30	40	50
Upper Mississippi	80	140	210	390	550
Lower Mississippi	700	900	2,100	3,050	4,150
Souris-Red-Rainy	10	15	90	240	250
Missouri	6,600	7,400	8,050	9,000	9,600
Arkansas-White-Red	3,100	3,800	5,600	6,400	6,850
Texas-Gulf	5,100	5,500	5,500	5,500	5,500
Rio Grande	1,950	2,000	2,050	2,050	2,050
Upper Colorado	1,370	1,440	1,800	2,000	2,000
Lower Colorado	1,520	1,660	1,750	1,800	1,800
Great Basin	1,700	1,860	1,950	2,000	2,000
Columbia-North Pacific	5,450	6,250	7,700	9,500	11,200
California	8,420	8,850	10,150	10,750	11,100
Total	37,240	41,840	49,480	56,550	62,280

Note: Areas in thousands of acres for the conterminous U.S.; for regions see 1.1–7.

Source: U.S. Water Resources Council, 1968.

2.4.4 FARM ANIMALS

2.4-15 WATER REQUIREMENTS FOR FARM ANIMALS AND POULTRY

Usage	gpd
Horse, work	12
Mule	12
Cattle	
Holstein calves (liquid milk or dried milk and water supplied)	
4 weeks of age	1.2–1.4
8 weeks of age	1.6
12 weeks of age	2.2–2.4
16 weeks of age	3.0–3.4
20 weeks of age	3.8–4.3
26 weeks of age	4.0–5.8
Dairy heifers, pregnant	7.2–8.4
Steers	
Maintenance ration	4.2
Fattening ration	8.4
Range cattle	4.2–8.4
Jersey cows[a]	
Milk production 5–30 lb/day	7.2–12
Holstein cows[a]	
Milk production 20–50 lb/day	7.8–22
Milk production 80 lb/day	23
Dry	11
Pigs	
Body weight, 30 lb	0.6–1.2
Body weight, 60–80 lb	0.8
Body weight, 75–125 lb	1.9
Body weight, 200–380 lb	1.4–3.6
Pregnant sows	3.6–4.6
Lactating sows	4.8–6.0
Sheep	
On range or dry pasture	0.6–1.6
On range (salty feeds)	2.0
On rations of hay and grain or hay, roots and grain	0–0.7
On good pasture	Little, if any
Chickens (100 birds)	
1–3 weeks of age	0.4–2.0
3–6 weeks of age	1.4–3.0
6–10 weeks of age	3.0–4.0
9–13 weeks of age	4.0–5.0
Pullets	3.0–4.0
Nonlaying hens	5.0
Laying hens (moderate temperatures)	5.0–7.5
Laying hens (temperature 90°F.)	9.0
Turkeys (100 birds)	
1–3 weeks of age	1.1–2.6
4–7 weeks of age	3.7–8.4
9–13 weeks of age	9–14
15–19 weeks of age	17
21–26 weeks of age	14–15

[a] Allow 15–20 additional gallons per day for each cow for flushing stables and washing dairy utensils.

Source: U.S. Department of Agriculture.

2.5   QUALITY CRITERIA

2.5.1   BENEFICIAL USE

2.5-1   CONCEPT OF WATER QUALITY CONTROL
FOR BENEFICIAL USE

Beneficial use to be preserved in river water	Water quality criteria prescribed by control board	Possible conditions if pollution not controlled
Irrigation	700 ppm dissolved minerals, maximum 40 percent sodium ratio, maximum 0.5 ppm boron, maximum	High concentrations: crop damage and loss of production
Bathing	10 coliform bacteria per ml, maximum	High coliform counts; an actual hazard to public health (contamination)
Fishing	5 ppm of dissolved oxygen (DO), minimum	Low dissolved oxygen; reduced fisheries resources or fish kills

Source: V.W. Bacon and C.A. Sweet, *"Setting Water Quality Criteria in California"*, Am. Soc. Civ. Eng., May 1955. With permission.

2.5-2 RECOMMENDATIONS FOR WATER FOR CERTAIN BENEFICIAL USES TO BE PROTECTED

Prado Water Quality Objectives

Water use	Quality factors				
	TDS, mg/l	Chloride, mg/l	Nitrate, mg/l as NO_3	Bicarbonate, mg/l HCO_3	Hardness, mg/l $CaCO_3$
Industrial uses					
Boiler feed water					
High pressure	200			48	0
Intermediate pressure	500			120	0
Low pressure	700			170	20
Cooling water	500	500		24	130
Food processing	500	250	10	155	250
Soft drinks		500		53	
Pulp and paper					
Mechanical pulping		1,000			
Unbleached		200			100
Bleached		200			100
Chemical *et al.*					
Organic chem.				128	250
Synth. rubber				155	350
Gums, wood chem.	1,000	500		250	900
Petroleum	1,000	300			350
Agriculture					
Vegetable crops					
Beans	132				
Carrots	132				
Bell peppers	350				
Lettuce	350				
Corn	393				
Cabbage	393				
Broccoli	700				
Tomatoes	700				
Spinach	1,000				
Ornamental shrubs					
Roses	362				
Pineapple guava	362				
Viburnum	362				
Pyracantha	700				
Pittosporum	700				
Xylosma	700				
Texas privet	700				
Arbor vitae	1,092				
Spreading juniper	1,092				
Lantana	1,092				
Oleander	1,750				
Bottlebrush	1,750				
Fruit crops					
Rangpur lime, Cleopatra mandarin		437			
Rough lemon tangelo sour orange		262			
Sweet orange, citrange		175			
West Indian avocado rootstock		140			

2.5-2   RECOMMENDATIONS FOR WATER FOR CERTAIN BENEFICIAL
USES TO BE PROTECTED (continued)

	Quality factors				
Water use	TDS, mg/l	Chloride, mg/l	Nitrate, mg/l as NO_3	Bicarbonate, mg/l HCO_3	Hardness, mg/l $CaCO_3$
Agriculture (continued)					
Mexican avocado rootstock		88			
Thompson seedless grape		437			
Cardinal, Black Rose grape		175			
Boysenberry		175			
Olallie blackberry		175			
Indian Summer raspberry		88			
Lassen strawberry		140			
Shasta strawberry		88			

Source: Reprinted, with publisher's permission, from *J. Am. Water Works Assn.*, *62*(2):106, 1970, by the American Water Works Association, Inc.,

2.5.2 HEALTH

2.5-3 HEALTH AND AESTHETIC SIGNIFICANCE OF TRACE ELEMENTS AND COMPOUNDS

Occurrence in U.S. Water Supplies

Trace element or compound	Health or aesthetic effect	1962 USPHS drinking water stand- ards limit, mg/l	Occurrence in 2,595 distribution samples, 1969, mg/l	
			Avg	Range
Alkyl benzine sulfonate (ABS)	USPHS limit gives a safety factor of 15,000; not highly toxic	0.5[a]	0.05	0–0.41
Arsenic (As)	Serious systemic poison – 100 mg usually causes severe poisoning, is cumulative, and may cause chronic effects. Late evidence indicates tiny amounts may be beneficial	0.01[a] 0.05[b]	0.0001	<0.03–0.10
Chloride (Cl^-)	Limit set for taste reasons	250.[a]	27.6	<1.0–1,950
Copper (Cu)	Body needs copper at level of about a mg/ day for adults; not a health hazard except when large amounts are ingested	1.[a]	0.13	0–8.35
Carbon chloroform extract (CCE)	At limit stated, organics in water are not considered a health hazard	0.2[a]	0.11	0.008–0.56
Cyanide (CN^{-2})	Rapid fatal poison, but limit set provides safety factor of about 100	0.01[a] 0.2[b]	0.00009	<0.1–0.008
Fluoride (F^-)	Beneficial in small amounts; above 2,250 mg dose can cause death	0.7–1.2[a] 1.4–2.4[b]	0.32	<0.2–4.40
Nitrate (NO_3)	Excessive amounts can cause methemoglo- binemia (blue baby) in infants	45[a]	6.3	<0.1–127
Sulfate (SO_4)	Above 750 mg/l, usually has laxative effect	250[a]	46	<1.0–770
Zinc (Zn)	Zinc is beneficial in that a child needs 0.3 mg/ kg day; 675–2,280 mg/l may be an emetic	5[a]	0.19	0–13.0
Barium (Ba)	Fatal dose is 550–600 mg as the chloride; it is a muscle (including heart) stimulant	1.0[b]	0.034	0–1.55
Cadmium (Cd)	13–15 ppm in food has caused illness	0.01[b]	0.003	<0.2–3.94
Chromium (Cr^{+6})	Limit provides a safety factor. Carcinogenic when inhaled	0.05[b]	0.0023	0–0.079
Lead (Pb)	Serious, cumulative, body poison	0.05[a]	0.013	0–0.64
Selenium (Se)	Toxic to both humans and animals in large amounts. Late research suggests small amounts may be beneficial	0.01[b]	0.003	0–0.07
Silver (Ag)	Can produce irreversible, adverse cosmetic changes	0.05[b]	0.008	0–0.03
Radium (Ra-226)	A bone-seeking, internal alpha emitter that can destroy bone marrow	3 pCi/l[c]	2.2 pCi/l	0–135.9 pCi/l
Strontium (^{90}Sr)	A bone-seeking, internal beta emitter	10 pCi/l[c]	<1.0 pCi/l	0–2.0 pCi/l

[a] Not to be exceeded where other more suitable supplies are, or can be made, available.
[b] Excess constitutes grounds for rejection of the supply.
[c] Consult 1962 *USPHS Drinking Water Standards* for interpretation.

Source: Reprinted, with publisher's permission, from *J. Am. Water Works Assn.*, *63*(11):728, © 1971 by the American Water Works Association, Inc.

2.5-4   HEALTH AND AESTHETIC SIGNIFICANCE OF TRACE ELEMENTS AND COMPOUNDS IN FINISHED DRINKING WATER

Name of ingredient	Health or aesthetic effect	Unofficial limit, mg/l	Occurrence in U.S. water supplies, mg/l	
			Avg	Range
Substance				
Antimony (Sb)	Similar to arsenic but less acute. Recommended limit not to exceed 0.1 mg/l; routinely below 0.05 mg/l; over longtime periods below 0.01 mg/l	0.05		
Beryllium (Be)	Poisonous in some of its salts in occupational exposure	None		
Bismuth (Bi)	A heavy metal in the arsenic family – avoid in water supplies	None		
Boron (B)	Ingestion of large amounts can affect central nervous system, and protracted ingestion may cause borism	1.0	0.069	0–3.28[a]
Cobalt (Co)	Beneficial in small amounts; about 7 μg/day	None	0.002	0.0–0.019[a]
Molybdenum (Mo)	Necessary for plants and ruminants. Excessive intakes may be toxic to higher animals; acute or chronic effects not well known	None		
Mercury (Hg)	Continued ingestion of large amounts can damage brain and central nervous systems	0.005	0.000	0.000–0.033[b]
Nickel (Ni)	May cause dermatitis in sensitive people; doses of 30–73 mg of $NiSO_4 \cdot 6H_2O$ have produced toxic effects	None	0.005	0.0–0.072[a]
Sodium (Na)	A beneficial and needed body element, but can be harmful to people with certain diseases	20[c]	40 percent above 20 mg/l	0.4–1,900[d]
Tin (Sn)	Long used in food containers without known harmful effects	None		
Uryanyl ion (UO$_2$)	May cause damage to kidneys	5.0		
Vanadium (V)	Some evidence that vanadium may be beneficial with respect to heart disease	None		
Pesticides				
Aldrin	One or all of these complex organic compounds have severe, acute, adverse health effects when ingested in large amounts. Small amounts accumulate, and long-range effects are generally unknown	0.017[e]	160 samples in NCWSS showed lower values than the limits	
Chlordane		0.003[e]		
DDT		0.042[e]		
Dieldrin		0.017[e]		
Endrin		0.001[e]		
Heptachlor		0.018[e]		
Heptachlor epoxide		0.018[e]		
Lindane		0.056[e]		
Methoxychlor		0.035[e]		
Toxaphene		0.005[e]		
Organic phosphorus plus carbamates, such as Parathion, Malathion, Carbaryl, and others	The organic phosphorus and carbamate pesticides are severe acute poisons affecting the central nervous system; ingestion of small amounts over time can harm the central nervous system	0.1[f]	Same as for pesticides	
Herbicides				
Algicide	A group having toxic properties of a generally lower order than the above pesticides. However, they should not be used without great care and should not be found in drinking water	1.0		
Copper key – El		1.0 (Cu)		
Cuprose		1.0 (Cu)		
Cutrine		1.0 (Cu)		
Diquot		0.2		

2.5-4   HEALTH AND AESTHETIC SIGNIFICANCE OF TRACE ELEMENTS AND COMPOUNDS
IN FINISHED DRINKING WATER (continued)

Name of ingredient	Health or aesthetic effect	Unofficial limit, mg/l	Occurrence in U.S. water supplies, mg/l	
			Avg	Range
Dalapon		0.1		
2,4-D		1.0		
2,4,5-T should not be used				
Micro-gard PR		0		
Mogul algicide AG-470		1.0		
Radapon		0.1		
Silvex (2,4,5-TP)		0.2		
Tordon 10K and 101		0		
Twink		0.2		
Ureabor		1.0 (as B)		

Note: Table shows limits not found in 1962 *USPHS Drinking Water Standards.*

[a] National Community Water Supply Study (NCWSS). 2500 samples.
[b] Late data from BWH, USPHS.
[c] The amount at, and above, which people on strict (500 mgd) sodium-restricted diets must include in their daily sodium-intake calculations.
[d] Survey of over 2,000 U.S. municipalities by USPHS in 1963–66.
[e] Report of the Committee on Water Quality Criteria, FWPCA, U.S. Department of the Interior.
[f] As parathion equivalent in cholinesterase inhibition.

Source: Reprinted, with publisher's permission, from *J. Am. Water Works Assn.*, *63*(11):728, © 1971 by the American Water Works Association, Inc.

2.5-5   POTABLE WATER QUALITY GOALS[a]

American Water Works Association

Characteristic	Goal

Physical Factors

Turbidity	Less than 0.1 unit
Nonfilterable residue	Less than 0.1 mg/l
Macroscopic and nuisance organisms	No such organisms
Color	Less than 3 units
Odor	No odor
Taste	No taste objectionable

Chemical Factors, mg/l

Aluminum (Al)	Less than 0.05
Iron (Fe)	Less than 0.05
Manganese (Mn)	Less than 0.01
Copper (Cu)	Less than 0.2
Zinc (Zn)	Less than 1.0
Filterable residue	Less than 200.0
Carbon chloroform extract (CCE)	Less than 0.04
Carbon alcohol extract (CAE)	Less than 0.10
Methylene blue active substances (MBAS)	Less than 0.20

Corrosion and Scaling Factors

Hardness (as $CaCO_3$)	80 mg/l; a balance between deposition and corrosion characteristics is necessary; a level of 80 mg/l seems best, generally, considering all the quality factors; however, for some supplies, a goal of 90 or 100 mg/l may be deemed desirable
Alkalinity (as $CaCO_3$)	Change of not more than 1 mg/l (decrease or increase in distribution system, or after 12 hr at 130°F in a closed plastic bottle, followed by filtration)
Coupon tests (incrustation and loss by corrosion)	90 day tests (incrustation on stainless steel not to exceed 0.05 mg/cm², loss by corrosion of galvanized iron not to exceed 5.00 mg/cm²)

Bacteriologic Factors

Coliform organisms (by multiple fermentation techniques)	No coliform organisms
Coliform organisms (by membrane filter techniques)	No coliform organisms

[a] For all health related constituents not stated herein, these goals shall require complete compliance with all recommended and mandatory limits contained in current USPHS Drinking Water Standards. Unless other methods are indicated, analyses shall be made in conformance with the latest edition of *Standard Methods of Examination of Water and Wastewater.*

2.5-5 POTABLE WATER QUALITY GOALS[a] (continued)

Characteristic	Goal

Radiologic Factors

Gross beta activity	Less than 100 pCi/l

[a] For all health related constituents not stated herein, these goals shall require complete compliance with all recommended and mandatory limits contained in current USPHS Drinking Water Standards. Unless other methods are indicated, analyses shall be made in conformance with the latest edition of *Standard Methods of Examination of Water and Wastewater.*

Note: The data here are not a standard. They implement the areas covered by *Drinking Water Standards.* They cover areas not directly associated with health effects, which are recognized as the prerogative of health agencies.

Source: Reprinted, with publisher's permission, from *J. Am. Water Works Assn.,* *60*(12):1317, © 1968 by the American Water Works Association, Inc.

2.5-6 DETECTABLE CONCENTRATIONS OF ASBESTOS IN WATER (10^6 FIBERS/l)[a]

Province/ state	City	Chrysotile asbestos fiber conc				Amphibole asbestos fiber conc		
		Raw water	Treated	Distribution	Filtration	Raw	Treated	Distribution
Alberta	Lethbridge	83	0—0.5	0—0.5	Yes			
	Medicine Hat	6.5	0—0.5	0.5—1.5	Yes			
British	Cassiar	25	—[b]	0.5—1.5	No			
Columbia	Kamloops	11	4	6—18	No	1.6		0.3
	Vancouver	0—0.5[c]	1—5.5	0—12	No			1.0
Manitoba	Flin Flon			0.1			0.2	
	Portage la Prairie	36	0—0.5	0—0.5	Yes			
	Selkirk	31	0—0.5	0—0.5	Yes			
	Thompson	190	1	0—1	Yes	13		
	Winnipeg	0—0.5	0.5—1.5	1—6.5	No			
New Brunswick	Fredericton			1.0				0.9
Newfoundland	Baie Verte	400	480	260—1800	No			12
	Gander	2	—[b]	1.5—7.0	No			0.5
	Labrador City	—[d]	5.6	0.8—15	No		1.6	
	La Scie	7	—[b]	8.5—15	No			
Northwest Territory	Yellowknife	3	31	1—2.5	No			0.9
Ontario	Hearst	—[c]	—[c]	11—22	No			
	Kirkland Lake			1.3				0.3
	Matheson	7.5	1	0—1.5	No			0.4
	Sault Ste. Marie			<0.1—0.3				0.2—0.3
	Thunder Bay	2.1				0.5		
	Tilbury			0.5				0.5
Quebec	Asbestos	170	9.5	3—8.9	Yes			
	Beaulac	—[c]	24	54—59	No			1.6
	Disraeli	220	—[b]	200—1200	No	9.1		7.0
	Sherbrooke	73	26	80—220	No		0.5	
	Thetford Mines	—[d]	140	110—150	No			
Saskatchewan	Prince Albert	8.6	0—0.5	0—0.5	Yes			
	Swift Current			0.6				0.6
Yukon	Dawson City	13	—[d]	0—0.5	No			
	Whitehorse	270	38	33—130	No	3.9		

[a] Compiled from: P. Toft *et al., CRC Crit. Rev. Environ. Control, 14*:151, 1984. Taken from: Health and Welfare Canada, A National Survey for Asbestos Fibers in Canadian Drinking Water Supplies, Rep. 70-END-34, Environmental Health Directorate, Health and Welfare Canada, Ottawa, 1979.

[b] No treatment.

[c] Sample not analyzed.

[d] Sample not available.

2.5-7   MIGRATION OF BACTERIA IN THE SUBSURFACE

Microorganism	Medium	Maximum distance traveled (m)		Ref.
		Vertical	Horizontal	
Bacillus stearothermophilus	Fractured rock		29	1
Bacteria	Fine sand		457	2
	Medium to coarse sand		21	3
	Alluvial gravel		90	4
	Pea gravel + sand		30	5
	Coarse gravels		457	6
	Gravel		920	7
	Sandy clay		15.25	8
	Fine to coarse sand		30.5	9
	Fine to medium sand		6.1	10
Clostridium welchii	Fine + medium sand		15.5	11
Coliforms	Loam + sandy loam			
	Sand + gravel	10—12	850	12
	Fine sandy loam	4	1.2	13
	Fine sand	4	2	14
	Pebbles		850	14
	Weathered limestone		1000	14
	Stone clay + sand	0.91		15
	Stone + clay	0.61		15
	Firm clay	0.3		15
	Coarse sand + gravel		55	16
	Sandy clay laom	2	6.1	17
	Sandy clay loam	4.3	13.5	17
	Sandy loam	0.64	28	17
Escherichia coli	Sand		3.1	18
	Fine + coarse sand	4	24.4	19
	Fine + medium sand	0.15		20
	Fine + medium sand		3.1	11
	Sand + sandy clay		10.7	21
	Silt loam		3	22
	Silty clay loam		1.5	22
	Medium sandy gravel		125	23
	Fine sandy gravel with cobbles		50	23
	Silty clay loam	1	15	24
	Fine sand		19.8	25
	Fine sand	0.3	70.7	26
Fecal coliforms	Fine loamy sand + gravel		9.1	27
	Stony silt loam		900	28
	Fine to medium sand		2.4	29
	Gravel with sand + clay		9	30
	Saturated gravels		42	30
	Sandy clay + clay	0.85		31
	Sandy clay	1.2		32
Salmonella enteriditis	Clay	0.1		33
S. typhi	Limestone		457	34
Streptococcus faecalis	Silty clay loam		0.5	22
	Silt loam		5	22
Strept. zymogenes	Sandy gravel	0.15	15.2	35

Source: M.V. Yates and S.R. Yates, *CRC Crit Rev. Environ. Control, 17*:307, 1987.

2.5-7 MIGRATION OF BACTERIA IN THE SUBSURFACE (continued)

REFERENCES

1. M.J. Allen and S.M. Morrison, *Ground Water, 11*:6, 1973.
2. A.F. Dappert, *Water Works Sewerage, 79*:265, 1932.
3. F. Ditthorn and A. Luersson, *Eng. Rec., 60*:642, 1909.
4. F. Marti, G.D. Valle, V. Krech, R.A. Gees, and E. Baumgartner, *Alimenta, 18*:135, 1979.
5. P.H. McGauhey and R.B. Krone, Report on the investigation of travel of pollution, Publ. No. 11, State Water Pollution Control Board, State of California, Sacramento, 1954.
6. J.C. Merrell Jr., The Santee Recreation Project, Santee, Calif., Water Pollut. Res. Ser. Publ. No. WP-20-7, Federal Water Pollution Control Administration, Cincinnati, Ohio, 1967.
7. L.W. Sinton, Investigations into the use of bacterial species *Bacillus stearothermophilus* and *Escherichia coli* (H_2S positive) as tracers of ground water movement, Water Soil Tech. Publ. No. 17, Water Soil Division, Ministry of Works and Development, Wellington, N.Z., 1980.
8. T. Viraraghavan, *Water Air Soil Pollut., 9*:355, 1978.
9. G.M. Wesner and D.C. Baier, *J. Am. Water Works Assn., 62*:203, 1970.
10. R.H.F. Young, *J. Water Pollut. Control Fed., 46*:1296, 1973.
11. E.L. Caldwell, *J. Infect. Dis., 62*:225, 1938.
12. N.I. Anan'ev and N.D. Demin, *Hyg. Sanit., 36*:292, 1971.
13. R.G. Butler, G.T. Orlob, and P.H. McGauhey, *J. Am. Water Works Assn., 46*:97, 1954.
14. B.M. Kudryavtseva, *J. Hyg. Epidemiol. Microbiol. Immunol., 16*:503, 1972.
15. A. Maline and J. Snellgrove, *Lab. Pract., 7*:219, 1958.
16. A.D. Randall, *J. Am. Water Works Assn., 62*:716, 1970.
17. R.B. Reneau and D. E. Pettry, *J. Environ. Qual., 4*:41, 1975.
18. J.K. Baars, *Bull. World Health Org., 16*:727, 1957.
19. E.L. Caldwell, *J. Infect. Dis., 62*:270, 1938.
20. E.L. Caldwell, *J. Infect. Dis., 62*:272, 1938.
21. E.L. Caldwell and L.W. Parr, *J. Infect. Dis., 61*:148, 1937.
22. C. Hagedorn, D.T. Hansen, and G.H. Simonson, *J. Environ. Qual., 7*:55, 1978.
23. B.H. Pyle and H.R. Thorpe, Evaluation of the potential for microbiological contamination of an aquifer using a bacterial tracer, in *Proc. Groundwater Pollut. Conf. Perth, West Australia*, Australian Resources Council, Australian Government Publication Service, Canberra, 1981, p. 213.
24. T.M. Rahe, C. Hagedorn, and E.L. McCoy, *Water Air Soil Pollut., 11*:93, 1979.
25. C.W. Stiles and H.R. Crohurst, *Public Health Rep., 38*:1350, 1923.
26. L.F. Warrick and O.J. Muegge, *J. Am. Water Works Assn., 22*:516, 1930.
27. H. Bouwer, J.C. Lance, and M.S. Riggs, *J. Water Pollut. Control Fed., 46*:844, 1974.
28. G.N. Martin and M.J. Noonon, Effects of domestic wastewater disposal by land irrigation on groundwater quality of central Canterbury plains, Water Soil Tech. Publ. No. 7, Water and Soil Division, Ministry of Works and Development, Wellington, N.Z., 1977.
29. F.C. McMichael and J.E. McKee, Wastewater reclamation at Whittier Narrows, Pub. No. 33, State Water Quality Control Board, State of California, Sacramento, 1965.
30. L.W. Sinton, *Water Air Soil Pollut., 28*:407, 1986.
31. K.W. Brown, J.F. Slowey, and H.W. Wolf, The movement of salts, nutrients, fecal coliforms, and virus below septic leach fields in three soils, in *Proc. 2nd Natl. Home Sewage Treatment Symp.*, American Society of Agricultural Engineers, St. Joseph, Mich., 1978.
32. K.W. Brown, H.W. Wolf, K.C. Donnelly, and J.F. Slowey, *J. Environ. Qual., 8*:121, 1979.
33. R.W. Weaver, Transport and fate — bacterial pathogens in soil, in Microbial Health Considerations of Soil Disposal of Domestic Wastewaters, Pub. No. EPA-600/9-83-017, U.S. Environmental Protection Agency, Washington, D.C., 1983, p. 123.
34. S.P. Kingston, *J. Am. Water Works Assn., 35*:1450, 1943.
35. H.J. Fournelle, E.K. Day, and W.B. Page, *Public Health Rep., 72*:203, 1957.

2.5-8   MIGRATION OF VIRUSES IN THE SUBSURFACE

Microorganism	Medium	Maximum distance traveled (m)		Ref.
		Vertical	Horizontal	
Bacteriophage	Sand	45.7	400	1
	Sandy clay	1.2		2
	Clay	0.85		3
	Boulder clay		510	4
	Sandstone		570	4
Coliphage f2	Silty sand	29	183	5
Coliphage T4	Karst		1600	6
Coxsackievirus B3	Fine loamy sand	18.3		7
	Sand	22.8	408	8
Echovirus	Coarse sand + fine gravel	11.3	45.7	9
Enterovirus	Sandy loam	3.5	14.5	10
Poliovirus	Loamy sand	0.4		11
	Medium sand		0.6	12
	Loamy sand	1.6		13
	Sand	0.2		14
	Silt loam		46.2	15
	Medium to fine sand		9	15
	Loamy medium sand		6	15
	Sand	9.1		8
	Coarse sand + fine gravel	10.6	3.0	9
	Coarse sand + fine gravel	7.62		16
Viruses	Sand	6		17
	Sandy clay	3		18
	Sand	38		18
	Sand + coarse gravel	16.8	250	19

Source: M.V. Yates and S.R. Yates, *CRC Crit. Rev. Environ. Control, 17*:307, 1987.

REFERENCES

1. D.B. Aulenbach, Long-Term Recharge of Trickling Filter Effluent into Sand, Pub. No. EPA-600/2-79-068, U.S. Environmental Protection Agency, Washington, D.C., 1979.
2. P.H. McGauhey and R.B. Krone, Report on the Investigation of Travel of Pollution, Publ. No. 11, State Water Pollution Control Board, State of California, Sacramento, 1954.
3. J.C. Merrell, Jr., The Santee Recreation Project, Santee, Calif., Water Pollut. Res. Ser. Publ. No. WP-20-7, Federal Water Pollution Control Administration, Cincinnati, Ohio, 1967.
4. R. Martin and A. Thomas, *J. Hydrol., 23*:73, 1974.
5. S.A. Shaub and C.A. Sorber, *Appl. Environ. Microbiol., 33*:609, 1977.
6. M.W. Fletcher and R.L. Myers, *Abstr. Annu. Mtg. Am. Soc. Microbiol., 74*:52, 1974.
7. B.H. Keswick and C.P. Gerba, *Environ. Sci. Technol., 14*:1290, 1980.
8. J.M. Vaughn and E.F. Landry, Data Report: An Assessment of the Occurrence of Human Viruses in Long Island Aquatic Systems, Department of Energy and Environment, Brookhaven National Laboratory, Upton, N.Y., 1977.
9. J.M. Vaughn, E.F. Landry, L.J. Baranosky, C.A. Beckwith, M.C. Dahl, and N.C. Delihas, *Appl. Environ. Microbiol., 36*:47, 1978.
10. K.E. Hain and R.T. O'Brien, The Survival of Enteric Viruses in Septic Tanks and Septic Tank Drainfields, Rep. No. 108, New Mexico Water Resources Research Institute, Las Cruces, N. Mex., 1979.
11. C.P. Gerba and J.C. Lance, *Appl. Environ. Microbiol., 36*:247, 1977.
12. K.M. Green and D.O. Cliver, Removal of virus from septic tank effluent by sand columns, in *Home Sewage Disposal, Proc. Natl. Home Sewage Disposal Symp.,* American Society of Agricultural Engineers, St. Joseph, Mich., 1974, p. 137.

2.5-8   MIGRATION OF VIRUSES IN THE SUBSURFACE (continued)

13. J.C. Lance and C.P. Gerba, *J. Environ. Qual.*, 9:31, 1980.
14. E. Lefler and Y. Kott, Virus retention and survival in sand, in *Virus Survival in Water and Wastewater Systems*, J.F. Malina and B.P. Sagik, Eds., Center for Research in Water Resources, Austin, Tex., 1974, 84.
15. S.L. Stramer, Fates of Poliovirus and Enteric Indicator Bacteria during Treatment in a Septic Tank System Including Septage Disinfection, Ph.D. dissertation, University of Wisconsin, Madison, 1984.
16. J.M. Vaughn, E.F. Landry, C.A. Beckwith, and M.Z. Thomas, *Appl. Environ. Microbiol.*, *41*:139, 1981.
17. F.M. Wellings, A.L. Lewis, and C.W. Mountain, Virus survival following wastewater spray irrigation of sandy soils, in *Virus Survival in Water and Wastewater Systems,* Malina, J.F. and B.P. Sagik, Eds., Center for Research in Water Resources, Austin, Tex., 1974, p. 253.
18. F.M. Wellings, A.C. Lewis, C.W. Mountain, and L.V. Pierce, *Appl. Environ. Microbiol.*, *29*:751, 1975.
19. E.L. Koerner and D.A. Haws, Long-Term Effects of Land Application of Domestic Wastewater Vineland, N.J., Rapid Infiltration Site, Pub. No. EPA-600/2-79-072, U.S. Environmental Protection Agency, Washington, D.C., 1979.

2.5-9 SUMMATION OF SUGGESTED NO-ADVERSE-RESPONSE LEVELS (SNARLs) AND RISK ESTIMATES FOR CHEMICALS

Compound	Estimated SNARLs (µg/l)		Upper 95% confidence estimate of lifetime cancer risk
	Adult	**Child**	
Disinfectants			
Chlorine	—[a]	—[a]	
Ozone	—[a]	—[a]	
Chlorine dioxide	210	60	
Chloramine	581	166	
Disinfectant byproducts			
Chlorate	24	7	
Chlorite	24	7	
Trihalomethanes			
Chloroform			8.9×10^{-8} [b]
			1.9×10^{-6} [c]
Chlorodibromo-methane			8.3×10^{-7} [c]
Haloacids			
Dichloroacetic acid	420	175	
Trichloroacetic acid	120	50	
Haloaldehydes			
Chloroacetaldehyde	—[d]	—[d]	
Dichloroacetaldehyde	—[d]	—[d]	
Trichloroacetaldehyde	—[d]	—[d]	
Haloketones			
1,1,1-Trichloroacetone	—[d]	—[d]	
1,1,3,3-Tetrachloroacetone	—[d]	—[d]	
Hexachloroacetone	—[d]	—[d]	
Haloacetonitriles			
Dichloroacetonitrile	56[e]		
Dibromoacetonitrile	161	23	
Bromochloroacetonitrile	—[d]	—[d]	
Trichloroacetonitrile	—[d]	—[d]	
Chloropicrin	40	12	
Chlorophenols			
2,4-Dichlorophenol	7000	2000	
2,4,6-Trichlorophenol	—[d]	—[d]	
2-Hydroxylchlorophenol	—[d]	—[d]	

[a] Not calculated.

[b] Tumor data for risk assessment calculation from drinking water animal study.

[c] Tumor data for risk assessment calculation from corn oil gavage animal study.

[d] Insufficient data for calculation.

[e] Not calculated; the adult value was calculated for comparison purposes; it is not recommended by the committee.

Source: NRC (National Research Council), *Drinking Water and Health,* Vol. 7, National Academy Press, Washington, D.C., 1987, 207 pp.

2.5.3   INDUSTRIAL

2.5-10   INDUSTRIAL WATER QUALITY TOLERANCES

Table A

Type of industrial use of water	Turbidity, mg/l	Color, mg/l	Oxygen consumed, mg/l	Dissolved oxygen, mg/l	Taste and odor	pH	Hardness as $CaCO_3$, mg/l	Alkalinity as $CaCO_3$, mg/l	Calcium, mg/l	Chlorides, mg/l
Air conditioning	10	10								
Baking[a,b]					None		a		a	
Boiler feedwater										
0–150 psi	20	80	15	1.4		8.0+	80			
150–250 psi	10	40	10	0.1		8.4+	40			
250–400 psi	5	5	4	0		9.0+	10			
over 400 psi	1	2	3	0		9.6+	2			
Brewing										
Light beer[c]	10	10			None	6.5–7.0		75–80	100–200	
Dark beer	10	10			None	6.5–7.0		80–150	200–500	
Carbonated beverages[b]	2	10	1.5	None			50–170		250	
Chemical process industries	Requirements vary widely within industry and process used; sometimes distilled water must be used.									
Concrete mixing[b,d]										
Confectionery	50				Low	7.0+	50			
Cooling water						6.5–7.5	Low			
Electroplating										
Fermentation	Same general requirements as brewing. Requirements for gin and spirits equal light beer, and requirements for whiskey equal dark beer.									
Food canning and freezing[b,f]	10				None					
Food equipment washing[b]	1	20			None	7.5+	10	30–250		250
Food processing – general[b]	10	10			Low		250	250		
Ice manufacture[b]	5	5	3		None		70	30–50		
Laundering						6.0–6.8	50	60		
Oil well flooding	0					7.0+	100	Low		
Photographic processing	2	2								
Plastics – clear, uncolored	2	2								
Paper and pulp										
Ground woodpulp[g]	50	20–30[c]					182–200[c]	150		25–75
Soda and sulfate pulp[g,h]	25	5					100	75		75

2.5-10   INDUSTRIAL WATER QUALITY TOLERANCES (continued)

Table A

Type of industrial use of water	Turbidity, mg/l	Color, mg/l	Oxygen consumed, mg/l	Dissolved oxygen, mg/l	Taste and odor	pH	Hardness as $CaCO_3$, mg/l	Alkalinity as $CaCO_3$, mg/l	Calcium, mg/l	Chlorides, mg/l
Kraft										
Bleached	40	25					100	75		200
Unbleached	100	100					200	150		200
Fine paper	10	5					100	40–75		
Rayon (viscose)										
Pulp production	5	5								
Manufacture	0.3					7.8–8.3	8	50		
Steel manufacturing						6.8–7.0	55	50–75		
Sugar manufacturing[e]							50			175
Tanning	20	10–100				6.8–8.0	50–513	128–135	20	20
Textile manufacturing										
General[f]	0.3–25	5–70	8				0–50		10	100
Dyeing	5	5–20					20			
Wool scouring	70						20			
Cotton bandage	5	5			Low		20			

[a] Calcium is necessary for some yeast actions, so some hardness is desirable. Too much hardness, however, retards fermentation and causes soggy bread.

[b] Except as specified, should conform to Public Health Service drinking water standards.

[c] Although the total dissolved solids limit is 500–1,000 ppm, there is a limit of 300 ppm for any one substance.

[d] Water should be free of acids, alkalies, oil, and decayed vegetable matter. Sugar should not exceed 3,400 ppm; SO_3 should not exceed 25 ppm.

[e] Magnesium in any form should not exceed 10 ppm.

[f] No chlorophenols. Hardness varies with product: general canning and freezing 50–85 ppm; peas 200–400 ppm; fruits and vegetables 100–200 ppm; legumes 25–75 ppm.

[g] Turbidity as SiO_2 must not be gritty.

[h] Calcium hardness as $CaCO_3$ should not exceed 50 ppm. Magnesium hardness as $CaCO_3$ should not exceed 50 ppm.

2.5-10   INDUSTRIAL WATER QUALITY TOLERANCES (continued)
Table B

Type of industrial use of water	Calcium chloride, mg/l	Magnesium chloride, mg/l	Sodium chloride, mg/l	Sulfates, mg/l	Magnesium sulfates, mg/l	Calcium sulfates, mg/l	Sodium sulfate, mg/l	Total dissolved solids, mg/l	Total solids, mg/l	Iron mg/l
Air conditioning										0.5
Baking[a,b]										0.2
Boiler feedwater										
0–150 psi									500–3,000	
150–250 psi									500–2,500	
250–400 psi									100–1,500	
over 400 psi									50	
Brewing										
Light beer[c]	100–200	100–200	275–500		100–200	100–500	100	500–1,500[j3]		0.1
Dark beer	100–200	100–200	275–500		100–200	100–500	100	850		0.1
Carbonated beverages[b]				250						0.1
Chemical process industries										
Concrete mixing[b,d]				Low	Low		Low			
Confectionery								100		0.2
Cooling water				Low						0.5
Electroplating				Low						Low
Fermentation			1,500							
Food canning and freezing[b,f]										
Food equipment washing[b]								850		0.2
Food processing – general[b]								850		0.2
Ice manufacture[b]	300	170–300	300		130–300	300	300	170–1,300		0.2
Laundering										0.2
Oil well flooding										0.1
Photographic processing						<100		Low	200	0.02
Plastics – clear, uncolored									200	0.02

2.5-10 INDUSTRIAL WATER QUALITY TOLERANCES (continued)
Table B

Type of industrial use of water	Calcium chloride, mg/l	Magnesium chloride, mg/l	Sodium chloride, mg/l	Sulfates, mg/l	Magnesium sulfates, mg/l	Calcium sulfates, mg/l	Sodium sulfate, mg/l	Total dissolved solids, mg/l	Total solids, mg/l	Iron mg/l
Paper and pulp										
Ground woodpulp[g]								500		0.3
Soda and sulfate pulp[g,h]								250		0.1
Kraft										
Bleached								300		0.2
Unbleached								500		1.0
Fine paper								200		0.1
Rayon (viscose)										
Pulp production									100	0.05
Manufacture										0.0
Steel manufacturing										
Sugar manufacturing[e]		e		20	e					0.1
Tanning										0.1
Textile manufacturing										
General[f]		i		100	i					0.1–1.0
Dyeing										0.25
Wool scouring										1.0
Cotton bandage										0.2

a Calcium is necessary for some yeast actions, so some hardness is desirable. Too much hardness, however, retards fermentation and causes soggy bread.
b Except as specified, should conform to Public Health Service drinking water standards.
c Although the total dissolved solids limit is 500–1,000 ppm, there is a limit of 300 ppm for any one substance.
d Water should be free of acids, alkalies, oil, and decayed vegetable matter. Sugar should not exceed 3,400 ppm; SO_3 should not exceed 25 ppm.
e Magnesium in any form should not exceed 10 ppm.
f No chlorophenols. Hardness varies with product: general canning and freezing 50–85 ppm; peas 200–400 ppm; fruits and vegetables 100–200 ppm; legumes 25–75 ppm.
g Turbidity as SiO_2 must not be gritty.
h Calcium hardness as $CaCO_3$ should not exceed 50 ppm. Magnesium hardness as $CaCO_3$ should not exceed 50 ppm.
i Magnesium should not exceed 5 ppm.
j Not more than 300 mg/l of any one substance.

2.5-10 INDUSTRIAL WATER QUALITY TOLERANCES (continued)
Table C

Type of industrial use of water	Manganese, mg/l	Iron and manganese together, mg/l	Aluminum oxide, mg/l	Silica mg/l	Copper, mg/l	Fluoride, mg/l	Carbonate, mg/l	Sodium carbonate, mg/l	Bicarbonate, mg/l	Nitrate, mg/l
Air conditioning	0.5	0.5								
Baking[a,b]	0.2	0.2								
Boiler feedwater										
0–150 psi			5.0	40			200	50		
150–250 psi			0.5	20			100	30		
250–400 psi			0.05	5			40	5		
over 400 psi			0.01	1			20	0		
Brewing	0.1	0.1		50		1.0	50–68	100		30
Light beer[c]	0.1	0.1		50		1.0	50–68	100		30
Dark beer	0.2	0.2				0.2–1.0				
Carbonated beverages[b]										
Chemical process industries										
Concrete mixing[b,d]										Low
Confectionery	0.2	0.2								
Cooling water	0.5	0.5								
Electroplating							Low			
Fermentation										
Food canning and freezing[b,f]	0.2	0.2				1.0				
Food equipment washing[b]		0.2				1.0				
Food processing – general[b]	0.2	0.3				1.0				
Ice manufacture[b]	0.2	0.2		10		1.0				
Laundering	0.2	0.2								
Oil well flooding									50	
Photographic processing	0.02	0.02							Low	
Plastics – clear, uncolored	0.02	0.02								

2.5-10 INDUSTRIAL WATER QUALITY TOLERANCES (continued)

Table C

Type of industrial use of water	Manganese, mg/l	Iron and manganese together, mg/l	Aluminum oxide, mg/l	Silica mg/l	Copper, mg/l	Fluoride, mg/l	Carbonate, mg/l	Sodium carbonate, mg/l	Bicarbonate, mg/l	Nitrate, mg/l
Paper and pulp										
Ground woodpulp[g]	0.1			50						
Soda and sulfate pulp[g,h]	0.05			20						
Kraft										
Bleached	0.1			50						
Unbleached	0.5			100						
Fine paper	0.05			20						
Rayon (viscose)										
Pulp production	0.03	0.05	8.0	<25	<5					
Manufacture	0.0	0.0								
Steel manufacturing									100	
Sugar manufacturing[e]									Low	
Tanning	0.1	0.2								
Textile and manufacturing										
General[f]	0.05–1.0	0.1–1.0							200	Low
Dyeing	0.25	0.25								
Wool scouring	1.0	1.0								
Cotton bandage	0.2	0.2								

a Calcium is necessary for some yeast actions, so some hardness is desirable. Too much hardness, however, retards fermentation and causes soggy bread.

b Except as specified, should conform to Public Health Service drinking water standards.

c Although the total dissolved solids limit is 500–1,000 ppm, there is a limit of 300 ppm for any one substance.

d Water should be free of acids, alkalies, oil, and decayed vegetable matter. Sugar should not exceed 3,400 ppm; SO_3 should not exceed 25 ppm.

e Magnesium in any form should not exceed 10 ppm.

f No chlorophenols. Hardness varies with product: general canning and freezing 50–85 ppm; peas 200–400 ppm; fruits and vegetables 100–200 ppm; legumes 25–75 ppm.

g Turbidity as SiO_2 must not be gritty.

h Calcium hardness as $CaCO_3$ should not exceed 50 ppm. Magnesium hardness as $CaCO_3$ should not exceed 50 ppm.

2.5-10   INDUSTRIAL WATER QUALITY TOLERANCES (continued)
Table D

Type of industrial use of water	Hydroxide, mg/l	Hydrogen sulfide, mg/l	Algae protozoa	Corrosion tendencies	Scale-forming tendencies	Slime-forming tendencies	Free carbon dioxide, mg/l
Air conditioning		None	None	No		No	
Baking[a,b]		0.2					
Boiler feedwater							
0–150 psi	50	5.0		No	No		
250–400 psi	40	3.0		No	No		
over 400 psi	30	0		No	No		
over psi	15	0		No	No		
Brewing							
Light beer[c]		0.2					
Dark beer		0.2					
Carbonated beverages[b]		0.2	None				
Chemical process industries							
Concrete mixing[b,d]				No			20
Confectionery		0.2					
Cooling water			Low	No	**No**	No	15
Electroplating							
Fermentation							
Food canning and freezing[b,f]		1.0					
Food equipment washing[b]							
Food processing – general[b]							
Ice manufacture[b]							
Laundering							
Oil well flooding			None				
Photographic processing			None				
Plastics – clear, uncolored							

2.5-10 INDUSTRIAL WATER QUALITY TOLERANCES (continued)
Table D

Type of industrial use of water	Hydroxide, mg/l	Hydrogen sulfide, mg/l	Algae protozoa	Corrosion tendencies	Scale-forming tendencies	Slime-forming tendencies	Free carbon dioxide, mg/l
Paper and pulp							
Ground woodpulp[g]							10
Soda and sulfate pulp[g,h]							10
Kraft							
Bleached							10
Unbleached							10
Fine paper							10
Rayon (viscose)							
Pulp production							
Manufacture				No			
Steel manufacturing							
Sugar manufacturing[e]							
Tanning							Low
Textile manufacturing							
General[f]							
Dyeing							
Wool scouring							
Cotton bandage							

[a] Calcium is necessary for some yeast actions, so some hardness is desirable. Too much hardness, however, retards fermentation and causes soggy bread.

[b] Except as specified, should conform to Public Health Service drinking water standards.

[c] Although the total dissolved solids limit is 500—1,000 ppm, there is a limit of 300 ppm for any one substance.

[d] Water should be free of acids, alkalies, oil, and decayed vegetable matter. Sugar should not exceed 3,400 ppm; SO^3 should not exceed 25 ppm.

[e] Magnesium in any form should not exceed 10 ppm.

[f] No chlorophenols. Hardness varies with product: general canning and freezing 50—85 ppm; peas 200—400 ppm; fruits and vegetables 100—200 ppm; legumes 25—75 ppm.

[g] Turbidity as SiO^2 must not be gritty.

[h] Calcium hardness as $CaCO^3$ should not exceed 50 ppm. Magnesium hardness as $CaCO^3$ should not exceed 50 ppm.

Note: All entries are maximum values unless indicated by plus sign; where range is indicated, both extremes are shown.

Source: *Report of the U.S. Study Commission — Texas,* March 1962.

REFERENCES

1. J. Eller *et al.,* *J. Am. Water Works Assn.,* 62(3):149, 1970.
2. T.R. Camp, *Water and Its Impurities,* Reinhold Publishing Corp., 1963, p. 140.
3. Am. Water Works Assn., *Water Quality and Treatment,* 3rd ed., McGraw-Hill, New York, 1971, p. 35.

2.5.4 AGRICULTURAL

2.5-11 TOLERANCE OF CROPS TO HERBICIDES USED IN AND AROUND WATER

Herbicide	Site of use	Formulation	Treatment rate	Concentration that may occur in irrigation water[a]	Crop injury threshold in irrigation water, mg/l[b]
Acrolein	Irrigation canals	Liquid	15 mg/l for 4 hr	10–0.1 mg/l	Flood or furrow: beans-60, corn-60, cotton-80, soybeans-20, sugar beets-60 Sprinkler: corn-60, soybeans-15, sugar beets-15
			0.6 mg/l for 8 hr	0.4–0.02 mg/l	
			0.1 mg/l for 48 hr	0.05–0.1 mg/l	
Aromatic solvents (xylene)	Flowing water in canals or drains	Emulsifiable liquid	5–10 gal/ft³ sec (350–750 mg/l) applied in 30–60 min	700 mg/l or less	Alfalfa >1,600, beans-1,200, carrots-1,600, corn-3,000, cotton-1,600, grain sorghum >800, oats-2,400, potatoes-1,300, wheat >1,200
Copper sulfate	Canals or reservoirs	Pentahydrate crystals	Continuous treatment 0.5–3.0 mg/l, slug treatment: 1/3–1 lb (0.15–0.45 kg)/ft³ sec water flow	0.04–0.8 mg/l during first 10 mi, 0.08–9.0 mg/l during first 10–20 mi	Threshold is above these levels
Dalapon	Banks of canals and ditches	Water soluble salt	15–30 lb/A or 17–34 kg/ha	Less than 0.2 mg/l	Beets >7.0, corn >0.35
Diquat	Injected into water or sprayed over surface	Liquid	3–5 mg/l, 1–1.5 lb/A, or 1.2–1.7 kg/ha	Usually less than 0.1 mg/l	Beans-5.0, corn-125
Diuron	Banks and bottoms of small dry powder ditches	Wettable powder	Up to 64 lb/A or 72 kg/ha	No data	No data

Herbicide	Use location	Formulation	Concentration	Concentration found in water	Crop tolerance
Dichlobenil	Bottoms of dry canals	Granules or wettable powder	7–10 lb/A or 7.9–12.6 kg/ha	No data	Alfalfa-10, corn >10, soybeans-1.0, sugar beets-1.0–10
Endothall	Ponds and reservoirs	Water soluble Na or K salts	1–4 mg/l	Absent or only traces	Corn-25, field beans-1.0, alfalfa >10.0
Entothall amine salts	Reservoirs and static water canals	Liquid or granules	0.5–2.5 mg/l	Absent or only traces	Corn >25, soybeans >25, sugar beets-25
Fenac	Bottoms of dry canals	Liquid or granules	10–20 lb/A or 12.6–25.2 kg/ha	Absent or only traces	Alfalfa-1.0, corn-10, soybeans-0.1, sugar beets-0.1–10
Monuron	Banks and bottoms of small dry powder ditches	Wettable powder	Up to 64 lb/A or 72 kg/ha	No data	No data
Silvex	Woody plants and brambles on floodways, along canal, stream, or reservoir banks	Esters in liquid form	2–4 lb/A or 2.2–4.4 kg/ha	No data. Probably well under 0.1 mg/l	Corn >5.0, sugar beets and soybeans >0.02
	Floating and emersed weeds in southern waterways	Esters in liquid form	2–8 lb/A or 2.2–8.8 kg/ha	0.01–1.6 mg/l 1 day after application	

a Herbicide concentrations given in this column are the highest concentrations that have been found in irrigation water, but these levels seldom remain in the water when it reaches crop land.

b Unless indicated otherwise, all crop tolerance data were obtained by flood or furrow irrigation. Threshold of injury is the lowest concentration causing temporary or permanent injury to crop plants even though, in many instances, neither crop yield nor quality was affected.

2.5-11   TOLERANCE OF CROPS TO HERBICIDES USED IN AND AROUND WATER (continued)

Herbicide	Site of use	Formulation	Treatment rate	Concentration that may occur in irrigation water[a]	Crop injury threshold in irrigation water, mg/l[b]
2,4-D amine	On banks of canals and ditches	Liquid	1–4 lb/A or 1.1–4.4 kg/ha	0.01–0.10 mg/l	Field beans >1.0, grapes->0.7, sugar beets >0.2, soybeans >0.02, corn-10, cucumbers, potatoes, sorghum, alfalfa, peppers >1.0
	Floating and emersed weeds in southern canals and ditches	Liquid	2–4 lb/A or 2.2–4.4 kg/ha	No data. Probably less than 0.1 mg/l	
Picloram	For control of brush on water-sheds	Liquid or granules	1–3 lb/A or 1.1–3.3 kg/ha	No data	Corn >10, field beans-0.1, sugar beets >1.0

[a] Herbicide concentrations given in this column are the highest concentrations that have been found in irrigation water, but these levels seldom remain in the water when it reaches crop land.

[b] Unless indicated otherwise, all crop tolerance data were obtained by flood or furrow irrigation. Threshold of injury is the lowest concentration causing temporary or permanent injury to crop plants even though, in many instances, neither crop yield nor quality was affected.

Source: *Water Quality Criteria, 1972.* National Academy of Sciences, U.S. Government Printing Office, Washington, D.C., 1973.

REFERENCES

1.   H.F. Arle, *Proc. West. Weed Control Conf.*, 12:58, 1950.
2.   H.F. Arle and G.N. McRae, *West. Weed Control Conf. Res. Progr. Report*, 1959, p. 72.
3.   H.F. Arle and G.N. McRae, *West. Weed Control Conf. Res. Progr. Report*, 1960, p. 61.
4.   V.F. Bruns, *Weeds, 3*:359, 1954.
5.   V.F. Bruns, *Weeds, 5*:250, 1957.
6.   V.F. Bruns, *Calif. Weed Conf. Proc., 16*:40, 1964.
7.   V.F. Bruns, *Proc. Wash. State Weed Conf.*, Agr. Exp.Serv., Washington State University, Pullman, 1969, p. 33.
8.   V.F. Bruns and W.J. Clore, *Weeds, 6*:187, 1958.
9.   V.F. Bruns and J.H. Dawson, *Weeds, 7*:333, 1959.
10.   V.F. Bruns et al., *The Use of Aromatic Solvents for Control of Submersed Aquatic Weeds in Irrigation Channels*, U.S. Department of Agriculture, 1955.
11.   V.F. Bruns, R.R. Yeo, and H.F. Arle, *Tolerance of Certain Crops to Several Aquatic Herbicides in Irrigation Water*, U.S. Department of Agriculture, 1964.
12.   P.A. Frank, R.J. Demint, and R.D. Comes, *Weed Sci., 18*(6):687, 1970.
13.   U.S. Department of Agriculture, Agriculture Research Service, Suggested Guide for Weed Control, Agricultural Handbook No. 332, 1969.
14.   R.R. Yeo, *Response of Field Crops to Acrolein*, Research Committee, West. Weed Control Conf., Salt Lake City, Utah, Research Progress Report, 1959, p. 71.
15.   V.F. Bruns, J.M. Hodgson, and N.F. Arle, Response of Several Crops to Six Herbicides in Irrigation Water, 1971, unpublished data.

2.5-12   KEY WATER QUALITY CRITERIA FOR FARMSTEAD USES

	Recommendations (at point of use)	
Characteristic	**General farmstead uses**	**Additional special use requirements**
Taste and odor	Substantially free	
Color	Substantially free	
pH	6.0–8.5	6.8–8.5 dairy sanitation
Total dissolved inorganic solids	500 mg/l (under certain circumstances, higher levels are acceptable)	
Dissolved organic compounds	No recommendations for total organics	
	The concentration of persistent chlorinated organic pesticides should not exceed the following:	

Compound, μg/l

Endrin	1
Aldrin	17
Dieldrin	17
Lindane	56
Toxaphene	5
Heptachlor	18
Heptachlor epoxide	18
DDT	42
Chlordane	3
Methoxychlor	35

Turbidity	Substantially free	
Hazardous trace elements	Levels in excess of those shown are grounds for rejection of a supply:	

Substances, mg/l

Arsenic	0.05
Barium	1.00
Cadmium	0.01
Chromium	0.05
Cyanides	0.2
Lead	0.05
Selenium	0.01
Silver	0.05

Other trace elements	Levels shown below should not be exceeded if alternate sources are available:	

Substances, mg/l

Manganese	0.05	In dairy sanitation, water should
Iron	0.3	contain <20 mg/l potassium
Copper	1.0	and <0.1 mg/l iron and copper
Zinc	5.0	In tooth formation, water should
Fluoride	0.7–1.2	contain <2.5 mg/l fluorides
Nitrate	45.0	

2.5-12   KEY WATER QUALITY CRITERIA FOR FARMSTEAD USES (continued)

Characteristic	Recommendations (at point of use)	
	General farmstead uses	Additional special use requirements
	pCi/l	
Radionuclides	Strontium-90 10 Radium-226 3 In absence of above radionu- clides, 1,000 pCi/l gross β activity	
Nonpathogenic microorganisms	To conform to USPHS drinking water standards	For dairy sanitation, water should not contain more than 20 orga- nisms/ml and contain not more than 5 lypolytic and/or proteolytic organisms/ml

Source: *Water Quality Criteria*, U.S. Department of the Interior, Fed. Water Pollut. Control Adm., April 1968.

2.5-13   KEY WATER QUALITY CRITERIA FOR LIVESTOCK USE

Characteristic	Recommendations
Total dissolved solids (TDS)	< 10,000 mg/l, depending upon animal species and ionic composition of the water
Hazardous trace elements, maximum allowable concentrations	
Arsenic	< 0.05 mg/l
Cadmium	< 0.01 mg/l
Chromium	< 0.05 mg/l
Fluorine	< 2.40 mg/l
Lead	< 0.05 mg/l
Selenium	< 0.01 mg/l
Organic substances	
Algae (water bloom)	Avoid abnormally heavy growth of blue-green algae
Parasites and pathogens	Conform to epidemiological evidence
Dissolved organic compounds	Biological accumulation from environmental sources, including water, shall not exceed established, legal limits in livestock products
Radionuclides	Conform to recommendations for farmstead water supplies

Source: *Water Quality Criteria*, U.S. Department of the Interior, Fed. Water Pollut. Control Adm., April 1968.

2.5.5   FISH AND WILDLIFE

2.5-14   MEDIAN TOLERANCE LIMITS OF CERTAIN CHEMICALS FOR VARIOUS ANIMALS IN DIFFERENT DILUTION MEDIA

Compound and animal tested	Medium	TL_m, mg/l			
		24 hr	48 hr	72 hr	96 hr
Acetic acid					
Culex sp. larvae	RDW	1,500	1,500	–	–
D. magna	SRW	47	–	–	–
L. macrochirus	RDW	>100<1,000	–	–	–
Acetone					
D. magna	RDW	10	10	–	–
Adipic acid					
L. macrochirus	RDW	<330	–	–	–
Ammonium chloride					
C. carassius	SRW	640	–	–	–
D. magna	ULW	202	161	67	50
D. magna	SRW	–	–	–	139[a]
L. macrochirus	SRW	725	725	725	725
Lymnaeae sp. egg	ULW	241	173	73	70
Ammonium hydroxide					
D. magna	SRW	60[a]	32[a]	–	20[a]
Ammonium sulfate					
D. magna	SRW	423[a]	433[a]	–	292[a]
Ammonium sulfite					
D. magna	SRW	299[a]	273[a]	–	203[a]
Barium chloride					
Rana sp. eggs	ULW	244.3 mg/l – 91.67% development			
		2,443.0 mg/l – 89.55% development			
		24,430.0 mg/l – 0.00% development			
Butyric acid					
D. magna	SRW	–	61	–	–
L. macrochirus	RDW	200	–	–	–
Calcium chloride					
D. magna	SRW	3,526[a]	3,005[a]	–	–
D. magna	ULW	1,838	759	759	649[a]
L. macrochirus	SRW	8,350	–	–	–
Lymnaea sp. eggs	ULW	4,485	3,094	3,308	2,573
Capric acid					
L. macrochirus	RDW	I			
Caproic acid					
L. macrochirus	RDW	>150<200	–	–	–
Caprylic acid					
L. macrochirus	RDW	I	–	–	–
Chromic sulfate					
D. magna	ULW	0.1	0.03	–	–
Chromic sulfate plus sodium dichromate					
Lymnaea sp. eggs	ULW	0.17–0.22	0.17–0.22	–	–
Ferric chloride					
D. magna	SRW	36[a]	21[a]	–	15[a]
Formaldehyde					
D. magna	RDW	>100<1,000	–	–	–
Formic acid					
L. macrochirus	RDW	175	–	–	–
Glutaric acid					
L. macrochirus	RDW	330	–	–	–

2.5-14   MEDIAN TOLERANCE LIMITS OF CERTAIN CHEMICALS FOR VARIOUS ANIMALS
IN DIFFERENT DILUTION MEDIA (continued)

Compound and animal tested	Medium	TLm, mg/l			
		24 hr	48 hr	72 hr	96 hr
Magnesium chloride					
D. magna	SRW	3,391[a]	3,699[a]	–	3,484[a]
Magnesium sulfate					
D. magna	ULW	963	929	861	788
D. magna	SRW	–	–	–	3,803
L. macrochirus	SRW	19,000	–	–	–
Lymnaea sp. eggs	ULW	10,530	6,525	6,300	6,250
Malonic acid					
L. macrochirus	RDW	150	–	–	–
Phenol					
D. magna	RDW	100	100	–	–
D. magna young	SRW	17[a]	7[a]	–	–
D. magna adult	SRW	61[a]	21[a]	–	–
L. macrochirus	SRW	>10<15[a]	–	–	–
M. latipinna	SRW	63[a]	22[a]	–	–
Potassium chloride					
D. magna	SRW	–	–	–	679[a]
D. magna	ULW	343	337	117	29
L. macrochirus	SRW	5,500	–	–	–
Lymnaea sp. eggs	ULW	1,941	1,492	1,018	1,100
Potassium cyanide					
D. magna	ULW	2	2	0.7	0.4
Lymnaea sp. eggs	ULW	796	796	147	130
Potassium dichromate					
C. carassius	SRW	705	–	–	–
D. magna	SRW	–	–	–	0.4[a]
L. macrochirus	SRW	739	–	–	–
Potassium ferricyanide					
D. magna	ULW	905	549	0.6	0.1
Potassium nitrate					
D. magna	ULW	490	490	226	39
D. magna	SRW	–	–	–	900[a]
L. macrochirus	SRW	5,500	–	–	–
Lymnaea sp. eggs	ULW	1,941	1,492	1,018	1,100
Propionic acid					
Culex sp. larvae	RDW	>1,000	>1,000	–	–
D. magna	SRW	–	50	–	–
L. macrochirus	RDW	188	–	–	–
Pyridine					
D. magna	RDW	2,114	944	–	–
Sodium acetate					
L. macrochirus	RDW	5,000	–	–	–
Culex sp. larvae	RDW	7,500	7,425	–	–
Sodium anthraquinone alpha-sulfonate					
D. magna	ULW	186	186	186	50
D. magna	SRW	–	–	–	12[a]
Lymnaea sp. eggs	ULW	186	186	186	186
Sodium bicarbonate					
Culex sp. larvae	RDW	2,000	2,000	–	–

2.5-14   MEDIAN TOLERANCE LIMITS OF CERTAIN CHEMICALS FOR VARIOUS ANIMALS IN DIFFERENT DILUTION MEDIA (continued)

Compound and animal tested	Medium	TL_m, mg/l			
		24 hr	48 hr	72 hr	96 hr
Sodium bisulfate					
Culex sp. larvae	RDW	300	300	–	–
Sodium bisulfite					
D. magna young	SRW	116[a]	81[a]	–	–
D. magna adult	SRW	–	–	–	102
D. magna adult	ULW	171	119	97	82
Dugesia sp.	ULW	179	179	179	179
Lymnaea sp. eggs	ULW	179	59	59	59
M. latipinna	SRW	241[a]	220[a]	–	–
Sodium bisulfite plus sodium carbonate					
D. magna	ULW	1,484–665	1,484–665	1,060–475	99–45
D. magna	SRW	–	–	–	436–85
Sodium bisulfite plus sodium chromate					
D. magna	SRW	–	–	–	67–0.278[a]
Sodium bisulfite plus sodium silicate					
D. magna	SRW	14,210–950	11,723–784	222–15	222–15
Sodium bisulfite plus sodium sulfate					
D. magna	SRW	–	–	–	82–3,654
Sodium bisulfite plus sodium carbonate and sodium chromate					
D. magna	SRW	–	–	–	86–441–0.354[a]
Sodium bisulfite plus sodium carbonate and sodium silicate					
D. magna	SRW	–	–	–	39–198–93[a]
Sodium bisulfite plus sodium carbonate and sodium sulfate					
D. magna	SRW	–	–	–	57–296–2,569[a]
Sodium bisulfite plus sodium chromate and sodium silicate					
D. magna	SRW	–	–	–	224–0.86–506[a]
Sodium bisulfite plus sodium chromate and sodium sulfate					
D. magna	SRW	–	–	–	78–0.32–3,443
Sodium bisulfite plus sodium silicate and sodium sulfate					
D. magna	SRW	–	–	–	52–126–2.326[a]
Sodium *p*-bromobenzene sulfonate					
D. magna	ULW	2,347	1,943	971	809
D. magna	SRW	–	–	–	523[a]
L. macrochirus	SRW	<1,560	–	–	–
Lymnaea sp. eggs	ULW	2,590	2,590	2,590	2,590

2.5-14   MEDIAN TOLERANCE LIMITS OF CERTAIN CHEMICALS FOR VARIOUS ANIMALS
IN DIFFERENT DILUTION MEDIA (continued)

Compound and animal tested	Medium	TL$_m$, mg/l			
		24 hr	48 hr	72 hr	96 hr
Sodium butyl sulfonate					
D. magna	ULW	8,000	8,000	5,400	2,700
Sodium butyrate					
L. macrochirus	RDW	5,000	–	–	–
Sodium carbonate					
Amphipoda	ULW	360	176	67	67
Culex sp. larvae	RDW	1,820	600	–	–
D. magna	ULW	347	265	–	–
D. magna	SRW	607[a]	565	–	524
Dugesia sp.	ULW	384	360	360	341
L. macrochirus	ULW	385	–	–	–
Lymnaea sp. eggs	ULW	403	403	395	411
M. latipinna	SRW	405[a]	297[a]	–	–
Sodium carbonate plus sodium chromate					
D. magna	SRW	–	–	–	420–0.34[a]
Sodium carbonate plus sodium chromate and sodium silicate					
D. magna	SRW	–	–	–	187–0.15–88
Sodium carbonate plus sodium chromate and sodium sulfate					
D. magna	SRW	–	–	–	240–0.19–2,078
Sodium carbonate plus sodium silicate					
D. magna	ULW	265–1,776	265–1,776	265–1,776	265–1,776
Sodium carbonate plus sodium silicate and sodium sulfate					
D. magna	SRW	–	–	–	161–76–1,396[a]
Sodium carbonate plus sodium sulfate					
D. magna	ULW	198–666	172–577	172–577	66–222
Sodium chloride					
C. carassius	SRW	13,750	–	–	–
Culex sp. larvae	RDW	10,500	10,200	–	–
D. magna	SRW	6,447[a]	5,874[a]	–	3,114[a]
D. magna	ULW	3,412	3,310	–	–
L. macrochirus	SRW	14,125	–	–	–
Lymnaea sp. eggs	ULW	3,412	3,388	–	–
M. latipinna	SRW	18,735[a]	16,595	–	–
Sodium p-chlorobenzene sulfonate					
D. magna	ULW	8,600	7,659	3,964	2,150
D. magna	SRW	–	–	–	2,394[a]
L. macrochirus	SRW	<3,219	–	–	–
Lymnaea sp. eggs	ULW	8,600	7,633	6,343	5,053
Sodium 2-chlorotoluene-4-sulfonate					
L. macrochirus	SRW	<1,374	–	–	–

2.5-14   MEDIAN TOLERANCE LIMITS OF CERTAIN CHEMICALS FOR VARIOUS ANIMALS
IN DIFFERENT DILUTION MEDIA (continued)

Compound and animal tested	Medium	TL$_m$, mg/l			
		24 hr	48 hr	72 hr	96 hr
Sodium 2-chlorotoluene-5-sulfonate					
D. magna young	SRW	0.8[a]	0.6[a]	–	0.4[a]
D. magna adult	SRW	3.3[a]	1.3[a]	–	–
Lymnaea sp. eggs	SRW	30[a]	–	–	–
M. latipinna	SRW	115.2[a]	66.1[a]	–	–
Sodium chromate plus sodium silicate					
D. magna	SRW	–	–	–	0.21–130[a]
Sodium chromate plus sodium silicate and sodium sulfate					
D. magna	SRW	–	–	–	0.28–122–2,255[a]
Sodium chromate plus sodium sulfate					
D. magna	SRW	–	–	–	0.28–3,044[a]
Sodium 2,5-dichlorobenzene sulfonate					
D. magna	ULW	4,931	4,931	2,490	938
D. magna	SRW	–	–	–	1,468[a]
L. macrochirus	SRW	<3,750	–	–	–
Lymnaea sp. eggs	ULW	4,981	4,513	3,984	3,144
Sodium dichromate					
D. magna	ULW	22	10	–	–
Sodium formate					
L. macrochirus	RDW	5,000	–	–	–
Sodium nitrate					
C. carassius	SRW	12,150	–	–	–
D. magna	ULW	5,980	3,581	2,125	665
D. magna	SRW	–	–	–	4,206
L. macrochirus	SRW	12,800	–	–	–
Lymnaea sp. eggs	ULW	6,375	6,460	5,950	3,251
Sodium m-nitrobenzene sulfonate					
D. magna	ULW	8,665	8,665	6,017	5,067
D. magna	SRW	–	–	–	2,235[a]
L. macrochirus	SRW	<1,350	–	–	–
Sodium 4-nitrochlorobenzene-2-sulfonate					
D. magna	ULW	4,698	3,483	948	948
D. magna	SRW	–	–	–	1,474
L. macrochirus	SRW	6,375	–	–	6,375[a]
Lymnaea sp. eggs	ULW	3,532	4,439	3,736	3,208
Sodium 4-nitrotoluene-2-sulfonate					
L. macrochirus	SRW	<1,440	–	–	–
Sodium oxalate					
L. macrochirus	RDW (without CaCl$_2$)	4,000	–	–	–
Sodium p-phenolsulfonate					
D. magna	ULW	13,510	13,510	3,494	1,471
L. macrochirus	SRW	–	–	–	19,616[a]
Lymnaea sp. eggs	ULW	10,700	9,122	8,828	8,828

2.5-14   MEDIAN TOLERANCE LIMITS OF CERTAIN CHEMICALS FOR VARIOUS ANIMALS
IN   DIFFERENT   DILUTION   MEDIA   (continued)

| Compound and animal tested | Medium | TL_m, mg/l | | | |
		24 hr	48 hr	72 hr	96 hr
Sodium mono-hydrogen phosphate					
D. magna	SRW	1,154[a]	1,089[a]	–	426[a]
Sodium mono-hydrogen phosphate plus sodium pyrophosphate					
D. magna	ULW	3,580–433	3,580–391	–	–
Lymnaea sp. eggs	ULW	2,685–63	2,954–223	2,828–60	2,685–248
Sodium phosphate					
D. magna	SRW	237[a]	177[a]	–	126[a]
Sodium propionate					
Culex sp. larvae	RDW	–	2,320	–	–
L. macrochirus	RDW	5,000	–	–	–
Sodium pyrophosphate					
D. magna	ULW	433	391	–	–
Sodium silicate					
Amphipoda	ULW	895	263	261	160
D. magna	ULW	575	494	413	216
D. magna	SRW	–	–	–	247
Lymnaea sp. eggs	ULW	632	630	630	632
Sodium sulfate					
Amphipoda	ULW	2,380	1,110	880	880
Culex sp. larvae	RDW	11,430	13,350	–	–
D. magna	ULW	8,384	2,564	725	630
D. magna adult	SRW	–	–	–	4,547
D. magna young	SRW	6,800	6,100	–	–
L. macrochirus	SRW	17,500	–	–	–
Lymnaea sp. eggs	ULW	5,401	5,400	5,400	3,553
M. latipinna	SRW	20,040	15,996	–	–
Sodium sulfide					
D. magna	SRW	16[a]	13[a]	–	9[a]
Sodium sulfite					
D. magna	SRW	299[a]	273[a]	–	203[a]
Sodium thiosulfate					
D. magna	SRW	2,245[a]	1,334[a]	–	805[a]
Sodium valerate					
L. macrochirus	RDW	5,000	–	–	–
Valeric acid					
D. magna	SRW	–	45	–	–
Xylene					
D. magna	RDW	>100<1,000	–	–	–
Zinc stearate					
L. macrochirus	RDW	I			

[a] Denotes 25 hr when placed after a value in the 24 hr column, 50 hr when in the 48 hr column, and 100 hr when in the 96 column.

Note: TL_m mg/l – median tolerance limit.
SRW – Standard Reference Water.[1]
RDW – Reference Dilution Water.[2]
ULW – glass-wool filtered University Lake water.
I – chemical to insoluble in water to be toxic.

Source: B.F. Dowden and H.J. Bennett, *J. Water Pollut. Control Fed.*, *37*(9):1310, 1965.

REFERENCES

1. L. Freeman, *Sew. Ind. Wastes*, 25(7):845, 1953.
2. B.F. Dowden, *Proc. Nat. Acad. Sci.*, 23:77, 1960.

2.5-15 48 HOUR TL$_m$ VALUES OF PESTICIDES FROM STATIC BIOASSAY FOR VARIOUS TYPES OF FRESH WATER ORGANISMS

o provide reasonably safe concentrations of these materials in receiving waters, application factors ranging from 1/10–1/100 ld be used with these values depending on the characteristic of the pesticide in question and used as specified in *Water ity Criteria,* U.S. Government Printing Office, 1968, p. 58. Concentrations thus derived tentatively may be considered safe r the environmental conditions recommended.

mg/l

Insecticides

Pesticide	Stream invertebrate[a] Species	TL$_m$	Cladocerans[b] Species	TL$_m$	Fish[c] Species	TL$_m$	Gammarus lacustris,[d] TL$_m$
Abate	Pteronarcys californica	100			Brook trout	1,500	640
Aldrin[e]	P. californica	8	Daphnia pulex	28	Rainbow trout	,3	12,000
Allethrin	P. californica	28	D. pulex	21	Rainbow trout	19	20
Azodrin					Rainbow trout	7,000	
Aramite			D. magna	345	Bluegill	35	100
Baygon[e]	P. californica	100			Fathead	25	50
Baytex[e]	P. californica	130	Simocephalus serrulatus	3.1	Brown trout	80	70
Benzene hexachloride (lindane)	P. californica	8	D. pulex	460	Rainbow trout	18	88
Bidrin	P. californica	1,900	D. pulex	600	Rainbow trout	8,000	790
Carbaryl (sevin)	P. californica	1.3	D. pulex	6.4	Brown trout	1,500	22
Carbophenothion (trithion)			D. magna	0.09	Bluegill	225	28
Chlordane[e]	P. californica	55	S. serrulatus	20	Rainbow trout	10	80
Chlorobenzilate			S. serrulatus	550	Rainbow trout	710	
Chlorthion		4.5	D. magna	4.5			
Coumaphos			D. magna	1			0.14
Cryolite			D. pulex	5,000	Rainbow trout	47,000	
Cyclethrin			D. magna	55			
DDD (TDE)[e]	P. californica	1,100	D. pulex	3.2	Rainbow trout	9	1.8
DDT[e]	P. californica	19	D. pulex	0.36	Bass	2.1	2.1
Delnav (dioxathion)					Bluegill	14	690
Delmeton (systex)				14	Bluegill	81	
Diazinon[e]	P. californica	60	D. pulex	0.9	Bluegill	30	500
Dibrom (naled)	P. californica	16	D. pulex	3.5	Brook trout	78	160
Dieldrin[e]	P. californica	1.3	D. pulex	240	Bluegill	3.4	1,000
Dilan			D. magna	21	Bluegill	16	600
Dimethoate (cygon)	P. californica	140	D. magna	2,500	Bluegill	9,600	400
Dimethrin					Rainbow trout	700	
Dichlorvos (DDVP)[e]	P. californica	10	D. pulex	0.07	Bluegill	700	1
Disulfoton (di-syston)	P. californica	18			Bluegill	40	70
Dursban	Peteronareella badia	1.8			Rainbow trout	20	0.4
Endosulfan (thiodan)	P. californica	5.6	D. magna	240	Rainbow trout	1.2	64
Endrin[e]	P. californica	0.8	D. pulex	20	Bluegill	0.2	4.7
EPH			D. magna	0.1	Bluegill	17	36
Ethion ion	P. californica	14	D. magna	0.01	Bluegill	230	3.2
Ethyl guthion[e]			D. pulex		Rainbow trout		

2.5-15   48 HOUR TL$_m$ VALUES OF PESTICIDES FROM STATIC BIOASSAY FOR VARIOUS TYPES OF FRESH WATER ORGANISMS (continued)

Insecticides

Pesticide	Stream invertebrate[a] Species	TL$_m$	Cladocerans[b] Species	TL$_m$	Fish[c] Species	TL$_m$	*Gammarus lacustris*,[d] TL$_m$
Fenthion	P. californica	39	D. pulex	4			
Guthion[e]	P. californica	8	D. magna	0.2	Rainbow trout	10	0.3
Heptachlor[e]	P. badia	4	D. pulex	42	Rainbow trout	9	100
Kelthane (dicofol)	P. californica	3,000	D. magna	390	Rainbow trout	100	
Kepone					Rainbow trout	37.5	
Malathion[e]	P. badia	6	D. pulex	1.8	Brook trout	19.5	1.8
Methoxychlor[e]	P. californica	8	D. pulex	0.8	Rainbow trout	7.2	1.3
Methyl parathion[e]			D. magna	4.8	Bluegill	8,000	
Morestan	P. californica	40			Bluegill	96	
Ovex	P. californica	1,500			Bluegill	700	
Paradichlorobenzene					Rainbow trout	880	
Parathion[e]	P. californica	11	D. pulex	0.4	Bluegill	47	6
Perthane			D. magna	9.4	Rainbow trout	7	
Phosdrin[e]	P. californica	9	D. pulex	0.16	Rainbow trout	17	310
Phosphamidon	P. californica	460	D. magna	4	Rainbow trout	8,000	3.8
Pyrethrins	P. californica	64	D. pulex	25	Rainbow trout	54	18
Rotenone	P. californica	900	D. pulex	10	Bluegill	22	350
Strobane[e]	P. californica	7			Rainbow trout	2.5	
Tetradifon (tedion)					Bluegill	1,100	140
TEPP[e]					Fathead	390	52
Thanite			D. magna	450			
Thimet					Bluegill	5.5	70
Toxaphene	P. californica	7	D. pulex	15	Rainbow trout	2.8	70
Trichlorfon (dipterex)[e]	P. badia	22	D. magna	8.1	Rainbow trout	160	60
Zectran	P. californica	16	D. pulex	10	Rainbow trout	8,000	76

Herbicides, Fungicides, Defoliants, Algicides

Pesticide	Stream invertebrate[a] Species	TL$_m$	Cladocerans[b] Species	TL$_m$	Fish[c] Species	TL$_m$	*Gammarus lacustris*,[d] TL$_m$
Ametryne					Rainbow trout	3,400	
Aminotriazole							
Aquathol					Bluegill	257	
Atrazine			Daphnia magna	3,600	Rainbow trout	12,600	
Azide, potassium					Bluegill	1,400	10,000
Azide, sodium					Bluegill	980	9,000
Copper chloride					Bluegill	1,100	
Copper sulfate					Bluegill	150	
Dichlobenil	Pteronarcys californica	44,000	Daphnia pulex	3,700	Bluegill	20,000	1,500
2,4-D, PGBEE			D. pulex	3,200	Rainbow trout	960	1,800
2,4-D, BEE	P. californica	1,800			Bluegill	2,100	760
2,4-D, isopropyl					Bluegill	800	
2,4-D, butyl ester					Bluegill	1,300	
2,4-D, butyl + ispropyl ester					Bluegill	1,500	
2,4,5-T isooctyl ester					Bluegill	16,700	
2,4,5-T isopropyl ester					Bluegill	1,700	
2,4,5-T PGBE					Bluegill	560	
2(2,4-DP) BEE					Bluegill	1,100	

2.5-15   48 HOUR TL$_m$ VALUES OF PESTICIDES FROM STATIC BIOASSAY FOR VARIOUS TYPES OF FRESH WATER ORGANISMS (continued)

Herbicides, Fungicides, Defoliants, Algicides

Pesticide	Stream invertebrate[a] Species	TL$_m$	Cladocerans[b] Species	TL$_m$	Fish[c] Species	TL$_m$	Gammarus lacustris,[d] TL$_m$
Dalapon	*P. californica* Very low toxicity		*D. magna*	6,000	Very low toxicity		
Dead-X	*P. californica*	5,000	*D. pulex*	3,700	Rainbow trout	9,400	5,600
DEF	*P. californica*	2,300			Bluegill	36	230
Dexon	*P. californica*	42,000			Bluegill	23,000	6,000
Dicamba					non-toxic		5,800
Dichlone			*D. magna*	26	Rainbow trout	48	11,500
Difolatan	*P. californica*	150			Channel cat	31	6,500
Dinitrocresol	*P. californica*	560			Rainbow trout	210	
Diquat					Rainbow trout	12,300	
Diuron	*P. californica*	2,800	*D. pulex*	1,400	Rainbow trout	4,300	380
Du-ter					Bluegill	33	
Dyrene			*D. magna*	490		15	
Endothal, copper					Rainbow trout	290	
Endothal, dimethylamine					Rainbow trout	1,150	
Fenac, acid	*P. californica*	70,000			Rainbow trout	16,500	
Fenac, sodium	*P. californica*	80,000	*D. pulex*	4,500	Rainbow trout	7,500	18,000
Hydram (molinate)	*P. californica*	3,500			Rainbow trout	290	
Hydrothol 191					Rainbow trout	690	1,000
Lanstan (korax)					Rainbow trout	100	5,500
LFN isopropyl ester					Rainbow trout	79	
Paraquat	*P. californica* Very low toxicity		*D. pulex*	3,700	Very low toxicity		18,000
Propazine					Rainbow trout	7,800	
Silvex, PGBEE			*D. pulex*	2,000	Rainbow trout	650	
Silvex, isooctyl					Bluegill	1,400	
Silvex, BEE					Bluegill	1,200	
Simazine	*P. californica*	50,000			Rainbow trout	5,000	21,000
Sodium arsenite	*P. californica* Very low toxicity		*Simocephalus serrulatus*	1,400	Rainbow trout	36,500	
Tordon (picloram)					Rainbow trout	2,500	48,000
Trifuralin	*P. californica*	4,200	*D. pulex*	240	Rainbow trout	11	5,600
Vernam (vernolate)[e]					Rainbow trout	5,900	25,000

[a] Stonefly bioassay was done at Denver, Colorado, and at Salt Lake City, Utah. Denver tests were in soft water (35 mg/l TDS), nonaerated, 60°F. Salt Lake City tests were in hard water (150 mg/l TDS), aerated, 48—50°F. Response was death.

[b] *Daphnia pulex* and *Simocephalus serrulatus* bioassay were done at Denver, Colorado, in soft water (35 mg/l TDS), nonaerated, 60°F. *Daphnia magna* bioassay was done at Pennsylvania State University in hard water (146 mg/l TDS), nonaerated, 68°F. Response was immobilization.

[c] Fish bioassay was done at Denver, Colorado, and at Rome, N.Y. Denver tests were with 2 in. fish in soft water (35 mg/l TDS), nonaerated; trout at 55°F; other species at 65°F. Rome tests were with 2—2 1/2 in. fish in soft water (6 mg/l TA:pH 5.85—6.4), 60°F. Response was death.

[d] *Gammarus* bioassay was done at Denver, Colorado, in soft water (35 mg/l TDS), nonaerated, 60°F. Response was death.

[e] Becomes bound to soil when used according to directions, but highly toxic (reflected in numbers) when added directly to water.

Source: *Handbook of Analytical Toxicology,* I. Sunshine, Ed., Chemical Rubber Co., Cleveland, 1969, p. 582.

2.5-16 COMPARATIVE TOXICITIES OF INSECTICIDES TO FISH AND MAMMALS AND SOLUBILITY OF INSECTICIDES IN WATER

Compound	96-hr TL_m[a] flathead minnows, mg/l[b,1,5]	LD_{50} for male white rats,[c] mg/kg[1]	LD_{50} for rats/kg body weight, mg[2,4]	Rats LD_{50} mg/kg[f,3]	Bluegills TL_m 96 hr mg/l[g,3]	Solubility in water, mg/l Ref. 1	Solubility in water, mg/l Ref. 3[h]
Chlorinated hydrocarbons							
Endrin	0.0013	17.8	5–45[d]	7.5	0.0006	0.1	0.19–0.26
Toxaphene	0.0051	90	40	–	–	Nil	
Dieldrin	0.016	46	60	46	0.0079	0.1	0.14–0.18
Aldrin	0.028	67	50	–	–	0.1	
DDT	0.034	113	250[e]	118	0.016	0.1	0.016–0.040[i]
Methoxychlor	0.035	6,000	5,000	–	–	Nil	
Heptachlor	0.056	100	90	–	–	Nil	
Lindane	0.056	125	125	91	0.077	10	0.5–6.6
Chlordane	0.069	335	500	–	–	Nil	
BHC	2.0	1,000	–	–	–	5–10	
Organic phosphorus							
EPN	0.20	12–40	13.6–42[d]	–	–	Slight	
Para-Oxon	0.33	3–3.5	3.5	–	–	20	
Parathion	1.4	13	6–15[d]	3.6	0.700	20	18–31
TEPP	1.7	2	1.2–2.0	–	–	High	
Chlorothion	3.2	880	1,500	–	–	40	
Systox	3.6	6.2	6–12	–	–	20	
Methyl parathion	8.3	14–42	9–25	–	–	50	
Malathion	12.5	1,375	1,156	–	–	145	
OMPA	121	10	–	–	–	High	
Dipterex	180	630	450	–	–	130,000	

a Median tolerance dose, or concentration at which 50% of organisms die.
b As 100% active compound.
c Oral feeding. The lethal dose at which 50% of the rats die.
d Males are more tolerant than females.
e LD_{50} man.
f Acute oral toxicity to female rats, in mg pesticide/kg body weight.
g 96 hr median tolerance limits, in mg of pesticide/l water.
h Apparent solubility in distilled water, varies with particle size.
i Technical grade DDT includes o,p' and p,p' isomers + DDE.

REFERENCES

1. R.T. Skrinde et al., *J. Am. Water Works Assn.*, 54(11):1407, 1962.
2. Compiled by H.J. Webb, *J. Am. Water Works Assn.*, 54(1):83, 1962.
3. G.G. Robeck et al., *J. Am. Water Works Assn.*, 57(2):181, 1965.
4. Pesticide Chemicals Official Compendium, Assn. Am. Pest. Control Off., College Park, Md.
5. C. Henderson et al., *The Toxicity of Organic Phosphorous and Chlorinated Hydrocarbon Insecticides to Fish*, a paper presented at the second seminar on Biological Problems in Water Pollution, Cincinnati, Ohio, 1959.

2.5-17   TL$_m$ VALUES FOR SHRIMP

Organochloride pesticides	μg/l
Aldrin	0.04
HC	2.0
Chlordane	2.0
Endrin	0.2
Heptachlor	0.2
Lindane	0.2
DDT	0.6
Dieldrin	0.3
Endosulfan	0.2
Methoxychlor	4.0
Perthane	3.0
DDE	3.0
Toxaphene	3.0

Organophosphorus pesticides	μg/l
Coumaphos	2.0
Dursban	3.0
Fenthion	0.03
Naled	3.0
Parathion	1.0
Ronnel	5.0

Note: The 48 hour TL$_m$ is listed for each chemical in μg/l.

Source: *Water Quality Criteria,* U.S. Department of the Interior, Fed. Water Pollut. Control Adm., April 1968.

2.5-18   EXAMPLES OF FISH AS INDICATORS OF WATER SAFETY FOR LIVESTOCK

Material	Toxic levels, mg/l for fish	Toxic effects on animals
Aldrin	0.02	3 mg/kg food (poultry)
Chlordane	1.0 (sunfish)	91 mg/kg body weight in food (cattle)
Dieldrin	0.025 (trout)	25 mg/kg food (rats)
Dipterex	50.0	10.0 mg/kg body weight in food (calves)
Endrin	0.003 (bass)	3.5 mg/kg body weight in food (chicks)
Ferban, fermate	1.0–4.0	
Methoxychlor	0.2 (bass)	14 mg/kg alfalfa hay, not toxic (cattle)
Parathion	2.0 (goldfish)	75 mg/kg body weight in food (cattle)
Pentachlorophenol	0.35 (bluegill)	60 mg/l drinking water not toxic (cattle)
Pyrethrum (allethrin)	2.0–10.0	1,400–2,800 mg/kg body weight in food (rats)
Silvex	5.0	500–2,000 mg/kg body weight in food (chicks)
Toxaphene	0.1 (bass)	35–110 mg/kg body weight in food (cattle)

Source: *Water Quality Criteria,* U.S. Department of the Interior, Fed. Water Pollut. Control Adm., April 1968. Compiled from data in *Water Quality Criteria,* Cal. State Water Qual. Control Board, 1963.

2.6   TREATMENT

2.6-1   WATER TREATMENT PROCESSES NOW BEING CONSIDERED FOR BAT[a]

Conventional processes	Advanced processes
Coagulation, sedimentation, filtration	Activated alumina
	Adsorption
Direct filtration	GAC
Diatomaceous earth filtration	Powdered activated carbon
Slow sand filtration	Resins
Lime softening	Aeration
Ion exchange	Packed column
Oxidation-disinfection	Diffused air
Chlorination	Spray
Chlorine dioxide	Slat tray
Chloramines	Mechanical
Ozone	Cartridge filtration
Bromine	Electrodialysis
Others	Reverse osmosis
	Ultrafiltration
	Ultraviolet light (UV)
	UV with other oxidants

[a]   BAT, best available technology.

Source: J.F. Dyksen, D.J. Hiltebrand, and R.F. Raczko, *Jour. Am. Water Works Assn., 80*:1, 30, 1988.

2.6-2 CONVENTIONAL TREATMENT PROCESSES

Communities of 25,000 or More

Process	1958[1]		1960[2]		1962[3]		1964[4]	
	No. plants	Estimated population served	No. plants	Estimated population served	No. plants	Estimated population served	No. plants	Estimated population served
Filtration								
Rapid sand (gravity)	316	47,835,760	338	49,711,716	358	48,911,243	474	58,845,020
Rapid sand (pressure)	19	2,712,970	22	1,864,390	21	1,902,572	33	2,152,295
Slow sand	23	5,752,875	2	97,265	19	5,386,790	25	4,152,490
Anthracite	45	5,208,330	49	5,517,605	65	6,704,995	87	6,839,710
Diatomaceous earth	—	—	—	—	—	—	2	362,000
Aeration	154	12,242,585	157	12,782,050	185	14,721,204	233	15,250,335
Softening								
Lime soda	43	6,235,350	42	6,020,867	42	6,168,298	67	10,015,730
Lime	36	6,052,150	46	7,294,978	58	7,497,542	66	8,456,320
Cation exchange	12	1,295,260	16	1,529,700	15	1,333,659	22	3,260,620
Taste and odor control	226[a]	34,294,615	249[a]	35,518,613	285[a]	36,514,982[a]	348	43,985,135
Activated carbon	182	28,060,625	198	29,650,550	232	30,876,855		
Chlorine dioxide	37	5,586,130	39	5,643,374	45	5,303,549		
Sulfur dioxide	—	—	—	—	5	672,208		
Others and not stated	38	6,637,760	40	4,858,720	40	4,308,083		
Corrosion control	236[a]	32,924,240	285[a]	34,791,144	333[a]	36,614,370[a]	421	39,464,595
Phosphate compounds	95	9,502,270	112	10,790,574	138	11,508,938		
Alkali for pH adjustment	161	24,781,475	195	25,392,858	216	26,608,720		
Chlorine gas	11	792,960	12	929,680	19	977,586		
Hypochlorites	—	—	—	—	3	16,225		
Sodium silicate	4	622,500	5	687,000	5	850,964		

a Figures entered under "Taste and odor control" and "Corrosion control" are not additive, because more than one chemical may be used.

REFERENCES

1. K.H. Jenkins, *J. Am. Water Works Assn.*, 53(1):31, 1961.
2. K.H. Jenkins and H.W. Owen, *J. Am. Water Works Assn.*, 55(1):35, 1963.
3. K.H. Jenkins, *J. Am. Water Works Assn.*, 55(12):1485, 1963.
4. K.H. Jenkins, *J. Am. Water Works Assn.*, 57:1360, 1965.

2.6-3   DESIGN CHARACTERISTICS OF WATER TREATMENT PROCESSES

	Design characteristic				
Treatment method	Loading rate, gpm/ft^2	Site area per mgd, ft^2	Head loss, ft	Proportion of wash water to design capacity, %	Cost[a] per mgd, $1,000
Filtration					
Rapid sand	3	375	6	1.5	15
Anthracite	3	375	6	1	16
Sand-anthracite	6	260	6	1.5	12.5
Diatomaceous earth	1.5	150	20	1	25
Microstraining	15	75	1.5	1	20
Sedimentation					
Conventional rate	0.85	1,050	1	0.5	12.5
High rate	1.25	630	1	0.5	8

[a] Estimated for 40 mgd capacity facility.

Source: Reprinted, with publisher's permission, from *J. Am. Water Works Assn.*, *56*(5): 535, © 1964 by the American Water Works Association, Inc.

2.6.1   CHEMICALS USED

2.6-4   PHASES OF TREATMENT IN WHICH VARIOUS CHEMICALS ARE USED

Chemical	Coagulant or aid	Alkalinity and pH adjustment	Disinfection	Dechlorination	Mineral oxidation	Taste and odor control	Algae control	Corrosion control	Softening	Fluoridation[a]
Activated carbon				x		x	x			
Aluminum sulfate	x									
Ammonia, anhydrous			x	x						
Ammonium hydroxide			x	x						
Ammonium sulfate	x		x							
Bentonite	x									
Calcium carbonate	x	x								
Calcium hydroxide		x						x	x	
Calcium oxide		x						x	x	
Carbon dioxide		x							x	
Chlorine			x		x	x	x			
Chlorine dioxide			x		x	x	x			
Copper sulfate							x			
Ferric chloride	x									
Ferric sulfate	x									
Ferrous sulfate	x									
Fluosilicic acid										x
Hydrochloric acid		x								
Hypochlorite (Ca)			x							
Hypochlorite (Na)			x							
Ozone			x		x	x				
Potassium permanganate						x				
Sodium aluminate	x									
Sodium bisulfite[b]				x						
Sodium carbonate		x						x	x	
Sodium chloride									x	
Sodium chlorite			x		x					
Sodium fluoride										x
Sodium hexametaphosphate[b]								x	x	
Sodium hydroxide[c]		x						x	x	
Sodium silicate	x							x		
Sodium silicofluoride										x
Sodium sulfite[b]				x						
Sodium tripolyphosphate								x		
Sulfuric acid		x							x	
Sulfur dioxide[c]				x						

[a] Chemicals utilized in processes for removal of fluoride include activated alumina, calcium hydroxide, magnesium oxide and sodium bisulfate.
[b] Also used for treatment of boiler water.
[c] Also used for cleaning filter sand.

Source: E.S. Hopkins and E.L. Bean, *Water Purification Control,* 4th ed., Copyright © 1966, Williams & Wilkins Co., Baltimore, 1966, p. 310. With permission.

2.6-5 PROPERTIES OF CHEMICALS MOST EXTENSIVELY USED IN WATER TREATMENT AND WASTEWATER PLANTS

Chemical	Formula	Trade name	Use	Available forms and appearances	Density, lb/ft² and commercial strength	Solubility in water, g/100 cc
Activated carbon	C	—	Adsorbent, taste and odor control, removal of trace organics	Black powder or granules (4–6 mesh), highly porous	15–20, 95% C	0
Aluminum sulfate	$Al_2(SO_4)_3 \cdot 14 H_2O$	Filter alum	Coagulation, sludge conditioning	Light tan to gray-green lumps, granules, or powder Liquid alum also available	60 to 75 (Powder is lighter) 17% Al_2O_3 (Minimum) Liquid–80 lb/ft², 8.3% Al_2O_3 49% dry aluminum sulfate	60.0 at 0°C, 65.3 at 10°C, 71.0 at 20°C, 78.8 at 30°C
Anhydrous ammonia	NH_3	—	Nitrogen feed for bacteria, pH adjustment	Liquefied gas	38.5 at 60°F, 99% NH_3	44 at 0°F, 32 at 70°F, 26 at 100°F
Ammonium hydroxide	NH_4OH	Aqua ammonia	Nitrogen feed for bacteria, pH adjustment	Liquid, colorless	56, 29.4% NH_3	NH_4OH solutions completely miscible
Bentonite (clay)	$(Al_2O_3 \cdot Fe_2O_3 \cdot 3MgO \cdot 4SiO_2 \cdot x H_2O)$	—	Coagulant aid	Pellets or powder	60	Forms colloidal sol
Calcium carbonate	$CaCO_3$	Limestone, calcite		White to gray powder, 100 mesh	100–135, 96–99% $CaCO_3$	

Name	Formula	Common name	Use	Description	Commercial strength	Solubility
Calcium hydroxide	$Ca(OH)_2$	Slaked lime	Coagulation, pH adjustment, sludge conditioning	White powder	25–35 $Ca(OH)_2$, 97–99%[a]	0.18 at 0°C, 0.16 at 20°C, 0.15 at 30°C
Calcium hypochlorite	$Ca(OCl)_2$	—	Disinfection, corrosion caused by biological growths, odor control	Lumpy white or yellowish-white powder, granules, or pellets	50–55, 70% available Cl_2	21.88 at 0°C, 22.7 at 20°C, 23.4 at 40°C
Calcium oxide	CaO	Quicklime	Coagulation, pH adjustment, sludge conditioning	White powder or porous white to light gray lumps	CaO 71–74% 55–60, 90–96% CaO, Below 88% is poor quality	Reacts to form $Ca(OH)_2$. See solubility of calcium hydroxide
Carbon dioxide	CO_2	Carbonic acid gas, solid (dry ice)	Recarbonation, pH control	Colorless, odorless gas; subliming white solid	Solid, 97.5; 100% CO_2	3.6 at 0°C, 1.8 at 20°C
Chlorinated ferrous sulfate	$Fe_2(SO_4)_3 \cdot FeCl_3$	Chlorinated copperas	Coagulation sludge, conditioning	Yellow solution		
Chlorine	Cl_2	—	Odor control, disinfection, corrosion control	Liquefied compressed gas	Gas–0.201, Liquid–91.70, 99.8% Cl_2	1.46 at 0°C, 0.98 at 10°C, 0.716 at 20°C, 0.57 at 30°C
Chlorine dioxide	ClO_2	—	Disinfection	Greenish-yellow gas, condenses to red liquid	—	>1.0

2.6-5 PROPERTIES OF CHEMICALS MOST EXTENSIVELY USED IN WATER TREATMENT AND WASTEWATER PLANTS (continued)

Chemical	Formula	Trade name	Use	Available forms and appearances	Density, lb/ft² and commercial strength	Solubility in water, g/100 cc
Copper sulfate	$CuSO_4 \cdot 5H_2O$	Blue vitriol	Algae control	Blue crystals or powder	75–93, 25.2% Cu	19.2 at 0°C, 31.2 at 30°C
Dolomitic hydrated lime	$Ca(OH)_2 \cdot Mg(OH)_3$	—	pH adjustments, sludge conditioning	White powder	35–40, 46.3% CaO, 33.7% MgO, Typical assay	See $Ca(OH)_2$
Dolomitic lime	$CaO \cdot MgO$	Quicklime	pH adjustment, coagulant, sludge conditioning	White (light gray, tan) pebble, crushed, lump, ground, pulverized	55–60, Average 57, CaO 55–57.5% MgO 37.6–40.5%	See $Ca(OH)_2$, Reacts to form $Ca(OH)_2 \cdot Mg(OH)_2$
Ferric chloride, $FeCl_3$ solution	$FeCl_3 \cdot 6H_2O$—crystal, $FeCl_3$—anhydrous	Liquid ferric chloride, crystals, ferric chloride	Coagulation, sludge conditioning, odor control (H_2S)	Dark brown solution Yellow, brown crystals Green, black granules	Solution—11.6 to 12.7 lb/gal 37–47% $FeCl_3$ Crystal—60–64 lb/ft² 60% $FeCl_2$ Anhydrous—65 to 70 lb/ft² 96–97% $FeCl_3$	Solution—completely Crystal—64.4 at 10°C 91.1 at 20°C Anhydrous—74.4 at 0°C
Ferric sulfate	$Fe_2(SO_4)_3 \cdot 2$-$3H_2O$	Partly hydrated ferric sulfate	Coagulation	Reddish tan-gray granules	70–72 $Fe(SO_4)_3 \cdot 3H_2O$ 68% $Fe_2(SO_4)_3$ 18.5% Fe $Fe(SO_4)_3 \cdot 2H_2O$ 76% $Fe_2(SO_4)_3$ 21% Fe	Very soluble
Ferrous sulfate	$FeSO_4 \cdot 7H_2O$	Copperas, sugar sulfate	Coagulation	Green to brownish yellow granules, crystals, powder, lumps	63–66 55% Fe_2SO_4 20% Fe	28.7 at 0°C, 37.5 at 10°C, 48.5 at 20°C, 60.2 at 30°C

Name	Formula	Common name	Use	Physical description	Composition/strength	Solubility
Fluosilicic acid	H₂SiF₆	Liquid fluoride	Fluoridation	Colorless to straw liquid	76–84, 24–36% H₂SiF₆	Completely miscible
Ozone	O₃	—	Disinfection	Colorless gas, sweetish odor	—	<0.5
Potassium permanganate	KMnO₄	—	Taste and odor control, removal of trace organics	Purple granules	90, 97–99% KMnO₄	3.0 at 0°C, 6.0 at 20°C
Sodium aluminate	NaAlO₂	Soda alum	Coagulation	White to gray crystals, colorless liquid,	45–60 (solid) 76 (liquid) 32% Al₂O₃ · Na₂O	3.6 at 20°C
Sodium bisulfate	Na₂S₂O₃	Meta-bisulfite	Deaeration	White crystalline powder, sulfur odor	74–85; 93–99%	5.4 at 20°C
Sodium carbonate	Na₂CO₃	Soda ash	Softening	White powder, free flowing	Extra light, 23; light, 35; dense, 65 99%Na₂CO₃	7.2 at 0°C, 10.8 at 10°C, 17.9 at 20°C
Sodium chloride	NaCl	Salt	Ion-exchange softening regenerant	Grayish crystals or rock	62–68 (crystals), 70–75 (rock) 98% NaCl	3.46 at 0°C, 3.6 at 20°C
Sodium hypochlorite	NaOCl	Chlorine bleach	Disinfection	Light yellow liquid	117–120, 10–15% available chlorine	Completely miscible
Sodium fluoride	NaF	—	Fluoridation	White crystals or powder	60–75 (powder), 75–90 (crystals) 97% NaF	<4.2

2.6-5 PROPERTIES OF CHEMICALS MOST EXTENSIVELY USED IN WATER TREATMENT AND WASTEWATER PLANTS (continued)

Chemical	Formula	Trade name	Use	Available forms and appearances	Density, lb/ft^2 and commercial strength	Solubility in water, g/100 cc
Sodium hexameta-phosphate	$Na_6(PO_3)_6$	Glassy phosphate	Sequestrant	Glassy flakes, pellets, granules or powder	Variable 65–67.5% P_2O_5	>24
Sodium hydroxide	$NaOH$	Caustic soda, lye	pH control	White crystalline, flakes, liquid solution	Variable 98.9% NaOH; solution 50% NaOH	29 at 0°C, 53 at 20°C
Sodium silicate	$2Na_2O \cdot$ x SiO_2 + H_2O; x = 1–7.5	Water glass		Syrupy opaque liquid	84–87	20–55
Sodium fluo-silicate	Na_2SiF_6	Sodium silico-fluoride	Fluori-dation	White to yellow powder, free-flowing	75–95; 59.7% F	0.43 at 1.7°C; 0.75 at 23.9°C
Sodium sulfite	Na_2SO_3	Sulfite	Deaeration	White crystals or powder	80–91; 93–98% Na_2SO_3	12–24
Sodium tripoly-phosphate	$Na_5P_3O_{10}$		Sequestrant	White granules or powder	60–70, 57.5% P_2O_5	<15 at 20°C
Sulfuric acid	H_2SO_4	Oil of vitriol	Sludge con-ditioning, pH adjust-ment	Colorless syrupy liquid	60° Bé' 106.4 lb/ft^2, 66° Bé' 114.5 lb/ft^2, 60° Bé' 77.7% H_2SO_4, 66° Bé' 93.2% H_2SO_4	Completely mis-cible

REFERENCES

1. *Manual of Practice No. 8*, Water Pollut. Control Fed., 1967.
2. E.S. Hopkins and E.L. Bean, *Water Purification Control*, Williams and Wilkins, Baltimore, 1966.

2.6.2   COAGULATION AND FILTRATION

2.6-6   MODES OF DESTABILIZATION AND THEIR CHARACTERISTICS

Phenomena	Physical double layer theory (coagulation)	Chemical bridging model (flocculation)	Aggregation by hydro-lyzed metal ions
Electrostatic interaction	Predominant	Subordinate	Important
Chemical interactions and adsorption	Absent	Predominant	Important
Zeta potential for optimum aggregation	Near zero	Usually not zero	Not necessarily zero
Addition of an excess of destabilizing species	No effect	Restabilization due to complete surface coverage	Restabilization usually accompanied by charge reversal; may be blurred by hydroxide precipitation
Fraction of surface coverage (θ) by destabilization for optimum aggregation	Negligible	$\theta = 0.5$ (La Mer (9)) in general, $0 < \theta < 1$	$0 < \theta < 1$, θ not necessarily $= 0.5$
Relationship between optimum dosage of destabilizing species and the concentration of colloid (or concentration of colloidal surface)	CCC virtually independent of colloid concentration	Stoichiometry, a linear relationship between flocculent dose and surface area	Stoichiometry possible, but does not always occur
Physical properties of the aggregates which are produced	Dense, great shear strength, but poor filtrability in cake filtration	Flocs of 3 dimensional structure; low shear strength, but excellent filtrability in cake filtration	Flocs of widely varying shear strength and density

Source: Reprinted with publisher's permission from *J. Am. Water Works Assn.*, *60*(5):514, © 1968, by the American Water Works Association Inc.,

2.6-7   COAGULATION SLUDGE CHARACTERISTICS

Plant	Treatment	BOD (5 day), mg/l	COD, mg/l	pH	Total solids, mg/l	Volatile solids, mg/l	Total suspended solids, mg/l	Volatile suspended solids, mg/l
A	Alum Coagulation and sedimentation	41 72[a] 144[b]	540	7.1	1,159	571	1,110	620
B	Alum Coagulation, clarifier	90	2,100	7.1	10,016	3,656	5,105	2,285
C[c]	Alum Coagulation, clarifier	108	15,500	6.0	16,830	10,166	19,044	10,722
D	Alum Coagulation, clarifier	44	–	6.0	–	–	15,790	4,130
E	Diatomite Filter	105	340	7.6	7,466	275	7,560	260
F[1]	Alum Coagulation, sedimentation	380	1,320 (1,162–1,580)[e]	6.6 (6.5–6.7)	17,350 (4,380–28,580)	4,120 (1,390–6,330)		
G[2]	Alum Coagulation, sedimentation[d]	–	–	7.0	8,050	3,320	7,750	3,200
H[f,3]	Sedimentation basin sludge	57	669	7	4,350	2,310	3,650	
I[g,3]	Sedimentation basin sludge	45	1,048	7	4,260	1,020	3,580	
J[h,3]	Sedimentation basin sludge	–	–	7	14,386	3,450	13,640	
K[i,4]	Sedimentation basin sludge	36–77	500–1,000	7	4,300		3,610	

[a] BOD after 7 days.
[b] BOD after 27 days.
[c] Activated carbon in sample.
[d] Milwaukee, Howard Avenue Plant.
[e] Range of values.
[f] Shoremont Plant, Rochester, N.Y.
[g] Wolcott, N.Y.
[h] Ithaca, N.Y.
[i] Average data from Shoremont and Wolcott.

Source: H.B. Russelmann, *Water Sewage Works, 115*(R.N.):R64, 1968. Courtesy of Scranton Gillette Communications, Inc.

REFERENCES

1. H.M. Bugg, *et al., J. Am. Water Works Assn., 62*(12):792, 1970.
2. A.E. Albrecht, *J. Am. Water Works Assn., 64*(1):46, 1972.
3. E.R. Sutherland, *J. Am. Water Works Assn., 61*(4):186, 1969.
4. W.K. Neubauer, *J. Am. Water Works Assn., 60*(7):819, 1968.

2.6-8   FILTER BACKWASH CHARACTERISTICS

Plant	Treatment	BOD (5 day), mg/l	COD, mg/l	pH	Total solids, mg/l	Volatile solids, mg/l	Total suspended solids, mg/l	Volatile suspended solids, mg/l
A	Alum Coagulation and sedimentation	4.2	28	7.8	121	44	47	31
B	Alum Coagulation, clarifier	3.7	75	7.2	378	115	104	53
C	Alum Coagulation, clarifier	2.8	160	7.8	166	45	75	40
D	Alum Coagulation, clarifier	1.8	–	–	–	–	100	60
F	Iron and manganese removal	–	–	6.9	1,487	343	1,370	400
H[1,2,a]					360			
I[1,2,b]					340			
K[1,3,d]					390			
L[1,4,e]							800[f]	
M[1,4,g]							120[f]	
N[1,5,h]					370			

[a] Shoremont Plant, Rochester, N.Y.
[b] Wolcott, N.Y.
[c] Ithaca, N.Y.
[d] Colorado Springs, Colo. (Plant No. 1)
[e] Detroit, Mich. (Water Works Park)
[f] Coliforms MPN/100 ml = 0
[g] Detroit, Mich. (Springwells)
[h] Cornell, N.Y.

Source: H.B. Russelmann, *Water Sewage Works, 115*(R.N.):R64, 1968. Courtesy of Scranton Gillette Communications, Inc.

REFERENCES

1. E.R. Sutherland, *J. Am. Water Works Assn., 61*(4):186, 1969.
2. *Waste Alum Sludge Characteristics and Treatment,* O'Brien and Gere, Consulting Engineers, Research Report No. 15, New York State Department of Health, Albany, 1966.
3. D.P. Proudfit, *J. Am. Water Works Assn., 60*(6):674, 1968.
4. W.W. Aultman *et al., J. Am. Water Works Assn., 58*(9):1102, 1966.
5. C.D. Gates and R.F. McDermott, *Characteristics of and Methodology for Measuring Water Filtration Plant Wastes,* Research Report No. 14, New York State Department of Health, Albany, 1966.

2.6-9   REMOVAL OF ALGAE BY WATER TREATMENT PROCESSES

Type of algae	Sedimentation with alum, softening[a]						Sedimentation with lime[b]			All operations[c]		
	Raw water, organisms/ml	Sedimentation basin effluent[a], organisms/ml	Removal, %	Softener effluent, organisms/ml	Removal, %	Overall removal, %	Raw water, organisms/ml	Sedimentation basin effluent[a], organisms/ml	Removal, %	Plant influent, organisms/ml	Plant effluent, organisms/ml	Removal, %
Blue-green algae												
Aphanizomenon	96	67	28	5	93	95	27	7	73	24	0	100
Oscillatoria	32	5	84	4	20	88	5	1	80	6	1	83
Diatoms												
Cyclotella	530	310	42	7	98	99	9,230	187	98	3,407	3	>99
Fragilaria	91	52	43	5	90	95	580	42	93	235	1	>99
Melosira	295	190	36	30	84	90	501	2	99	300	0.3	>99
Navicula	20	5	75	0	100	100	37	2	95	32	1	97
Stephanodiscus	24	20	17	0	100	100	575	0	100	230	0	100
Others	7	5	29	2	60	72	43	1	98	95	0	100
Green algae												
Ankistrodesmus	11	25	127 increase	1	96	91	21	1	95	13	0.2	>98
Chlorella	249	106	57	26	75	90	2,300	78	97	920	5	>99
Scenedesmus	4	9	125 increase	1	89	75	69	7	90	34	0.3	>99
Others	5	7	40 increase	0	100	100	43	2	95	30	0.4	>98
Flagellates												
Chlamydomonas	160	162	1 increase	1	99	99	85	11	88	62	0.6	99
Chloromonad	28	21	25	0	100	100						
Euglena	10	13	30 increase	2	85	80	75	4	95	40	2	95
Others	40	20	50	0	100	100	5	0	100	36	0	100
Totals	1,599	1,017	37	84	92	95	13,596	345	97	5,466	14.9	>99

[a] Four samplings averaged.
[b] Three samplings averaged.
[c] Ten samplings averaged.

REFERENCE

1. R.R. Speedy, N.B. Fisher, and D.B. McDonald, *J. Am. Water Works Assn.*, 61:289, 1969.

2.6.3   *IRON AND MANGANESE REMOVAL*

2.6-10   IRON AND MANGANESE SLUDGE TREATMENT

Process	Sludge form	Sludge disposal	Comments
Oxidation with or without chemical assist. Slow sand filtration	Retained on filter media	Physical-mechanical removal from filter media, hauled to disposal site	Primarily used in older plants. Considered ultimate disposal; no further study needed
Coagulation and clarification. Rapid sand filtration	Clarifier sludge; retained on filter media	Removed with clarifier or filter backwash sludge	Usually combined with coagulation or softening sludges. Additional disposal study needed
Oxidation with chemicals. Pressure or gravity filtration	Retained on filter media	Removed with filter backwash sludge	Disposal problem same as for other filter backwash sludge
Aeration with pH adjustment. With or without chemicals. Detention. Pressure or gravity filtration	Retained on filter media	Removed with filter backwash sludge	Disposal problem same as for other filter backwash sludge
Manganese zeolite	Precipitation on filter, generally anthracite	Removed with backwash sludge. Regenerant brines removed with rinse water	Additional study needed on effects of brines on receiving vehicle
Ion exchange	Insignificant, predominantly and regenerant brines	Removed with rinse water	Additional study needed on effect of brines on receiving vehicle

Source: Reprinted, with publisher's permission, from *J. Am. Water Works Assn., 61*(10):542, © 1969, by the American Water Works Association, Inc.

2.6.4   REMOVAL OF COLOR, ODOR, TASTE

2.6-11   EFFECT OF ACTIVATED CARBON ON HERBICIDES AND INSECTICIDES

| Pesticide | Concentration treated with carbon, ppm | Carbon dosage used, ppm | Threshold odor values | | Threshold odor units removed per ppm carbon | Concentration after treatment | |
			Before	After		Calculated from odor reduction, units	Determined by chemical analysis, ppm
Parathion	10	10	50	4	4.6	0.8	2.6
BHC-37, gamma	25	5	70	6	12.6	0.22	0.08
Malathion, 50%	2	10	50	4	4.6	0.08	0.25
2,4-D, 23.5%	6	20	50	3	2.4	0.085	1.38
2,4-D, 11.7%	1	10	70	2	6.8	0.005	L[a]
Chlordane, 6%	50	10	50	1.4	4.9	0.084	L[a]
DDT, 50%	5	2	70	4	33	0.15	L[a]

[a] Concentrations were too low to determine by available test methods.

Source: Reprinted, with publisher's permission, from *J. Am. Water Works Assn.*, 57(8):1016, © 1965, by the American Water Works Association, Inc.

2.6.5   DISINFECTION

2.6-12   SUMMARY OF MAJOR POSSIBLE DISINFECTION METHODS FOR DRINKING WATER[a]

Disinfection agent[c]	Technological status	Efficacy in demand-free systems[b]			Persistence of residual in distribution system
		Bacteria	Viruses	Protozoan cysts	
Chlorine[d]	Widespread use in U.S. drinking water				Good
As hypochlorous acid (HOCl)		++++	+++++	++	
As hypochlorite ion (OCl⁻)		+++	++	NDR[e]	
Ozone[c]	Widespread use in drinking water outside U.S., particularly in France, Switzerland, and Quebec Province	++++	++++	++++	No residual possible
Chlorine dioxide[d]	Widespread use for disinfection (both primary and for distribution system residual) in Europe; limited use in U.S. to counteract taste and odor problems and to disinfect drinking water	++++	++++	NDR[e]	Fair to good (but possible health effects)
Iodine	No reports of large-scale use in drinking water				Good (but possible health effects)
As diatomic iodine (I₂)		++++	+++	+++	
As hypoiodous acid (HOI)		++++	++++	+	
Bromine	Limited use for disinfection of drinking water	++++[f]	++++[f]	+++[f]	Fair
Chloramines	Limited present use on a large scale in U.S. drinking water	++	+	+	Excellent

[a] Data from NRC (National Research Council), *Drinking Water and Health,* Vol. 2, National Academy Press, Washington, D.C., 1980, pp. 114—115.

[b] Ratings: ++++, excellent biocidal activity; +++, biocidal activity; ++, moderate biocidal activity; +, low biocidal activity.

[c] The sequence in which these agents are listed does not constitute a ranking.

[d] By-product production and disinfection demand are reduced by removal of organics from raw water before disinfection.

[e] Either no data reported or only available data were not free from confounding factors, thus rendering them not comparable to other data.

[f] Poor in the presence of organic material.

2.6-13   STATUS OF POSSIBLE METHODS OF DRINKING WATER DISINFECTION[a]

Disinfection agent	Suitability as inactivating agent	Limitations	Suitability for drinking water disinfection[b]
Cl_2	Yes	Efficiency decreases with increasing pH; affected by NH_3 or organic N	Yes
O_3	Yes	On-site generation required, no residual; other disinfectant needed for residual	Yes
ClO_2	Yes	On-site generation required; interim maximum contaminant limit (MCL) 1.0 mg/l	Yes
I_2	Yes	Biocidal activity sensitive to pH	No
Br_2	Yes	Lack of technological experience; activity may be pH sensitive	No
Chloramines	No	Mediocre bactericide; poor virucide;	No[c]
Ferrate	Yes	Moderate bactericide; good virucide; residual unstable; lack of technological experience	No
High pH conditions	No	Poor biocide	No
H_2O_2	No	Poor biocide	No
Ionizing radiation	Yes	Lack of technological experience	No
$KMnO_4$	No	Poor biocide	No
Silver	No	Poor biocide; MCL 0.05 mg/l	No
UV light	Yes	Adequate biocide; no residual; use limited by equipment maintenance considerations	No

[a]   Data from NRC (National Research Council), *Drinking Water and Health,* Vol. 2, National Academy Press, Washington, D.C., 1980, p. 118.

[b]   This evaluation relates solely to the suitability for controlling infectious disease transmission.

[c]   Chloramines may have use as secondary disinfectants in the distribution system, in view of their persistence.

2.6-14   COMPARATIVE EFFICACY OF DISINFECTANTS IN THE INACTIVATION OF 99% OF MICROORGANISMS IN DEMAND-FREE SYSTEMS [a]

Disinfection agent	*Escherichia coli*			Poliovirus 1			*Endamoeba histolytica* cysts		
	pH	Temperature (°C)	C·t[a]	pH	Temperature (°C)	C·t[a]	pH	Temperature (°C)	C·t[a]
HOCl	6.0	5	0.04	6.0	0	1.0	7	340	20
				6.0	5	2.0			
OCl⁻	10.0	5	0.92	10.5	5	10.5		—[b]	
O₃	6.0	11	0.031	7.0	20	0.005	7.5—8.0	19	1.5[c]
	7.0	12	0.002	7.0	25	0.42			
ClO₂	6.5	20	0.18	7.0	15	1.32			
	6.5	15	0.38	7.0	25	1.90		—[b]	
I₂	6.5	20—25	0.38	7.0	26	30	7.0	30	80
	7.5	20—25	0.40						
Br₂		—[b]		7.0	20	0.06	7.0	30	18
NH₂Cl	9.0	15	64	9.0	15	900		—[b]	
	9.0	25	40	9.0	25	320			
NHCl₂	4.5	15	5.5	4.5	15	5,000		—[b]	

[a] Concentration of disinfectant (mg/l) times contact time (min).
[b] Either no data were reported or only available data were not free from confounding factors, thus rendering them not comparable to other data.
[c] This value was derived primarily from experiments that were conducted with tap water; however, some parallel studies with distilled water showed essentially no difference in inactivation rates.

Source: From NRC (National Research Council), *Drinking Water and Health,* Vol. 2, National Academy Press, Washington, D.C., 1980, p. 117.

2.6-15 C·t PRODUCTS FOR 99% INACTIVATION OF SELECTED ORGANISMS[a]

Disinfectant	Microorganism	Temp (°C)	pH	Disinfect. conc (mg/l)	Time (min)	Mean C·t	Re
Free	*Escherichia coli*	5	6	0.1	0.4	0.04	1
chlorine	Poliovirus	5	6	1.0	1.7	1.7	1
	Endamoeba histolytica cysts[b]	5	6	5.0	18	90	2
	Giardia lamblia cysts[c]	5	6	1.0	50	50	3
		5	6	2.0	40	80	3
		5	6	4.0	20	80	3
		5	6	8.0	9	72	3
	G. lamblia cysts[d]	5	6	2.5	30	75	4
	Giardia lamblia cysts[c]	5	6	2.5	100	250[b]	4
	G. muris cysts	5	6	2.5	100	250[b]	4
Ozone	*E. coli*	1	7.2	0.07	0.083	0.006	5
		1	7.2	0.065	0.33	0.022	5
	Poliovirus 1	5	7.2	0.15	1.47	0.22	6
	G. lamblia	5	7	0.1—0.5	1.0—5.0	0.6	7
	G. muris	5	7	0.2—0.7	2.5—9.6	1.9	7
	G. muris	15	7	0.1—0.3	1.7—7.7	0.4	8
	G. lamblia	25	7	0.03—0.2	1.0—5.5	0.18	7
	G. muris	25	7	0.03—0.2	1.3—8.2	0.3	7
Chlorine	*G. muris*	5	7	0.11—0.55	1.3—168	11.0	9
dioxide		25	7	0.22—1.13	3.3—28.8	5.0	9
		25	9	0.22—0.82	2.1—19.2	2.8	9
Free	*G. lamblia*	5	6	2.5	20—100	162	4
chlorine		5	6	1.0—8.0	6—47	65	3
		5	7	2.0—8.0	7—42	97	3
		5	8	2.0—8.0	72—164	110	3
		15	6	2.5—3.0	7	20	3
		15	7	2.5—3.0	6—18	32	5
		15	8	2.5—3.0	7—21	37	3
		25	6	1.5	<6	<9	3
		25	7	1.6	<7	<10	3
		25	8	1.5	<8	<12	3
Free	*G. muris*	1	7	1.3—2.2	597—1038	1330	9
chlorine		3	6.5	0.24—1.1	32—297	68	10
		3	7.5	0.24—1.0	150—770	140	10
		5	7	0.41—273	236—267	370	9
		25	5	4.4—13	3.9—16.3	66	11
		25	7	2.9—7.1	3.6—16.0	29	11
		25	9	11.6—72.6	3.0—15.6	206	9
Chloramine	*G. muris*	3	6.5	1.6—2.6	188—276	463	12
		3	7	1.5—2.4	236—276	496	12
		3	7.5	1.5—2.3	225—296	514	12
		10	7	1.3—2.7	122—227	327	12
		10	8.5	1.3—2.3	164—263	351	12
		18	6.5	1.1—2.1	58—225	197	12
		18	7	1.0—1.9	75—241	184	12
		18	7.5	1.1—2.0	68—256	217	12
		1	8	3.8	502	1880	9
		5	7	6.4—17.7	107—220	1720	9
		5	8	3.8	380	1430	9
		15	7	5.0—16.6	55.3—182	970	9
		15	8	3.2—8.4	58.3—133	530	9

[a] Compiled from NRC (National Research Council), *Drinking Water and Health,* Vol. 7, National Academy Press, Washington, D.C., 1987, pp. 16—19, 21, 22.

[b] Extrapolated data.

[c] Cysts from asymptomatic carriers.

[d] Cysts from symptomatic carriers.

2.6-15 C·t PRODUCTS FOR 99% INACTIVATION OF SELECTED ORGANISMS[a] (continued)

REFERENCES

1. P.V. Scarpino, G. Berg, S.L. Chang, D. Dahling, and M. Lucas, *Water Research, 6*:959, 1972.
2. W.B. Snow. *J. Am. Water Works Assn., 48*:1510, 1956.
3. E.L. Jarroll, A.K. Bingham, and E.A. Meyer, *Appl. Environ. Microbiol., 41*:483, 1981.
4. E.W. Rice, J.C. Hoff, and F.W. Schaeffer, III, *Appl. Environ. Microbiol., 43*:250, 1982.
5. E.B. Katzenelson, B. Klettner, and H.I. Shuval, *J. Am. Water Works Assn., 66*:725, 1974.
6. D. Roy, R.S. Englebrecht, and E.S.K. Chian, *J. Am. Water Works Assn., 74*:660, 1982.
7. G.B. Wickramanayake, A.J. Rubin, and O.J. Sproul, *J. Am. Water Works Assn., 77*(8):74, 1985.
8. G. B. Wickramanayake, A.J. Rubin, and O.J. Sproul, *J. Water Pollut. Control Fed., 56*:983, 1984.
9. A.J. Rubin, Ohio State University, Columbus, Ohio.
10. E.L. Jarroll, J.C. Hoff, and E.A. Meyer, in S.L. Erlandsen and E.D. Meyer, *Giardia and Giardiasis: Biology, Pathogenesis, and Epidemiology,* Plenum, New York, 1984, 311.
11. G. Leahy, Ohio State University, Columbus, Ohio, personal communication, 1986.
12. J. Glicker, Portland, Ore., personal communication, 1986.

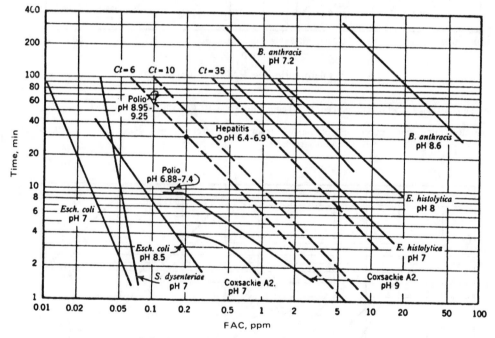

2.6-16   Disinfection vs. FAC[a] residuals.

[a] FAC–free available chlorine.

Note: Time scale is for 99.5–100 percent kill. Temperature was in the range of 20–29°C, with pH as indicated.

Source: Reprinted, with publisher's permission, from *J. Am. Water Works Assn., 54*(11):1379, © 1962, by the American Water Works Association, Inc.

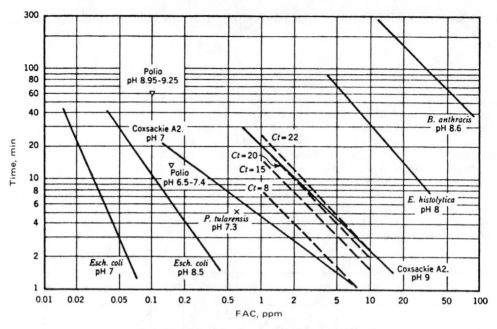

2.6-17   Disinfection vs. FAC residuals.

Note: Time scale is for 99.6–100 percent kill. Temperature was 10°C, with pH as indicated.

Source: Reprinted, with publisher's permission, from *J. Am. Water Works Assn.*, *54*(11):1379, © 1962, by the American Water Works Association, Inc.

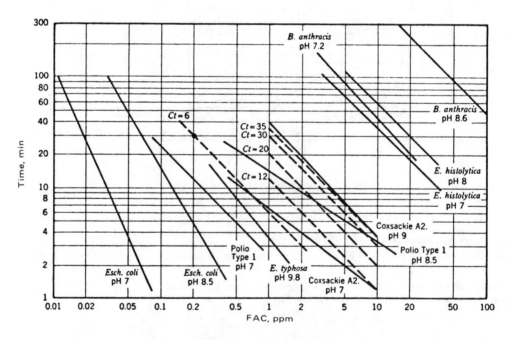

2.6-18   Disinfection vs. FAC residuals.

Note: Time scale is for 99.6–100 percent kill. Temperature was in the range of 0–5°C, with pH as indicated.

Source: Reprinted, with publisher's permission, from *J. Am. Water Works Assn.*, *54*(11):1379, © 1962, by the American Water Works Association, Inc.

2.6-19   INACTIVATION OF SELECTED VIRUSES BY CHLORINE

Test microorganism	Disinfectant	pH	Temperature (°C)	C·t	Ref.
E. coli	HOCl	6.0	5	0.04	1
	OCl⁻	10.0	5	0.92	1
	NH_2Cl	9.0	5	175.0	1
Poliovirus 1	HOCl	6.0	5	1—2	1
	OCl⁻	10.0	5	10.5	1
	NH_2Cl	9.0	15	900.0	1
Rotovirus	HOCl	—[a]	—[a]	0.25	1
	OCl⁻	—[a]	—[a]	1.4	1
Hepatitis A	HOCl	7.0	5	60[b]	1
	HOCl	6.0	—[a]	0.32[c]	3
	OCl⁻	10.0	—[a]	1.04[c]	3
Norwalk agent	HOCl/OCl	7.4	25	—[a]	4

[a]   Not reported.

[b]   C·t estimated from animal infectivity data.

[c]   C·t estimated from disinfection curves. Chlorine residual data suggested that the test mixtures contained significant demand. Concentration used for the calculation was the dose reported.

Source: NRC (National Research Council), *Drinking Water and Health,* Vol. 7, National Academy Press, Washington, D.C., 1987, 207 pp.

REFERENCES

1. NRC (National Research Council), *Drinking Water and Health,* Vol. 2, National Academy Press, Washington, D.C., 1980, 393 pp.
2. D.A. Peterson, T.R. Hurley, J.C. Hoff, and L.G. Wolfe, *Appl. Environ. Microbiol., 45*:223, 1983.
3. W.O.K. Grabow, V. Gauss-Müller, O.W. Prozesky, and F. Deinhardt, *Appl. Environ. Microbiol., 46*:619, 1983.
4. B.K. Keswick, T.K. Satterwhite, P.C. Johnson, H.L. DuPont, S.L. Secor, J.A. Bitsura, G.W. Gary, and J.C. Hoff, *Appl. Environ. Microbiol., 50*:261, 1985.

2.6-20   COEFFICIENTS OF ULTRAVIOLET ABSORPTION[1] FOR VARIOUS WATERS

$$(\lambda = 2,537 \text{ Å})$$

Source	a/cm (coefficient of absorption per centimeter)	
Distilled	0.008	
Swimming pool	0.031	for use in
Cleveland tap water (1944)	0.050	$I = I_o e^{-ax}$
Drilled well	0.056	I = effective solution intensity
Fish pool	0.070	I_o = initial intensity
Lake Erie (1944)	0.083	a = coefficient of absorption
Concrete cistern	0.297	x = distance between points where the intensities are I and I_o

REFERENCE

1. M. Luckiesh and L.L. Holladay, "Disinfecting Water by Means of Germicidal Lamps," *Gen. Elect. Rev., 47*:45, 1944.

2.6.6   SOFTENING AND BOILER WATER TREATMENT

2.6-21   WASTE BRINES FROM ION EXCHANGE SOFTENERS

Constituent	mg/l[1]	mg/l[2]
Calcium	1,720	3,000 - 6,000
Magnesium	600	1,000 - 2,000
Sodium and potassium	3,325	2,000 - 5,000[a]
Chloride	9,600	9,000 - 22,000
Total dissolved solids	15,656	
Total hardness	7,762	

[a] Sodium.

REFERENCES

1. H.A. Faber, *J. Am. Water Works Assn.*, *61*(10):542, 1969.
2. J.W. Krasaukas and L. Streicher, *J. Am. Water Works Assn.*, *61*(11):621, 1969.

2.6-22   AMOUNT OF SLUDGE PRODUCED BY LIME HARDNESS REMOVAL

City	Acre ft of 50% sludge/yr/mgd/ 100 ppm hardness removed
Cedar Rapids, Iowa	0.62
Columbus, Ohio	0.52
Sandusky, Ohio	0.66
Lansing, Mich.	0.65
Hinsdale, Ill.	0.45

Source: Reprinted, with publisher's permission, from *J. Am. Water Works Assn.*, *53*(9):1169, © 1961 by the American Water Works Association, Inc.

2.6.7 MISCELLANEOUS AND NONCONVENTIONAL PROCESSES

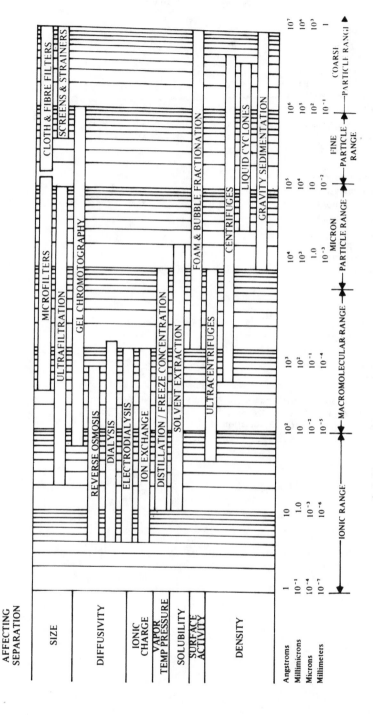

2.6-23   Useful ranges of various separation processes.

Courtesy of Dorr-Oliver, Incorporated, Stamford, Conn.

2.6-24   DESALTING PROCESSES

Process	Application	
	Sea water	Brackish
Distillation		
Multistage flash (MSF)	x	x
Multiple effect, long-tube vertical (LTV)	x	x
Combination MSF and LTV	x	x
Vapor compression	x	x
Vapor reheat	x	x
Solar	x	x
Membrane processes		
Electrodialysis	?	x
Reverse osmosis	?	x
Dynamic membrane	?	x
Freezing processes		
Vapor freezing–vapor compression	x	x
Direct contact	x	x
Hydrate	x	x
Others		
Ion exchange		x
Solvent extraction		x
Electrochemical demineralization		x

Source: Reprinted, with publisher's permission, from *J. Am. Water Works Assn., 60*(8):869, © 1968, by the American Water Works Association, Inc.

2.7   DISTRIBUTION

2.7.1   *PLUMBING SYSTEMS*

2.7-1   FIXTURES AND THEIR WATER SUPPLY PROTECTION

Fixtures	Type of protection[a]	Remarks
Aspirators		
Laboratory	Vacuum breaker	
Portable	Vacuum breaker	
Vacuum system	Vacuum breaker	
Bedpan		
Washers	Vacuum breaker	
Washer hose	Vacuum breaker	Locate 5 ft above floor
Boiling type sterilizer	Air gap	Not less than twice the effective opening of the water supply
Exhaust condenser	Vacuum breaker	
Flush floor drain	Vacuum breaker	
Hose connection	Vacuum breaker	Locate 6 ft above floor
Pressure instrument washer-sterilizer	Vacuum breaker	
Pressure sterilizer	Vacuum breaker	
Vacuum systems		
Cleaning	Air gap or vacuum breaker	
Fluid suction	Air gap or vacuum breaker	

[a] Where vacuum breakers are used, they shall be installed after the last control valve.

Source: *Report of Public Health Service Committee on Plumbing Standards,* U.S. Department of Health, Education, and Welfare, Public Health Service, 1963.

2.7-2   MINIMUM DISTANCES FROM SOURCE OF CONTAMINATION TO INDIVIDUAL WATER SUPPLIES AND PUMP SUCTION LINES

Source of contamination	Distance, ft[a]
Sewer	50
Septic tanks	50
Subsurface pits	50
Pasture	100
Sewer of cast iron (leaded or mechanical joints)	10
Subsurface disposal fields	100[b]
Seepage pits	100
Cesspools	150
Barnyards	100
Farm site	25
Pumphouse floor drain of cast iron draining to ground surface	2

[a] These distances constitute minimum separation and shall be increased in areas of creviced rock or limestone or where the direction of movement of the ground water is from source of contamination toward the well.

[b] When approved by the authority having jurisdiction and under special soil conditions, this distance may be reduced. However, a minimum separation of not less than 50 feet shall be maintained.

Source: *Report of Public Health Service Committee on Plumbing Standards,* U.S. Department of Health, Education, and Welfare, Public Health Service, 1963.

2.8   RECREATIONAL WATERS

2.8-1   AVERAGE NUMBER OF INDICATOR ORGANISMS IN POOL WATER

85 Samples

	Average	Range
Chlorine, ppm	0.53	0.1–1.0
Total count per 1 ml	8	0–300
Staphylococcus per 100 ml	47	1–658
Coliform per 100 ml	0.8	0–58
Strept. salvarius per 100 ml	1.9	0–52
Enterococci per 100 ml	0.7	0–32
Ps. aeruginosa per 100 ml	0	0
Bathing load per day	90	30–250
pH	7.7	7.4–8.0

Source: M.S. Favero, C.H. Drake, and G.B. Randal, *U.S. Public Health Rep., 79*:61, 1964.

2.8-2   POLLUTION CHARACTERISTICS OF SWIMMING POOL AND SALINE BATHING BEACHES

	Fresh water swimming pools	Saline bathing beaches
Water supply	Normally free from contamination	Normally contaminated by soil washings and possibly animal wastes
Primary source of pollution	Respiratory tract and body surfaces of swimmers	Soil washings, animal wastes, diluted raw and treated sewage
Secondary sources of pollution	Soil trackings from immediate vicinity of pool	Respiratory tract and body surfaces of bathers
Effect of dilution and currents	Relatively slight	Considerable
Natural purification	Slight, if any	May be considerable
Effect of artificial purification	Considerable	None
Bacterial findings	Measure of efficiency of filtration and disinfection	Measure of resultant of contamination plus purification processes
Possible bacterial indices	*Salmonella salivarius* *S. pyrogenes* *Neisseria catarrhalis* *N. sicca* Coliform *Escherichia coli* Enterococci *S. fecalis*	Coliform *E. coli* Enterococci *(S. fecalis)*

Source: M. Levine *et al., Proc. 1st Int. Conf. Waste Disp. Marine Environ.,* July 22–25, 1959, Pergamon Press, 1960. With permission.

2.8-3   EFFECTIVENESS OF IODINE AND FREE CHLORINE FOR DISINFECTING SWIMMING POOL WATER

	Iodine	Chlorine
Total number of samples	790	439
Desired residual range, ppm	0.6–0.8	0.4–0.6
Samples confirming for coliforms, percent	2	3
Samples confirming for fecal streptococci, percent	3	2
Average number of staphylococci per 100 ml	3.5[a]	46[b]
Median 48 hr plate count	210	5

[a] Based upon 60 samples.
[b] Based upon 54 samples.

Source: Reprinted, with publisher's permission, from *J. Am. Water Works Assn.*, *60*(1):69, © 1968, by the American Water Works Association, Inc.

2.8-4   BACTERIOLOGICAL TESTS ON WADING POOLS

Halogen	Plate count	Coli-aerogenes	*E. coli*	Staph.	Entero-cocci	Diphth-oids
0–0.19 ppm						
Chlorine	71800	54	23	3550	55	475
Iodine	81600	58	11	2360	18	20
Bromine	61000	66	30	4930	33	288
0.2–0.40 ppm						
Chlorine	5280	22	15	2750	8	134
Iodine	7700	10	2	380	3	17
Bromine	3270	29	11	1720	12	15
0.5–1.0 ppm						
Chlorine	1550	1	0	14	0	6
Iodine	2400	0	0	100	2	6
Bromine	1600	0	0	260	0	4

REFERENCE

1.   J.R. Brown, D.M. McLean, and M.C. Nixon, *Can. J. Publ. Health, 54*:121, 1963.

2.8-5   BACTERIOLOGICAL RESULTS AT STATIONS WITH VARYING INTENSITIES OF RECREATIONAL USE

Location	Total coliform No. samples	MPN/100 ml Min	Max	Med	Fecal coliform No. samples	MPN/100 ml Min	Max	Med	Fecal streptococci No. samples	MPN/100 ml Min	Max	Med
Folsom Reservoir, First Study, May 4–May 15, 1960[a]												
Low Use Area												
Intake	21	2.1	64	15	21	<2.1	2.1	<2.1	21	<2.1	2.1	<2.1
Boat refueling	11	8.6	>700	32	7	<2.1	32	2.1	9	<2.1	32	12
Docks	8	48	240	8.6	4	<2.1	32	6.7	5	<2.1	240	<2.1
Docks	9	7.4	32	15	4	<2.1	<8.6	2.1	4	<2.1	2.1	<2.1
Docks	8	8.6	700	20	3	2.1	32	<4.8	5	<2.1	2.1	<2.1
Docks	10	12	700	20	6	<2.1	32	3.4	9	<2.1	12	2.1
High Use Area												
Shore fishing	9	15	>700	64	5	2.1	64	32	7	2.1	64	15
Supervised swimming	12	12	700	185	6	<2.1	32	6.1	12	<2.1	700	4.5
	13	7.4	>700	210	10	<2.1	240	32	13	<2.1	700	2.1
	11	8.6	700	130	6	2.1	32	20.3	11	<2.1	700	4.8
Boat launching	8	32	700	240	4	4.8	64	12.4	5	<2.1	240	15
	8	15	>700	470	4	4.8	32	5.3	8	4.8	240	15
	9	2.1	700	130	4	4.8	64	18.4	9	<2.1	240	20
Fishing, picnicking	10	25	>700	700	5	2.1	32	32	10	4.8	240	20
Swimming	10	240	>700	700	5	8.6	240	32	10	20	>700	470
	10	240	>700	700	10	2.1	240	64	7	7.4	>700	32
Cackuma Reservoir, First Study, May 20–June 7, 1960[a]												
Low Use Area												
Dam intake	16	<4.5	700	47	14	<2.1	2.1	<2.1	16	<2.1	4.8	<2.1
Dam surface	19	7.4	2,100	64	17	<2.1	12	<2.1	19	<2.1	240	2.1
Near toilet	20	4.8	>700	68	17	<2.1	4.8	<2.1	21	<2.1	>700	2.1
Near park intake	29	4.8	>24,000	500	18	<2.1	23	6	29	<2.1	>700	240
Tower intake surface	19	15	>24,000	620	17	<2.1	1,300	<2.1	19	<2.1	>700	15

Cackuma Reservoir, First Study, May 20–June 7, 1960[a]

High Use Area

Shore fishing	23	20	>24,000	620	13	<2.1	62	23	23	<2.1	>7,000	64
	27	2.1	>24,000	240	16	<2.1	7,000	14	27	<2.1	>700	32
	19	15	7,000	620	18	<2.1	230	6	19	<2.1	>700	130
Boat refueling	29	4.8	>2,400	500	16	<2.1	62	<2.1	29	<2.1	700	15
Launching	31	4.8	7,000	230	18	<2.1	62	<2.1	31	<2.1	240	4.8
Docks	27	32	2,400	620	16	<2.1	64	6	27	<2.1	240	15
Near toilet	18	20	>24,000	620	17	<2.1	230	6	18	<2.1	>700	152

[a] Data are for the entire study period.

Source: Reprinted, with publisher's permission, from *J. Am. Water Works Assn., 61*(6):297.© 1969, by the American Water Works Association, Inc.

2.9   MONITORING

2.9-1   SUBSTANCES ANALYZED FOR WATER USED FOR SPECIFIC PURPOSES

Component	Public water supply	Industrial water	Fish and wildlife	Livestock	Irrigation
Alkalinity	x	x	x		
Acidity		x			
Hardness	x	x	x		
pH	x	x	x		x
Total dissolved solids	x	x			
Total suspended solids		x			
Temperature	x	x			
Bicarbonate		x			
Chloride	x	x			x
Fluoride	x	x		x	
Nitrate	x	x			
Phosphate	x	x			
Sulfate	x	x			
Organics (CCl$_4$ extractable)		x			
COD		x			
Hydrogen sulfide		x			
Dissolved oxygen	x		x		
Carbon dioxide			x		
Total dissolved materials			x	x	x
Salinity			x		
Turbidity	x		x		
Color	x		x		
Cyanide	x		x		
Pesticides	x		x		
Toxic substances			x		
Oil	x		x		
ABS			x		
Odor					
Nitrite	x				
Herbicides	x				
Phenols	x				
Grease	x				
MBAS (surfactants)	x				
Carbon chloroform extracts (CCE)	x				
Ammonia	x				
Radionuclides	x		x	x	x
Gross beta	x			x	
Radium 226	x			x	
Strontium 90	x			x	
Arsenic	x			x	x
Barium	x				x
Boron	x				
Cadmium	x			x	
Copper	x				
Iron	x				
Lead	x			x	
Manganese	x				
Selenium	x			x	
Silver	x				
Uranium	x				
Dissolved organic compounds				x	
Sodium adsorption ratio					x
Metals (aluminum to zinc)	x		x		x
Zinc					
Chromium	x			x	

Source: D.G. Ballinger, *Water Sewage Works,* Ref. No. 1968, p. R38. Courtesy of Scranton Gillette Communications, Inc.

MINIMUM NUMBER OF SAMPLES PER MONTH

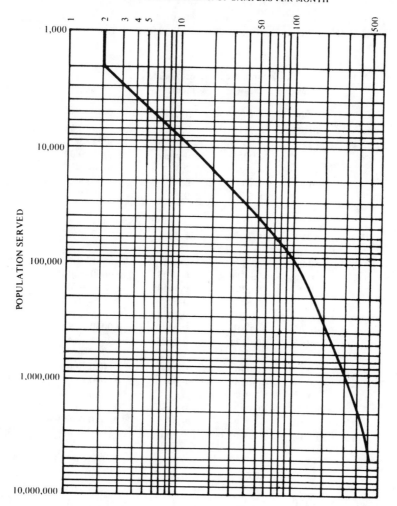

2.9-2   Recommended minimum monthly samples for bacteria counts for water supply systems.

Source: U.S. Public Health Service.

2.9-3 WATER QUALITY DATA

U.S. Public Health Service Study, 1961–1966

Component	Goal	Arithmetic mean	Approximate median	Range	No. samples	Less than values
Aluminum, mg/l	<0.05	0.1054	0.0583	0.0033–65[a]	91	
Arsenic, mg/l	<0.05	<0.0117	0.0104	0.0000–0.45	565	515
Barium, mg/l	<1	0.0403	0.0228	0.0005–3	558	
Cadmium, mg/l	<0.01	<0.0085	0.0064	0.0000–0.09	559	536
Color units	<3	4.2	2.3	0.0–70	218	
Turbidity, J.T.U.	<0.1	2.25	0.88	0.00–99.9	227	
Odor, units	0	0.5	0.56	0.0–8	112	
Carbon alcohol extract, mg/l	<0.1	<0.13	0.124	0.1–0.9	478	306
Chloride, mg/l	<250	40.7	21.3	0.7–605	273	
Carbon chloroform extract, mg/l	<0.04	<0.067	0.056	0.005–0.423	540	367
Chromium, mg/l	<0.05	<0.0041	0–003	0.0000–0.045	557	468
Copper, mg/l	<0.2	<0.0293	0.00549	0.00000–0.98	560	122
Fluoride, mg/l	0.6–1.2	0.61	0.52	0.00–10.1	445	
Cyanide, mg/l	<0.2	0.00375	0.00375			
Iron, mg/l	<0.05	0.0806	0.0279	0.0000–1.92	470	
Lead, mg/l	<0.05	<0.0171	0.0125	0.0001–0.4	559	468
Methylene blue active substances, mg/l	<0.2	<0.046	0.033	0.000–0.64	556	151
Manganese, mg/l	<0.01	<0.0369	0.01	0.0000–2	492	244
Nitrate, NO_3, mg/l	<45	2.3	0.69	0.0–99.8	260	
Selenium, mg/l	<0.01	<0.007	0.01	0.0000–0.01	556	503
Sulfate (SO_4), mg/l	<250	57.2	39.3	0.0–400.9	258	
Silver, mg/l	<0.08	0.00083	0.00025		189	
Total dissolved solids, mg/l	<200	271	205	4–1,739	271	
Zinc, mg/l	<1	<0.62	0.39	0.00–7	559	
Hardness, mg/l	80	136	144	0–920	341	
Beta activity (displaced), pCi/l	<100	<7.1	4.4	0.1–300	367	84
Beta activity (suspended), pCi/l	<100	<2.5	–	0.0–35	343	299
90 Si., pCi/l	<10	<1.7	1	0.0–10	97	50

[a] 1962 data only.

REFERENCES

1. F.B. Taylor, J.H. Eagen, and F.D. Maddox, *J. Am. Water Works Assn., 60*:764, 1968.
2. C.N. Durfor and E. Becker, *J. Am. Water Works Assn., 56*:231, 1964.

2.9-4   INCIDENTS OF WATER POLLUTION BY ORGANIC PESTICIDES

Location	Year	Pesticide	Source	Remarks	Ref.
Montebello, Cal.	1945	2,4-D	IW[a]	Well waters contaminated for several years with the phenol	1, 2
Saskatchewan River, Canada	1949–51	DDT	DA[b]	Blackfly larvae control	3
Tennessee River, Ala.	1950	toxaphene	R[c]	Fish kill	4
Detroit River, Mich., and Lake St. Clair, Ont.	1953	DDT	R	Recovered from raw and treated drinking water	5, 6
St. Lucie County, Fla.	1955	dieldrin	DA	Sandfly larvae control	7, 8
Snake, Missouri, and Mississippi rivers	1957	aldrin DDT	R	Isolated by carbon filter technique	5, 6
Great Lakes	1957	TFM	DA	Sea lamprey control	9
Lake Hopatcong, N.J.	1957–58	2,4-D 2,4,5-T	DA	Aquatic-plant control	10
Dickinson Reservoir, N.D.	1958	toxaphene, rotenone	DA	Trash fish control	11
Dutchess County, N.Y.	1958–60	2,4,5-T	DA	Aquatic-plant control	12
Denver, Colo.	1959	2,4-D	IW	Ground water contaminated	13, 14, 15
Clayton Lake, N.M.	1959	toxaphene	DA	Distribution study	16
Various lakes in Illinois	1960	endothal	DA	Aquatic-plant control	17
Farm pond, S.C.	1960	parathion	R	Insect control in peach orchard	18

[a] Industrial waste water.
[b] Direct application.
[c] Runoff.

Source: Reprinted, with publisher's permission, from *J. Am. Water Works Assn.*, *56*(3):267, © 1964, by the American Water Works Association, Inc.

REFERENCES

1.   H.A. Swenson, *Proc. Soc. Water Treat.*, *11*:84, 1962.
2.   A.N. Sayre and V.T. Stringfield, *J. Am. Water Works Assn.*, *40*(11):1152, 1948.
3.   F.J.H. Freeden *et al.*, *Can. J. Agric. Sci.*, *33*:379, 1953.
4.   L.A. Young and H.P. Nicholson, *Progr. Fish Cult.*, *13*:193, 1951.
5.   A.A. Rosen and F.M. Middleton, *Anal. Chem.*, *31*:1729, 1959.
6.   F.M. Middleton and J.J. Lichtenberg, *Ind. Eng. Chem.*, *52*:99A, 1960.
7.   H.P. Nicholson, *Am. Water Works Assn.*, *51*:981, 1959.
7.   H.P. Nicholson, *J. Am. Water Works Assn.*, *51*:981, 1959.
8.   R.W. Harrington and W.L. Bidlingmayer, *J. Wildlife Manage.*, *22*:176, 1958.
9.   M.A. Smith, *Anal. Chem.*, *32*:167, 1960.
10.   R.R. Younger, *Proc. Northeast Weed Control Conf.*, *13*:322, 1959.
11.   J.M. Cohen *et al.*, *J. Am. Water Works Assn.*, *53*:233, 1961.
12.   M. Pierce, *Proc. Northeast Weed Control Conf.*, *15*:545, 1961.
13.   T.R. Walker, Bull. 72, Geological Society of America, 1961.
14.   G. Walton, *Public Health Aspects of the Contamination of Ground Water in South Platte River Basin in Vicinity of Henderson, Colo.*, U.S. Public Health Service, 1959.
15.   G. Walton, *Public Health Aspects of the Contamination of Ground Water in the Vicinity of Derby, Colo.*, Robert A. Taft San. Eng. Center, Cincinnati, 1961.
16.   B.J. Kallman, *Trans. Am. Fisheries Soc.*, *91*:14, 1962.
17.   R.C. Hiltibran, *Weeds*, *10*:17, 1962.
18.   H.P. Nicholson *et al.*, *Trans. Am. Fisheries Soc.*, *91*:213, 1962.

2.9-5   MOVEMENT OF INSECTICIDES FROM SOIL OR WATER INTO PLANTS

| | | | Residues | | | |
Insecticide	Crop	Source of insecticide	Amount in source $\mu g/l$	Amount in plant mg/l	Concentration or dilution factor[a]	Re
Toxaphene	Aquatic plants	Water	0.41	0.21	512.0	1
DDT	Aquatic vegetation	Water	200.0	75.0	375.0	2
DDT	Aquatic vegetation	Water	20.0	31.0	1,550.0	3
Organo-chlorines	Aquatic plants	Water	0.45	1.0	2,220.0	4
Organo-chlorines	Aquatic plants	Water	0.35	1.1	3,171.0	4
Organo-chlorines	Aquatic plants	Water	0.23	0.8	3,478.0	4
Organo-chlorines	Aquatic plants	Water	0.30	30.3	100,000.0	4
DDT	Algae and moss	Water	0.33	0.01	33.0	5
DDT	Algae	Water	5.8	0.002	0.34	6
Chlordane	Algae	Water	6.6	0.013	1.97	6
Endrin	Algae	Water	10.5	0.007	0.66	6
DDT	Vascular plants	Water	5.8	0.003	0.52	6
Chlordane	Vascular plants	Water	6.6	0.003	0.45	6
Endrin	Vascular plants	Water	10.5	0.006	0.57	6
			mg/kg	mg/kg		
Aldrin	Carrot roots	Muck soil	8.36	0.01 (root)	0.0013	7
Aldrin	Carrot roots	Clay soil	0.48	0.01 (root)	0.021	7
Dieldrin	Carrot roots	Muck soil	3.9	0.02	0.0051	7
Dieldrin	Carrot roots	Clay soil	0.48	0.11	0.23	7
Heptachlor	Rutabagas roots	Soil	0.320	0.024	0.075	8
Heptachlor	Wheat foliage	Seed treated	543.0	0.015	0.036	9
Aldrin & dieldrin	Wheat foliage	Soil	1.8	0.014	0.0077	10
Aldrin & dieldrin	Cucumber fruit	Soil	3.7	0.113	0.031	11
Heptachlor	Cucumber fruit	Soil	3.8	0.091	0.024	11
Dieldrin	Cucumber fruit	Soil	1.4	0.043	0.031	11
Aldrin	Alfalfa	Soil	0.84	0.009	0.011	11
Heptachlor	Alfalfa	Soil	0.78	0.028	0.036	11
Aldrin	Carrot roots	Soil	0.94	0.32	0.34	12
Aldrin	Potato tuber	Soil	0.94	0.07	0.074	12
Heptachlor	Carrot roots	Soil	0.49	0.36	0.73	12
Heptachlor	Potato tuber	Soil	0.49	0.05	0.10	12
DDT	Alfalfa foliage	Soil	1.39	0.113	0.08	13

[a] Concentration factor = $\dfrac{\text{concentration in animal}}{\text{concentration in water}}$.

Source: C.E. Edwards, *Persistent Pesticides in the Environment*, CRC Press, Cleveland, 1971.

REFERENCES

1.   L.C. Terriere, *J. Agr. Food Chem.*, *14*:66, 1965.
2.   R.A. Crocker and A.J. Wilson, *Trans. Am. Fish Soc.*, *94*(2):152, 1965.
3.   W.R. Bridges, B.J. Kallman, and A.K. Andrews, *Trans. Am. Fish Soc.*, *92*:421, 1963.
4.   J.O. Keith, *J. Appl. Ecol. (Suppl. 3)*, 71, 1966.
5.   G.L. Mack *et al.*, *N.Y. Fish Game J.*, *11*(2):148, 1964.
6.   P.J. Godsil and W.C. Johnson, *Pest. Mon. J.*, *1*(4):21, 1968.
7.   H. Hurtig and C.R. Harris, *Conf. Pollut. Environ.*, Montreal, 1966, 16.
8.   J.G. Saha and W.W.A. Stewart, *Can. J. Plant Sci.*, *47*:79, 1967.
9.   R.H. Burrage and J.G. Saha, *Can. J. Plant Sci.*, *47*:114, 1967.
10. J.G. Saha and M. McDonald, *J. Agr. Food Chem.*, *15*:205, 1967.
11. E.P. Lichtenstein and K.R. Schulz, *J. Agr. Food Chem.*, *13*(1):57, 1965.
12. E.P. Lichtenstein, T.W. Fuhremann, and K.R. Schulz, *J. Agr. Food Chem.*, *16*(2):348, 1968.
13. G.W. Ware, B.J. Estesen, and W.P. Cahill, *Pest. Mon. J.*, *2*(3):129, 1968.

2.9-6 CONCENTRATION OF RESIDUES FROM WATER INTO AQUATIC INVERTEBRATES

Organism	Insecticide	Amount In water (pp 10^{-9})[a]	Amount In animals (pp 10^{-6})[b]	Concentration factor[c]	Ref.
Aquatic invertebrates	Toxaphene	0.63	1.43	2,270	1
Shrimp	DDT	0.5	0.14	2,800	2
Crayfish	DDT	20.0	1.47	73	3
Crayfish	DDT	20.0	2.0	100	4
Crayfish	DDT	24.0	2.32	97	5
Crayfish	DDT	20.0	0.33	16.5	6
Plankton	DDT	20.0	5.0	250	7
Plankton	DDT	0.3	5.0	16,666	8
Eastern oyster	DDT	10.0	151.0	15,100	9
Eastern oyster	DDT	1.0	30.0	30,000	
Eastern oyster	DDT	0.1	7.0	70,000	
Pacific oyster	DDT	1.0	20.0	20,000	
Hand clam	DDT	1.0	3.0- 9.0	3,000 – 9,000	
Eastern oyster	Dieldrin	1.0	3.5	3,500	
Commercial shrimp	DDT	0.5	0.14	280	
Sea squirt	DDT	100.0 10.0	20.0 10.0	200 1,000	
Sea squirt	DDT	0.1	20.0	200,000	
Sea squirt	DDT	0.01	10.0	1,000,000	
Sea hare	DDT	0.01	1.78	178,000	
Crabs	DDT	50.0	7.2	144	
Snails	DDT	50.0	74.0	1,480	
Oysters	BHC	0.33	0.006	18.2	10
	DDT	0.55	0.033	60.0	
	Dieldrin	0.44	0.006	13.6	
	Heptachlor	0.55	0.002	3.6	
Clams	DDT	5.8	0.008	1.4	11
	Chlordane	6.6	0.006	0.9	
	Endrin	10.5	0.013	1.2	
Common clam	DDT	0.1	7.0	70,000	12
Oyster	Dieldrin	1.0	3.5	3,500	13
Hooked mussel	DDT	1.0	24.0	24,000	
Eastern oyster	DDT	1.0	26.0	26,000	
Pacific oyster	DDT	1.0	20.0	20,000	
European oyster	DDT	1.0	15.0	15,000	
Crested oyster	DDT	1.0	23.0	23,000	
Northern quahogs	DDT	1.0	6.0	6,000	

[a] $\mu g/l$.
[b] mg/kg.
[c] Concentration factor = $\dfrac{\text{concentration in animal}}{\text{concentration in water}}$.

Source: C.E. Edwards, *Persistent Pesticides in the Environment,* CRC Press, Cleveland, 1971.

REFERENCES
1. L.C. Terriere *et al., J. Agr. Food Chem., 14*:66, 1965.
2. P.A. Butler, in *Effects of Pesticides on Fish and Wildlife,* U.S. Department of the Interior, *Fish Wildlife Circ., 226*:65, 1965.
3. O.B. Cope, *J. Appl. Ecol. (Suppl.), 3*:33, 1966.
4. W.R. Bridges, B.J. Kallman, and A.K. Andrews, *Trans. Am. Fish Soc., 92*:421, 1963.
5. H. Cole *et al., Bull. Environ. Contam. Toxicol.,* 2:127, 1967.
6. *A Review of Fish and Wildlife Service Investigations During the Calendar Year,* U.S. Department of the Interior, *Fish Wildlife Serv. Circ., 167*:1963, 109.

2.9-6   CONCENTRATION OF RESIDUES FROM WATER INTO AQUATIC
INVERTEBRATES ((continued)

7.   E.G. Hunt and A.I. Bischoff, *Calif. Fish Game, 46*:91, 1960.
8.   J.O. Keith, *U.S. Fish Wildlife Serv., Denver Wildlife Res. Center Ann. Progr. Rept.,* 1964,
     14.
9.   *A Review of Fish and Wildlife Service Investigations During the Calendar Year,* U.S.
     Department of the Interior, *Fish Wildlife Serv. Circ., 199*:1964, 129.
10. G.E. Burdick *etal., Trans. Am. Fish. Soc., 93*:127, 1964.
11. P.J. Godsil and W.C. Johnson, *Pest. Mon. J., 1*(4):21, 1968.
12. V.L. Loosanoff, in *Research in Pesticides,* Academic Press, New York, 1965, 135.
13. P.A. Butler, *J. Appl. Ecol. (Suppl.), 3*:253, 1966.

2.9-7   CONCENTRATION OF RESIDUES FROM WATER TO FISH

Organism	Insecticide	Amount of residue		Concentration factor[c]	Ref.
		In water (pp 10^{-9})[a]	In animal (pp 10^{-6})[b]		
Rainbow trout	DDT	20.0	4.15	207	1
Black bullhead	DDT	20.0	3.11	155	1
Bluegill	Heptachlor	50.0	15.7	314	1
Catfish	Aldrin and dieldrin	0.044	0.07	1,590	2
Catfish	Aldrin and dieldrin	0.009	0.04	4,444	2
Catfish	Aldrin and dieldrin	0.021	0.02	952	2
Catfish	Aldrin and dieldrin	0.007	0.01	1,428	2
Buffabfish	Aldrin and dieldrin	0.023	0.09	3,913	2
Buffabfish	Aldrin and dieldrin	0.007	0.21	30,000	2
Scaled sardine	DDT	0.1	0.11	1,100	3
Rainbow trout	Toxaphene	0.41	7.72	18,829	4
Fish	DDT	0.30	1.0 – 6.4	3,333 – 21,333	5
Fish	Endrin	0.10	7.0	70,000	6
Fish	Toxaphene	1.0 – 4.0	0.8 – 2.5	2,000–2,500	7
Fish (5 spp.)	DDT	30–40	4–58	130–1,450	8
Bullhead trout	DDT	20	2–4	100–200	9
Fathead minnows	Endrin	0.015	0.15	10,000	10
Croakers	DDT	0.1	2.0	20,000	11
Pinfish	DDT	1.0	12.0	12,000	11
Pinfish	DDT	0.1	4.0	40,000	11
Fish	Dieldrin and DDT	10.0	0.1–1.0	10–100	12
Trout	Dieldrin	2.3	7.7	3,300	13
Chubs	DDT	5.8	0.029	5	14
Chubs	Chlordane	6.6	0.008	1.2	14
Chubs	Endrin	10.5	0.050	4.7	14
Trout	DDT	20.0	4.0	200.0	15
Bluegills	Heptachlor	50.0	56.8	1,130.0	16
Fish	DDT	0.015	12.44	829,300	17

2.9-7   CONCENTRATION OF RESIDUES FROM WATER TO FISH (continued)

Organism	Insecticide	Amount of residue		Concentration factor[c]	Ref.
		In water (pp 10^{-9})[a]	In animal (pp 10^{-6})[b]		
Fish	DDT	0.11	3.85	35,000	17
White catfish	DDD	14.0	30.4–129.0	2,172–9,214	18
Largemouth bass	DDD	14.0	19.7–25.0	1,407–1,785	18
Brown bullhead	DDD	14.0	15.5–24.8	1,107–1,771	18
Black crappie	DDD	14.0	5.4–115.0	386–8,214	18
Bluegill	DDD	14.0	6.6–10.0	471– 714	18
Sacramento blackfish	DDD	14.0	10.9–17.6	778–1,257	18
Brook trout	DDT	24.0	17.3	710.0	19

[a] μg/l.
[b] mg/kg
[c] Concentration factor = $\dfrac{\text{Concentration in animal}}{\text{Concentration in water}}$.

Source: C.E. Edwards, *Persistent Pesticides in the Environment,* CRC Press, Cleveland, 1971.

REFERENCES

1.   O.B. Cope, *J. Appl. Ecol. (Suppl.)*, *3*:33, 1966.
2.   B.I. Sparr, *et al., Adv. Chem. Ser.*, *60*:146, 1966.
3.   P.A. Butler, in *Effects of Pesticides on Fish and Wildlife,* U.S. Department of the Interior, *Fish Wildlife Circ.*, *226*:65, 1965.
4.   L.C. Terriere, *et al., J. Agr. Food Chem.*, *14*:66, 1965.
5.   J.O. Keith, *U.S. Fish Wildlife Serv., Denver Wildlife Res. Center Ann. Progr. Rept.,* 1964, 14.
6.   E. Langer, *Science, 144*:35, 1964.
7.   B.J. Kallman, O.B. Cope, and R.J. Navarre, *Trans. Am. Fish Soc., 91*:14, 1962.
8.   R.A. Crocker and A.J. Wilson, *Trans. Am. Fish Soc., 94*(2):152, 1965.
9.   W.R. Bridges, B.J. Kallman, and A.K. Andrews, *Trans. Am. Fish Soc., 92*:421, 1963.
10.   D.I. Mount, and G.J. Putnicki, *Trans. 31st N. Amer. Wildlife Nat. Res. Conf.,* 177, 1966.
11.   D.J. Hansen, *Ann. Rept. Bur. Comm. Fish Circ., 247*:10, 1966.
12.   A.V. Holden and K. Marsden, *J. Proc. Sew. Purif., 4*:295, 1966.
13.   A.V. Holden, *J. Appl. Ecol. (Suppl.), 3*:45, 1966.
14.   P.J. Godsil and W.C. Johnson, *Pest. Mon. J., 1*(4):21, 1968.
15.   *A Review of Fish and Wildlife Service Investigations During the Calendar Year,* U.S. Department of the Interior, *Fish Wildlife Serv. Circ., 167,* 1963, 109.
16.   *A Review of Fish and Wildlife Service Investigations During the Calendar Year,* U.S. Department of the Interior, *Fish Wildlife Serv. Circ., 199,* 1964, 129.
17.   G.L. Mack *et al., N.Y. Fish Game J., 11*(2):148, 1964.
18.   E.G. Hunt and A.I. Bischoff, *Calif. Fish Game, 46*:91, 1960.
19.   H. Cole *et al., Bull. Environ. Contam. Toxicol., 2*:127, 1967.

Section 3
Wastewater

3.1 SOURCES

3.1.1 RUNOFF

3.1-1 SURFACE RUNOFF COEFFICIENTS

Character of surface	Runoff coefficients[a,b]
Pavement	
Bituminous streets	0.70−0.95
Concrete streets	0.80−0.95
Driveways, walks	0.75−0.85
Brick	0.70−0.85
Roofs	0.75−0.95
Lawns, sandy soil	
Flat, 2%	0.05−0.10
Average, 2−7%	0.10−0.15
Steep, 7%	0.15−0.20
Lawns, heavy soil	
Flat, 2%	0.13−0.17
Average, 2−7%	0.18−0.22
Steep, 7%	0.25−0.35

[a] Applicable to storms of 5−10 yr frequencies.
[b] For use with the rational formula which relates runoff to rainfall as follows:

$$Q = CiA$$

where

Q = peak runoff, cfs.
C = runoff coefficient (ratio of peak runoff rate to average rainfall rate for period known as the time of concentration).
i = average rainfall intensity, in./hr for a period equal to the time of concentration.
A = drainage area, acres.

REFERENCES

1. *Manual of Practice No. 9,* Water Pollut. Control Fed., 1969.
2. Design and construction of sanitary and storm sewers, *Manual of Engineering Practice, No. 37,* Am. Soc. Civ. Engineers, New York, 1960.

3.1-2 AREA RUNOFF COEFFICIENTS

Description of area	Runoff coefficients, C[a]	
	Ref. 1	Ref. 2
Downtown	0.70−0.95	
Neighborhood	0.50−0.70	
Residential		
Single family	0.30−0.50	
Multi-units, detached	0.40−0.60	
Multi-units, attached	0.60−0.75	
Compact	−	0.50
Suburban	0.25−0.40	0.40
Apartment	0.50−0.70	0.60
Industrial		
Light	0.50−0.80	0.60
Heavy	0.60−0.90	0.70
Parks, cemeteries	0.10−0.25	0.20
Playgrounds	0.20−0.35	
Railroad yard	0.20−0.35	0.20
Unimproved areas	0.10−0.30	
Densely built, paved areas		0.90

[a] Applicable to storms of 5−10 yr frequencies;
C − a coefficient based on

$$Q = CiA$$

where

Q = peak runoff, cfs.
C = runoff coefficient (ratio of peak runoff rate to average rainfall rate for period known as the time of concentration).
i = average rainfall intensity, in./hr for a period equal to the time of concentration.
A = drainage area, acres.

REFERENCES

1. *Manual of Practice No. 9,* Water Pollut. Control Fed., 1969.
2. D.B. Redfern, *J. Water Pollut. Control Fed., 34*(10): 1055, 1962.

3.1-3   RUNOFF RETARDANCE
COEFFICIENT C_r

Surface	C_r^a
Smooth asphalt	0.007
Concrete paving	0.012
Tar and gravel paving	0.017
Closely clipped sod	0.046
Dense bluegrass turf	0.060

[a] For use with Izzard's formula for concentration time for small experimental plots without developed channels:

$$t_c = \frac{41 \, b \, L_o^{1/3}}{(k \, i)^{2/3}}$$

where t_c = time of concentration, min

b = coefficient, $= \dfrac{0.0007i + C_r}{S_o^{1/3}}$

L_o = overland flow length, ft

k = rational runoff coefficient (see Table 1.1–14, coefficient, Surface Runoff)

i = rainfall intensity, in./hr during time t_c

S_o = surface slope

C_r = coefficient of retardance.

The equation is valid only for laminar flow conditions where $iL_o = {<}500$.

Source: C.F. Izzard, *Proc. Highway Res. Board,* 26:129, 1946. With permission.

3.1-4   INFILTRATION RATES TO SEWERS

Per mile of sewer	2,000–200,000 gpd
Per in. diameter mi. of sewer	500–2,000 gpd
Per capita	25–200 gpd
Per acre of sewered area	300–1,500 gpd

Source: *Manual of Practice No. 8,* Water Pollut. Control Fed., 1959. With permission.

3.1-5   INFILTRATION AT JOINTS[a]

Joint material	Clay pipe, 6 in. diam		Concrete pipe, 6 in. diam	
	Head above flow line, in.	Avg infiltration rate, gpd/in. diam mi	Head above flow line, in.	Avg infiltration rate, gpd/in. diam mi
Jute only	3	8,270	3	6,710
	9	71,050	9	52,800
	15	155,250	15	118,000
	21	258,000	21	205,500
	27	356,000	27	278,000
Cement	3	3,360	3	680
	9	15,000	9	4,950
	15	28,700	15	10,450
	21	41,200	21	16,500
	27	53,200	27	22,000
Hot pour	3	1,330	3	0
	9	1,660	9	107
	15	3,410	15	235
	21	4,720	21	419
	27	5,520	27	513
PVC compression joint	3	0		
	9	645		
	15	1,450		
	21	1,850		
	27	2,400		
Compression joint B	0–27	Negligible		
Cold mastic			3	990
			9	1,450
			15	3,210
			21	5,130
			27	7,810
Rubber gasket			0–27	Negligible

[a] Results obtained under laboratory conditions.

Source: I.W. Santry, Jr., *J. Water Pollut. Control Fed., 36*:1256, 1964. With permission.

3.1-6   OVERFLOW OCCURRENCES

Interceptor capacity (x dry weather flow)	Total time, %			
	Boston[1]	Detroit[2]	Washington, D. C.[3]	Kansas City[4]
1.5				3.72
2	5.7	2.5	–	–
2.5	–	–	–	3.55
3.0	3.7	2.1	2.7	–
3.5	–	–	–	3.22
4.5	–	–	–	2.83
5.0	2.2	1.5	2.1	–
5.5	–	–	–	2.50
6.5	–	–	–	2.30
10	1.2	–	1.1	–

REFERENCES

1. J.E. McKee, *J. Boston Soc. Civil Eng., 34*(2):55, 1957.
2. C.L. Palmer, *Sew. Ind. Waste, 22*(2):154, 1950.
3. C.F. Johnson, *Civ. Eng., 28*(2):428, 1958.
4. H.H. Benges, *J. Water Pollut. Control Fed., 33*(12):1252, 1961.

3.1.2   DOMESTIC

3.1-7   PER CAPITA SEWAGE FLOWS

Population range	Gal per capita per day			Liters per capita per day		
	1[a]	1[b]	2	1[a]	1[b]	2
<5,000	63—158	197	—	240—600	750	—
5,000—10,000	93—282	—	—	350—1,070	—	—
10,000—25,000	47—111	126—162	53—207	180—420	480—615	200—785
25,000—75,000	41—149	78—230	190—217	155—565	295—870	340—820
75,000—150,000	76—97	116—137	141—187	290—370	440—520	535—710
150,000—500,000	86—149	107—149	139—161	325—565	405—565	525—610
500,000—1,000,000	104	—	140—196	395	—	530—740
>1,000,000	—	145	144—164	—	550	545—620

[a]   Separate sewerage systems.
[b]   Combined sewerage systems.

REFERENCES

1. C.E. Keefer, *Sewage Treatment Works,* McGraw-Hill, N.Y., 1940.
2. H.E. Babbitt and E.R. Baumann, *Sewerage and Sewage Treatment,* 8th ed., John Wiley & Sons, N..Y., 1958.

3.1-8   DESIGN RATES OF FLOW FOR
SANITARY SEWERS

Sewer type	Flow, cfs/acre	Reference
Local sanitary	0.013	1
Local sanitary in light industrial and commercial areas	0.019	1
Commercial areas	0.0046–0.094[a]	2
Trunk sanitary	0.0093	1

[a] Office — 40–50 gpd/capita; 180–525 gpd/hotel room; stores, offices and small business — 12–25 gpd/capita (average); laterals — 400 gpd/capita; trunks — 250 gpd/capita.

REFERENCES

1. D.B. Redfern, *J. Water Pollut. Control Fed.*, *34*(10):1053, 1962.
2. R.E.L. Johnson, *J. Water Pollut. Control Fed.*, *37*(11): 1600, 1965.

3.1-9   DESIGN BASIS FOR SEWERS

Type of sewer	Design flow condition	Item	Value	cfs/acre
Main	Full	Residential flow Development		
		density, houses/acre	3	–
		Persons/house	3.33	–
		Per capita flow (gpd)[a]	250	0.00039
		Residential flow (gpd/acre)	2,500	0.0039
		Infiltration (gpd/acre)	2,000	0.0031
		Peak total flow (gpd/acre)	4,500	0.0070
		Commercial flow = 1.33 residential flow (gpd/acre)	6,000	0.0093
Interceptor	Full	Residential peak flow (gpd/acre)[b]	3,500	0.0054
		Commercial flow = 1.30 residential flow (gpd/acre)	4,500	0.0070

[a] Includes infiltration into house services.
[b] Based on diversity factor of approximately 80 percent.
[c] Calculated by editors.

Source: R.L. Brown and W.R. Condon, *J. Water Pollut. Control Fed.*, *39*(8):1374, 1967. With permission.

3.1.3 COMMERCIAL AND INDUSTRIAL

3.1-10 WASTE FLOWS–COMMERCIAL AND INDUSTRIAL ALLOWANCES

City	Year and source	Commercial allowance, gal/day acre	Industrial allowance, gal/day acre
Berkeley, Cal.	–	–	50,000
Grand Rapids, Mich.	–	Hospitals: 200 gpd/bed schools: 200–300 gpd/room	250,000
Hagerstown, Md.	–	Hospitals: 150 gpd/bed schools: 120–150 gpd/room	–
Houston, Texas	1960[2]	Peak flows: offices: 0.36 gpd/ft^2 retail: 0.20 gpd/ft^2 hotels: 0.93 gpd/ft^2	–
Las Vegas, Nev.		Schools: 15 gpd/capita	–
Los Angeles, Cal.[a]	1965	Commercial: 11,700 hospitals: 0.75 mgd schools: 0.12 mgd universities: 0.73 mgd	15,500
Los Angeles County Sanitation District, Cal.	1964	10,000 avg; 25,000 peak	–
Kansas City, Mo.	1958	5,000	10,000
Lincoln, Nebr.	1962	7,000	–
Memphis, Tenn.	–	2,000	2,000
St. Joseph, Mo.	1962	6,000	–
Santa Monica, Cal.		Commercial: 9,700	13,600

[a] Values used for future planning; individual studies made for specific projects.

REFERENCES

1. *Manual of Practice No. 9,* Water Pollut. Control Fed., 1969.
2. R.E.L. Johnson, *J. Water Pollut. Control Fed., 37*:1597, 1965.

3.1-11 AVERAGE COMMERCIAL FLOWS

Type of establishment	Average flow, gpd/cap
Stores, offices, and small businesses	12–25
Hotels	50–150
Motels	50–125
Drive-in theaters, 3 persons/car	8–10
Schools (no showers), 8 hr period	8–35
Schools (with showers), 8 hr period	17–25
Tourist and trailer camps	80–120
Recreational and summer camps	20–25

Source: *Manual of Practice No. 9,* Water Pollut. Control Fed., 1969.

3.1-12   SEWER DESIGN CAPACITY IN COMMERCIAL AREAS

City	Year	Design allowance gpd/acre	cfs/acre[a]	Remarks	Ref.
Baltimore, Md.	1949	6,750–13,500	0.0105–0.021		1, 2
Columbus, Ohio	1946	40,000	0.062	Added to domestic	1, 2
Cranston, R.I.	1943	25,000	0.039		1, 2
Dallas, Texas	1949	30,000	0.046	Added to domestic	1, 2
Los Angeles, Cal.	1948	9,700	0.015		1, 2
Milwaukee, Wisc.	1945	60,500	0.094		1, 2
Toledo, Ohio	1946	15,000–30,000	0.023–0.046		1, 2
Memphis, Tenn.	–	2,000	0.0031		2
Santa Monica, Cal.		9,700	0.015		2
Shreveport, La.		3,000	0.0046		2
		25,000	0.039	Average	3
		50,000	0.077	Assume peak 2x average	3
Buffalo, N.Y.		60,000	0.093		4
Detroit, Mich.		50,000	0.077		4
Cincinnati, Ohio		40,000	0.062		4
Grand Rapids, Mich.		40–50 gpd/capita		Office	2
		400–500 gpd/room		Hotel	2
Hagerstown, Md.		180–250 gpd/room		Hotel	2
Las Vegas, Nev.		310–525 gpd/room		Hotel	2
Los Angeles, Cal.	1948	450–500 gpd/capita		Hotel	2
		80–100 gpd/capita		Office	2
Santa Monica, Cal.		7,750 gpd/acre		Hotel	2
		12–25 gpd/cap average		Stores, offices, small business	2
		400 gpd/capita		Lateral includes infiltration	2
		250 gpd/capita		Trunks but not storm flows	2
Toronto, Canada		63,500	0.098	Downtown sewers	1
Dallas, Texas		30,000	0.046	Normal added to domestic	1
		60,000	0.093	Tunnel relief sewer	1
St. Louis, Mo.		90,000	0.14	Average flow	1
		165,000	0.266	Peak flow	1
Los Angeles, Cal.		22,600	0.035	Average flow	1
		45,200	0.070	Peak flow	1

[a] Calculated by the editors.

Source: R.E.L. Johnson, *J. Water Pollut. Control Fed.,* 37(11):1600, 1965. With permission.

REFERENCES

1. W.E. Stanley *et al., J. Boston Soc. Civil Eng., 40*:312, 1953.
2. *Manual of Practice No. 9,* Water Pollut. Control Fed., 1969.
3. G.M. Fair and J.C. Geyer, *Water Supply and Waste Disposal,* John Wiley and Sons, New York, 1954.
4. H.E. Babbitt and E.R. Bauman, *Sewerage and Sewage Treatment,* 6th ed., John Wiley and Sons, New York, 1947.

3.1-13   QUANTITIES OF WASTE FLOW

Type of establishment	gal/cap day	Type of establishment	gal/cap day
Small dwellings and cottages with seasonal occupancy	50	Day schools with cafeterias, but no gymnasiums or showers	20
Single family dwellings	75	Day schools with cafeterias, gymnasiums, and showers	25
Multiple family dwellings (apartments)	60	Boarding schools	75–100
Rooming houses	40	Day workers at schools and offices (per shift)	15
Boarding houses	50	Hospitals	105–250+
Additional kitchen wastes for nonresident boarders	10	Institutions other than hospitals	75–125
Hotels without private baths	50	Factories (gal/person/ shift, exclusive of industrial wastes)	15–35
Hotels with private baths (2 persons/room)	60	Picnic parks with bathhouses, showers, and flush toilets	10
Restaurants (toilet and kitchen wastes/patron)	7–10	Picnic parks (toilet wastes only), (gal/picnicker)	5
Restaurants (kitchen wastes/ meal served)	2.5–3	Swimming pools and bathhouses	10
Additional for bars and cocktail lounges	2	Luxury residences and estates	100–150
Tourist camps or trailer parks with central bathhouse	35	Country clubs (per resident member)	100
Tourist courts or mobile home parks with individual bath units	50	Country clubs (per nonresident member present)	25
Resort camps (night and day) with limited plumbing	50	Motels (per bed space)	40
Luxury camps	100–150	Motels with bath, toilet, and kitchen wastes	50
Work or construction camps (semi-permanent)	50	Drive-in theaters (per car space)	5
Day camps (no meals served)	15	Movie theaters (per auditorium seat)	5
Day schools without cafeterias, gymnasiums, or showers	15	Airports (per passenger)	3–5
		Self-service laundries (gal/ wash, i.e., per customer)	50
		Stores (per toilet room)	400
		Service stations (per vehicle served)	10

Source: *Manual of Septic Tank Practice,* U.S. Public Health Service, 1957.

3.1.4   AGRICULTURAL

3.1-14   AVERAGE CONCENTRATIONS OF POLLUTANTS IN FEEDLOT RUNOFF

	Concrete lot 7.5% slope	Dirt lot 3.5% slope
BOD	7,500	4,000
COD	15,000	9,500
Nitrate	50	40
NH$_3$-N	600	400
Organic-N	100	75
Alkalinity	2,500	1,500
Total solids	15,000	9,000
Suspended solids	6,000	4,000

Note: All values in mg/l.

Source: H. Bernard, Effect of agriculture on waste quality, Relationship of Agriculture to Soil and Water Pollution, Cornell Conference on Agricultural Waste Manufacture, 1970, p. 6.

3.1-15   FEEDLOT RUNOFF CHARACTERISTICS

Constituents	Concrete surfaced	Nonsurfaced
COD, mg/l	2,760–19,400	1,900–8,900
BOD, mg/l[2,b]	450–1,400[k]	250–750[l]
Kjeldahl-N, mg/l	94–1,000	50–540
Ammonia-N, mg/l	1.3–139[a]	1.0–62[b]
Nitrite-N, mg/l	1.0–6.0[c]	1.0–23[d]
Nitrate-N, mg/l	0.1–11[e]	0.1–6.0[f]
Suspended solids, mg/l	1,400–12,000[g]	1,500–10,500[h]
Chloride, mg/l	300[i] (200–415)[j]	270 (210–315)
Phosphate as PO$_4^{\equiv}$, mg/l	50 (20–80)	26 (14–45)
Lignin as tannic acid, mg/l	70 (41–150)	50 (20–90)
pH	7.9 (7.5–8.2)	8.2 (7.7–8.4)
Total coliform, mil org/100 ml	130 (33–348)	79 (22–348)
Fecal coliform, mil org/100 ml	130 (35–240)	33 (8–79)
Fecal streptococci, mil org/100 ml	79 (13–240)	24 (8–79)

[a] Winter 1.3–7.0; fall 20–77; and summer 50–139 mg/l.
[b] Winter 1.0–3.8; fall 13–45; and summer 26–62 mg/l.
[c] July–Aug 1.0–6.0; Oct–Nov 1.0–5.0 mg/l.
[d] July–Aug 1.0–7.0; Oct–Nov 1.0–23 mg/l.
[e] July–Aug 0.1–11; Oct–Nov 0.4–2.3 mg/l.
[f] July–Aug 0.1–6.0; Oct–Nov 0.5–2.6 mg/l.
[g] July–Aug; dry 1,400–3,000; moist 6,000; wet 3,000–12,000 mg/l; Oct–Nov, wet 2,000–2,500 mg/l.
[h] July–Aug, dry 1,500–2,000; moist 5,000; wet 3,000–10,500 mg/l; Oct–Nov, wet 1,800 mg/l.
[i] Median.
[j] Values in parentheses for chemical constituents represent 70% limits.
[k] Winter 300–600; spring and fall 750–1,050; summer 1,100–1,400.
[l] Winter 150–350; spring and fall 350–550; summer 550–750 mg/l.

REFERENCES

1. J.R. Miner *et al., J. Water Pollut. Control Fed., 38*(10):1582, 1966.
2. J.R. Miner, Water Pollution Potential of Cattle Feedlot Runoff, Ph.D. thesis, Kansas State, Manhattan, 1967.

3.1-16   RUNOFF FROM RURAL AREAS

Calculated Average Annual Loads

Watershed[b]	Cover	Total solids, lb/acre	BOD, lb/acre	COD, lb/acre	PO_4,[a] lb/acre	Total N, lb/acre
A	Corn	3,660	27.5	480	8.4	88
B	Corn	13,200	120.0	1,300	27.7	237
A	Wheat	480	3.7	64	1.1	11
B	Wheat	1,730	15.5	170	3.6	31
A	Meadow	trace	–	–	–	–
B	Meadow	trace	–	–	–	–
C	Meadow and apple orchard	18.5	3.7	27.8	3.9	0.81

a   Hydrolyzable PO_4.
b   Location of watershed:

A – watershed in Coshocton, Ohio with improved practice, i.e., contour tillage; liming to pH 6.8; 5-20-20 fertilizer – 180 lb/acre yr each on corn and wheat, and 6 ton/acre yr of manure on corn and 4 ton/acre yr for wheat; 200 lb/acre yr of fertilizer 0-20-20 on first year meadow; a clover-alfalfa-timothy mixture is used on the meadow.

B – watershed in Coshocton, Ohio with prevailing practice, i.e., straight-row tillage across the slope; liming to pH 5.4; 5-20-20 fertilizer – 50 lb/acre yr on corn and 100 lb/acre yr on wheat and 4 ton/acre yr of manure; a red clover-timothy mixture is used for the meadow.

C – Ripley, Ohio 5 acre apple orchard (300 trees) in meadow.

Source: R.B. Weidner *et al., J. Water Pollut. Control Fed., 41*(3):377, 1969. With permission.

3.2   WASTEWATER CHARACTERISTICS

3.2.1   RUNOFF

3.2-1   SEWERAGE SYSTEM OVERFLOWS – CHEMICAL VALUES

Arithmetic Mean Values, mg/l

Sewer Type	NH_3-N	Organic N-N	Phenols	Suspended solids	Total PO_4 as $PO_4 \equiv$
Combined[a]	3.25 (19)[c]	0.37 (20)	0.027 (15)	150 (31)	9.0 (31)
	(0.00–11.52)[d]	(0.08–2.98)	(0.011–0.125)	(23–1,398)	(1.8–25.0)
Separate[b]	0.48 (9)	0.36 (9)	(3)	1,280 (6)	2.9 (7)

[a] Conner Combined Sewerage System, Detroit, 1964.
[b] Allen Creek Drain, Separate System, Ann Arbor, 1964.
[c] Value in parentheses – number of samples analyzed.
[d] Range in values.

REFERENCE

1. W.J. Benzie *et al., J. Water Pollut. Control Fed., 38*(3):415, 1966.

3.2-2   BACTERIAL DENSITIES IN SEPARATE AND COMBINED STORMWATER SYSTEMS

Nature	Location	Organisms/100 ml			Ref.
		Coliform	Fecal coliform	Fecal streptococci	
Combined sewer overflows	Detroit	4.3 x 10⁶			2
Stormwater	Detroit	930,000			2
Surface runoff	Detroit	2,300–430,000			1
Street gutters	Seattle	up to 16,100			3
Streets and parks	Stockholm	4,000 (median value)			4
		200,000 (maximum)			4
Residential-commercial	Cincinnati	10% > 460,000	10% > 76,000	10% > 110,000	5
Combined sewer	Detroit	9.4 x 10⁶	2.7 x 10⁶	580,000	6
		(0.57 x 10⁶ to	(0.19 x 10⁶ to	(0.18 x 10⁶ to	
		12 x 10⁶)	20 x 10⁶)	1.5 x 10⁶)	
Separate sewage	Ann Arbor	1.2 x 10⁶	82,000	140,000	6
		(0.12 x 10⁶ to	(7.4 x 10³ to	(12 x 10³ to	
		34 x 10⁶)	750 x 10³)	670 x 10³	

REFERENCES

1. C.L. Palmer, *J. Water Pollut. Control Fed., 35*(2): 162, 1963.
2. C.L. Palmer, *Sewage Ind. Wastes, 22*(2): 154, 1950.
3. R.O. Sylvester, *An Engineering and Ecological Study for the Rehabilitation of Green Lake,* University of Washington, Seattle, 1960.
4. G. Akerlinch, *Nord. Hyg. Tidskr. (Denmark), 31:* 1, 1950.
5. S.R. Weibel *et al., J. Water Pollut. Control Fed., 36*(7): 914, 1964.
6. R.J. Burm *et al., J. Water Pollut. Control Fed., 38*(3): 409, 1966.

3.2-3 DISTRIBUTION OF PRIORITY POLLUTANTS FREQUENTLY DETECTED[a]

Pollutant	Frequency[b]		Range[c]	
	Liquid	Sediment	Liquid µg/l	Sediment µg/kg[d]
Phenols	6	—	0.3—0.9	—
Polynuclear aromatic hydrocarbons (PAHs)				
Fluoranthene	5	3	0—98	1,200—4,000
Pyrene	4	3	0.76	1,200—3,000
Anthracene	—	3	—	375—7,000
1,2-Benzanthracene	—	2	—	1,100—4,100
Benzo(*a*)pyrene	—	2	—	600—25,000
3,4-Benzofluoranthene + 11,12-benzofluoranthene	—	2	—	2,800—4,600
Chrysene	—	2	—	1,500—4,100
Fluorene	—	2	—	210—300
Phenanthrene	—	2	—	375—2,000
Esters				
Bis(2-ethylhexyl)phthalate	6	3	0—60	130—59,000
Di-*n*-butyl phthalate	6	3	0.5—3.4	300—25,000
Butylbenzyl phthalate	—	2	—	450—1,200
Heavy metals				
Arsenic	7	3	2—15	2.8—4.8
Beryllium	7	3	1—2	0.1—0.2
Cadmium	7	3	2—7	0.06—1
Chromium	7	3	9—36	8.10—28
Copper	7	3	11—49	7—140
Lead	7	3	30—400	24—370
Nickel	7	3	9—40	3—16
Silver	7	3	3—8	0.5—2.1
Zinc	7	3	90—330	37—380
Antimony	7	3	<20	3—3.3
Selenium	7	3	3—10	0.06—1.4
Thallium	7	3	2—10	<0.2
Mercury	7	3	0.2—0.56	2.1—2.4

[a] R. Field, *CRC Crit. Rev. Environ. Control, 16*:147, 1986.

[b] Reported when found four or more times out of seven liquid samples and two or more times out of three sediment samples.

[c] These are mostly grab samples and therefore cannot be accurately related to pounds of pollutants discharged. Also, the mass emission of pollutants will vary for each rainfall event.

[d] Heavy metals in mg/kg.

3.2-4   PATHOGENS IN WASTEWATER[a]

Organism	No. types known	Disease	No./100 ml	Ref.
Bacteria				
Salmonella		Typhoid, paratyphoid, salmonellosis	2.3—8,000	1
Shigella		Bacillary dysentery	1—1,000	1
Enteropathogenic				
E. coli		Gastroenteritis	—[b]	
Yersinia		Gastroenteritis	—[b]	
Campylobacter		Gastroenteritis	—[b]	
Vibrio		Cholera	10—10^4 [c]	2
Leptospira		Weil's disease	—[b]	3
Viruses				
Enterovirus				
Poliovirus	3	Paralysis	0—6,000	
		Aseptic meningitis	182—492,000[d]	
Coxsackievirus				
A	24	Herpangia		
		Aseptic meningitis		
		Respiratory illness		
		Paralysis		
B	6	Pleurodynia		
		Aseptic meningitis		
		Pericarditis		
		Myocarditis		
		Congenital heart disease		
		Anomalies		
		Nephritis		
Echovirus	34	Respiratory infection		
		Aseptic meningitis		
		Diarrhea		
		Pericarditis		
		Myocarditis		
		Fever, rash		
Reovirus	3	Respiratory disease		
		Gastroenteritis		
Adenovirus	41	Acute conjunctivitis		
		Diarrhea		
		Respiratory illness		
		Eye infection		
Hepatitis A virus	1	Infectious hepatitis		
Rotavirus	2	Infantile gastroenteritis	400—85,000[e]	4
Norwalk agent	1	Nonbacterial gastroenteritis		
Protozoa				
Giardia lamblia		Diarrhea, malabsorption	530—10^5	5, 6
Entamoeba coli		Mild diarrhea, colonic ulceration	28; 52	7, 8
Entamoeba histolytica		Amebic dysentery, liver abscess, colonic ulceration	4	7
Cryptosporidium		Gastroenteritis	4.1—1732 28.4[f]	9
Helminths				
Ascaris		Ascariasis	5—100	1
Ancylostoma		Anemia	6—188	1
Necator		Anemia	—[b]	
Trichuris		Diarrhea, anemia	10—20 41	10

[a] Compiled from J.B. Rose, *CRC Crit. Rev. Environ. Control, 16*:231, 1986.

[b] Unknown.

[c] During an epidemic.

3.2-4 PATHOGENS IN WASTEWATER[a] (continued)

[d] pfu — plaque-forming unit.

[e] iff — immunofluorescent foci.

[f] Geometric mean.

REFERENCES

1. R.G. Feachem, D.J. Bradley, H. Garelick, and D.D. Mara, *Sanitation and Disease Health Aspects of Excreta and Wastewater Management,* John Wiley & Sons, New York, 1983, 501.
2. Y. Kott and N. Betzer, *Israel J. Med. Sci., 8*:1912, 1972.
3. L.G. Irving, Viruses in wastewater effluents, in *Viruses and Disinfection of Water and Wastewater,* M. Butler, A.R. Medlen, and R. Morris, Eds., University of Surrey, Guildford, U.K., 1982, 7.
4. C.P. Gerba, S.N. Singh, and J.B. Rose, *CRC Crit. Rev. Environ. Control, 15*:213, 1984.
5. J.C. Fox and P.R. Fitzgerald, The presence of *Giardia lamblia* cysts in sewage and sewage sludges from the Chicago area, in Waterborne Transmission of Giardiasis, Rep. EPA-600-9-79-001, W. Jakubowski and H.C. Hoff, Eds., U.S. Environmental Protection Agency, Cincinnati, Ohio, 1979, p. 268.
6. W. Jakubowski and T.H. Ericksen, Methods for detection of *Giardia* cysts in water supplies, in Waterborne Transmission of Giardiasis, Rep. EPA-600-9-79-001, W. Jakubowski and H.C. Hoff, Eds., U.S. Environmental Protection Agency, Cincinnati, Ohio, 1979, p. 268.
7. H. Kott and Y. Kott, *Water Sewage Works, 140*:177, 1967.
8. W.-L.L. Wang and S.G. Dunlop, *Sewage Ind. Wastes, 26*:1020, 1954.
9. J.B. Rose, *J. Am. Water Works Assn., 80*:53, 1988.
10. H. Liebmann, *Berl. Muench. Tieraerztl. Wochenschr., 78*:106, 1965.
11. W.B. Rowan and A.L. Gram, *J. Parasitol., 45*:615, 1959.

3.2-5 POLLUTION FROM URBAN AND AGRICULTURAL RUNOFF

Constituent	Urban runoff[1] (storm water)	Agricultural runoff[2]
Suspended solids, mg/l	5–1,200	–
COD, mg/l	20–610	–
BOD, mg/l	1–173	–
Total phosphorus, mg/l	0.02–7.3	0.1–0.65
Nitrate nitrogen, mg/l	–	0.03–5.0
Total nitrogen, mg/l	0.3–7.5	0.5–6.5
Chlorides, mg/l	3–35	–

REFERENCES

1. S.R. Weibel *et al., J. Water Pollut. Control Fed., 36*(7):914, 1964.
2. Algae and metropolitan wastes, Transaction of seminar on algae and metropolitan wastes, held at Cincinnati, Ohio, April 27–29, 1960, under sponsorship of Division of Water Supply and Pollution Control and R.A. Taft Sanitary Engineering Center, Cincinnati, 1961.

3.2-6 NUTRIENT CONCENTRATIONS IN URBAN STORMWATER RUNOFF FROM 27 ACRE RESIDENTIAL AND LIGHT COMMERCIAL AREA

Cincinnati, Ohio, July 1962–July 1964[a]

Constituent	Ranges of discrete samples	Storm average
Total N	0.3–7.5	3.1
Inorganic N, mg/l	0.1–3.4	1.0
Total hydrolyzable P, mg/l	<0.007–2.4	0.37

[a] January and February 1963, not included.

Source: Reprinted, with publisher's permission, from *J. Am. Water Works Assn., 59*(43):355, 1967. © 1967, by the American Water Works Association, Inc.

REFERENCES

1. S.R. Weibel *et al., J. Am. Water Works Assn., 58*(8):1075, 1966.
2. S.R. Weibel *et al.,* Characterization, Treatment, and Disposal of Urban Stormwater, 3rd Int. Conf. Water Pollut. Res., Munich, 1966, unpublished.

3.2.2 DOMESTIC

3.2-7 WEIGHT OF HUMAN EXCRETA[a]

Wet weight

	Feces	Urine	Ref.
Men	150	1500	1
Women	45	1350	1
Boys	110	569	1
Girls	25	450	1
Average	82.5	967	1
Adults, mixed diet	60—250	600—1600	2
Young children	60—150	500—1000	2
Mixed population	90.7	1261	5
Infants	4—120	100—500	2

Dry weight

	Feces	Urine	Ref.
	$23(4.9—41)^b$	55—70	2
	35—70	50—70	3
	20.5	43.3	4
	20.9^c	46.6^d	5

[a] Grams per capita per day
[b] Range.
[c] Assuming 23% solids.
[d] Assuming 3.7% solids.

REFERENCES

1. C.E. Keefer, *Sewage Treatment Works*, McGraw-Hill, N.Y., 1940, 52.
2. K. Diem, Ed., *Scientific Tables*, 6th ed., Geigy Pharmaceuticals, Ardsley, N.Y., 1962.
3. H.B. Gotaas, *Composting*, World Health Organization, Geneva, 1956, 35.
4. L. Metcalf and H.P. Eddy, *Sewerage and Sewage Disposal, A Textbook*, McGraw-Hill, N.Y., 1930, 351.
5. A.P. Folwell, *Sewerage*, 6th ed., John Wiley & Sons, N.Y., 1912, 368.

3.2-8 CHEMICAL COMPOSITION OF HUMAN FECES AND URINE

Human feces without urine
Approximate quantity
0.3−0.6 lb (135−270 g)/ cap dry moist weight
0.08−016 lb (35−70 g)/ cap day dry weight

	Feces	Urine
Approximate composition, %		
Moisture content	66−80	93−96
Organic matter content[a] (dry basis)	88−97	65−85
Nitrogen	5.0−7.0	15−19
Phosphorus (as P_2O_5)	3.0−5.4	2.5−5
Potassium (as K_2O)	1.0−2.5	3.0−4.5
Carbon	40−55	11−17
Calcium (as CaO)	4−5	4.5−6
C/N ratio	5−10	

Human urine
Approximate quantity
Volume: 1¾−2¼ pt (1.0−1.3ll)/cap day
Dry solids: 0.12−0.16 lb (50−70 g)/cap day

[a] The quantities of organic matter are based on the loss in weight on ignition of the dry material.

Source: H.B. Gotaas, *Composting,* World Health Organization, Geneva, 1956.

3.2-9   TYPICAL SANITARY CHEMICAL ANALYSIS OF SEWAGE

Constituent (mg/l)	Strong		Medium		Weak	
	1	2	1	2	1	2
Solids, total	1,000	1,200	500	720	200	350
Volatile	700	600	350	365	120	185
Fixed	300	600	150	355	80	165
Suspended solids, total	500	350	300	220	100	100
Volatile	400	275	250	165	70	80
Fixed	100	75	50	55	30	20
Dissolved solids, total	500	850	200	500	100	250
Volatile	300	325	100	200	50	105
Fixed	200	525	100	300	50	145
Settleable solids	12	20	8	10	4	5
BOD, 5-day, 20°C	300	400	200	220	100	110
Oxygen consumed	150		75		30	
Dissolved ogygen	0		0		0	
Nitrogen, total as N	85	85	50	40	25	20
Organic	35	35	20	15	10	8
Free ammonia	50	50	30	25	15	12
Nitrites (RNO_2)	0.10	0	0.05	0	0	0
Nitrates (RNO_3)	0.40	0	0.20	0	0.10	0
Chlorides	175	100[a]	100	50[a]	15	30[a]
Alkalinity, as $CaCO_3$	200	200[a]	100	100[a]	50	50[a]
Fats	40	150[b]	20	100	0	50
Total organic carbon, TOC		290		160		80
Chemical oxygen demand, COD		1000		500		250
Phosphorus, total as P		15		8		4
Organic		5		3		1
Inorganic		10		5		3

[a]   Values should be increased by amount in domestic water supply.
[b]   Grease.

REFERENCES

1. H.E. Babbitt and E.R. Baumann, *Sewerage and Sewage Treatment,* 8th ed., John Wiley & Sons, N.Y., 1958
2. Metcalf and Eddy, Inc., *Wastewater Engineering: Treatment, Disposal, Reuse,* 2nd ed., revised by G. Tchobanoglous, McGraw-Hill, N.Y., 1979.

3.2-10   ORGANIC CONSTITUENTS OF MUNICIPAL SEWAGE

Constituent	Type	Concentration
Volatile acids	Formic, acetic, propionic, butyric, and valeric	8.5 – 20 mg/l
Nonvolatile soluble acids	Glutaric, glycolic, lactic, citric, benzoic, and phenyllactic	Any acid 0.1 – 1.0 mg/l
Higher fatty acids	Palmitic, stearic, and oleic	2/3 fatty acid content
Proteins and amino acids	At least 20 types	45 – 50% of the total nitrogen
Carbohydrates	Glucose, sucrose, lactose, some galactose, and fructose	—

Source: L. Walter, *Water Sewage Works, 108*:428, 1961. Courtesy of Scranton Gillette Communications, Inc.

3.2-11   MUNICIPAL SEWAGE COMPOSITION

Fraction, %	Total solids	Organic matter[a]	Nitrogenous matter[b]
Settleable (> 100 μm)	18	28 – 34	23
Supracolloidal (1 – 100 μm)	11 – 13	22 – 27	27 – 34
Colloidal (1 nm – 1 μm)	7	11 – 14	8 – 11
Soluble	62 – 64	25 – 37	22 – 42

Strength parameter	Particulates, %	Solubles, %
Total solids	34.7	65.3
Volatile solids	57.6	42.4
COD	77.3	22.7
Organic nitrogen	80.5	19.5

[a]   Volatile solids or organic carbon.
[b]   Organic nitrogen data.

REFERENCE

1.   J.V. Hunter and H. Heukelekian, *J. Water Pollut. Control Fed., 37*(8):1142, 1965.

3.2-12   ORGANIC COMPONENTS OF SEWAGE PARTICULATES

Component	Settleable fraction,[b] mg/l	Supracolloidal fraction,[c] mg/l	Colloidal fraction,[d] mg/l
Total grease	11.70	9.57	5.55
Alcohol-soluble matter[a]	4.62	2.01	1.40
Amino acids			
Alcohol-soluble	1.05	0.28	0.66
Alcohol-insoluble	7.54	12.56	4.71
Carbohydrates and lignin	18.05	10.60	6.09
Total organic matter	42.96	35.02	18.41
Volatile solids	52.25	41.29	20.64

[a] Excluding amino acids
[b] > about 100 μm
[c] Approximately 1–100 μm
[d] Approximately 1 nm–1 μm

REFERENCE

1.   J.V. Hunter and H. Heukelekian, *J. Water Pollut. Control Fed., 37*(8):1142, 1965.

3.2-13   CONTRIBUTIONS OF PHYSICALLY SEPARATED SEWAGE FRACTIONS TO WHOLE SEWAGE ORGANIC MATTER

Organic Constituent	Winter-Spring, 1959					Fall-Winter, 1959-60				
	Settleable,[a] mg/l	Supracolloidal,[b] mg/l	Colloidal,[c] mg/l	Soluble, mg/l	Constituent sum, mg/l	Settleable,[a] mg/l	Supracolloidal,[b] mg/l	Colloidal,[c] mg/l	Soluble, mg/l	Constituent sum, mg/l
Acids	0.89	1.64	1.48	22.56	26.57	1.14	1.66	1.82	33.95	38.57
Bases	–	–	–	3.24	3.24	–	–	–	3.55	3.55
Solvent soluble neutrals	9.45	6.64	3.58	13.59	33.26	8.99	5.64	3.54	17.31	35.48
ABS	0.08	0.14	0.10	3.94	4.26	0.11	0.13	0.09	4.02	4.35
Phenols	0.004	0.002	0.002	0.12	0.128	0.005	0.001	0.001	0.10	0.107
Cholesterol	0.4	0.02	0.03	0.03	0.12	0.02	0.03	0.04	0.05	0.14
Creatine-creatinine	–	–	–	0.20	0.20	–	–	–	0.17	0.17
Uric acid	–	–	–	0.33	0.33	–	–	–	0.34	0.34
Tannin-lignin	2.02	1.05	1.15	–	4.22	2.12	0.64	1.34	–	4.10
Cellulose	11.50	3.15	2.43	–	17.08	21.58	2.05	2.34	–	25.97
Hemi-cellulose	3.53	4.32	1.33	–	9.18	3.21	5.36	0.94	–	9.51
Pectin	1.48	2.16	1.35	–	4.99	1.30	1.32	1.35	–	3.97
Hexoses	0.26	0.13	0.10	9.77	10.26	0.24	0.14	0.10	8.65	9.13
Pentoses	–	–	–	0.77	0.77	–	–	–	0.66	0.66
Amino acids	8.59	12.84	5.37	9.05	35.85	12.41	12.26	4.76	9.01	38.44
Fraction sum	37.84	32.09	16.92	63.60	150.46	51.12	29.23	16.32	77.81	174.49
Volatile solids	52.25	41.29	20.64	72.19	186.37	63.89	36.86	18.79	87.86	207.40

Sampling period

[a] > about 100 μm
[b] Approximately 1-100 μm
[c] Approximately 1 nm – 1 μm

Source: J.V. Hunter, Ph.D. thesis, Rutgers University, New Brunswick, N.J., 1962.

3.2-14   AMINO ACID CONTENTS OF DOMESTIC WASTEWATER

Constituent amino acid	Concentration		Ref.
	Free, mg/l	Total, mg/l	
Cystine	0–trace	1.4–5.7	1,2,3,4
Lysine and histidine	Trace	5.1–9.7	2,3,4
Histidine	Present	Present	1,4
Lysine	Absent	Absent	1
	Present	Present	4
Arginine	Trace	4.6–11.0	1,2,3,4,5
Serine, glycine and aspartic acid	0.02–0.13	9.4–19.4	2,3,4,5
Threonine and glutamic acid	0.01–0.18	4.5–24.8	2,3,4,5
Alanine	0.02–0.09	5.1–11.9	2,3,4
Proline	0	0	2,3
Tyrosine	0.06–0.09	1.7–6.4	2,3,6
Methionine and valine	0.05–0.24	0.09–15.7	2,3,4,5
Phenylalanine	0.02–0.33	4.7–16.8	2,3
Leucine	0.06–0.28	4.2–13.1	2,3,4,5
Tryptophane	Present	Present	5,7

Source: Reprinted from S.J. Faust and J.V. Hunter, *Organic Compounds in Aquatic Environments,* Marcel Dekker, New York, 1971, by courtesy of Marcel Dekker, Inc.

REFERENCES

1. A. Buswell and S. Neave, Bulletin No. 3, Dept. of Registration and Education, State of Illinois, 1930.
2. C. Sastry, P. Subrahanyam, and S. Pillai, *Sewage Ind. Wastes, 30:*1241, 1958.
3. P. Subrahanyam *et al., J. Water Pollut. Control Fed., 32:*344, 1960.
4. L. Kahn and C. Wayman, *J. Water Pollut. Control Fed., 36:*1368, 1964.
5. M. Aurich, W. Dummler, and D. Mucke, *Wasserwirtech. Technol. (Germany), 8:*496, 1958.
6. W. Rudolfs and N.S. Chamberlin, *Ind. Eng. Chem., 24:*111, 1932.
7. W. Rudolfs and B. Heinemann, *Sewage Ind. Wastes, 11:*587, 1939.

3.2-15   NONAMINO ACID NITROGENOUS CONSTITUENTS OF DOMESTIC WASTEWATER

Compound	Concentration, mg/l	Ref.
Urea	2–16	1
Muramic acid	0.5	2
Amino sugars	1.2–2.2	2
Uric acid	0.2–1.0	3,4,5,6,7
Hippuric acid	Present	2
Xanthine	Trace	5
Indole	0.00025	8
Skatole	0.00025	8
Aliphatic amines	0.1	4
Creatine-creatinine	0.2–7.0	2,3,5
Organic bases	3.4	3
Thiamine	0.029	9
Riboflavin	0.022–0.044	9,10
Niacin	0.135	9
Cobalamin	0.0008	9
Biotin	0.0003	11
Pantothenic acid	Present	9
Folic acid	Present	9

Source: Reprinted from S.J. Faust and J.V. Hunter, *Organic Compounds in Aquatic Environments,* Marcel Dekker, New York, 1971, by courtesy of Marcel Dekker, Inc.

REFERENCES

1. A. Hanson and T. Flynn, *Proc. 19th Ind. Waste Conf., Purdue Univ.,* 1964, p. 32.
2. H.H. Painter and M. Viney, *J. Biochem. Microbiol. Technol. Eng., 1:*143, 1959.
3. J.V. Hunter and H. Heukelekian, *J. Water Pollut. Control Fed., 37:*1142, 1965.
4. H.H. Painter, M. Viney, and A. Bywaters, *J. Inst. Sewage Purif.,* 302, 1961.
5. A. Buswell and S. Neave, Bulletin No. 3, Dept. of Registration and Education, State of Illinois, 1930.
6. G. Kupchik and G. Edwards, *J. Water Pollut. Control Fed., 34:*410, 1965.
7. J. O'Shea and R. Bunch, *J. Water Pollut. Control Fed., 37:*1444, 1965.
8. W. Rudolfs and N.S. Chamberlin, *Ind. Eng. Chem., 24:*111, 1932.
9. E. Srinath and S. Pillai, *Curr. Sci., 35:*247, 1966.
10. L. Kraus, *Sewage Ind. Wastes, 14:*811, 1942.
11. H. Neujahr and J. Hartwig, *Acta Chem. Scand., 13:*954, 1961.

3.2-16   MISCELLANEOUS ORGANIC COMPOUNDS PRESENT IN DOMESTIC WASTEWATER

Compound	Concentration, mg/l	Ref.
Lignin	1.5–5.0	1,2
Aromatic hydrocarbons	1.3	2
Aliphatic hydrocarbons	4.0	2
Phenols	0.1–1.0	2,3
Nonionic surfactants	1–2	4
Cholesterol	0.3–0.26	2,5
Coprostanol	0.1–0.75	5
Ascorbic acid	Present	6

Source: Reprinted from S.J. Faust and J.V. Hunter, *Organic Compounds in Aquatic Environments*, Marcel Dekker, New York, 1971, by courtesy of Marcel Dekker, Inc.

REFERENCES

1. H.H. Painter and M. Viney, *J. Biochem. Microbiol. Technol. Eng.*, *1*:143, 1959.
2. J.V. Hunter and H. Heukelekian, *J. Water Pollut. Control Fed.*, *37*:1142, 1965.
3. H.H. Painter, M. Viney, and A. Bywaters, *J. Inst. Sewage Purif.*, 302, 1961.
4. Ministry of Technology (Brit.), *Notes on Water Pollution*, No. 34, 1966.
5. J. Murtaugh and R. Bunch, *J. Water Pollut. Control Fed.*, *37*:410, 1965.
6. W. Rudolfs and B. Heinemann, *Sewage Ind. Wastes*, *11*:587, 1939.

3.2-17   MINERAL PICKUP FROM DOMESTIC WATER USAGE

Constituent	Concentration, mg/liter		
	Palo Alto water[a]	Palo Alto effluent[b]	Increment range[c]
Anions			
Carbonate (CO_3)	2.4	0.0	
Bicarbonate (HCO_3)	45.0	251.0	
Chloride (Cl)	3.5	215.0[d]	20–50
Sulfate (SO_4)	5.8	47.5[e]	15–30
Nitrate (NO_3)	1.1	18.4	20–40[f]
Phosphate (PO_4)	0.0	21.4	20–40
Cations			
Sodium (Na)	0.5	155.0[e]	40–70
Potassium (K)	0.8	8.8	7–15
Calcium (Ca)	10.4	49.6[d]	15–40[g]
Magnesium (Mg)	9.8	32.6[d]	15–40[g]
Other data			
Silica (SiO_2)	5.8	14.5	
Fluoride (F)	0.8	3.8	
Manganese (Mn)	0.0	0.0	
Iron (Fe)	0.0	<0.1	
Aluminum (Al)	0.1	<3.0	
Boron (B)	0.1	0.93	0.1–0.4
Total dissolved solids (TDS)	63.8	693.0	100–300
Total alkalinity ($CaCO_3$)	39.0	206.0	100–150

[a] Provided by city of San Francisco from its Hetch Hetchy source in the Sierra.
[b] Effluent from the Palo Alto, Calif., wastewater treatment plant.
[c] Reported national average range of mineral pickup by domestic usage.
[d] Approximately 15 percent local well water used along with Hetch Hetchy water.
[e] High due to use of water softeners.
[f] Total nitrogen as N.
[g] Reported as $CaCO_3$.

Source: Copyright 1969 by the Dun-Donnelley Corp. Reprinted by special permission from the June 1969 issue of *Water and Wastes Engineering.*

3.2-18 INCREMENTS OF COMMON IONS IN DOMESTIC WATER AFTER ONE CYCLE OF WATER USE

mg/l

Component	California[1][g]			Chanute, Kansas[2]	NEIWPCC[3][h]		Sanks and Kaufman[4] Use increment			Wells and Gloyna[5]		Bunch et al.[6]	
	Min	Max	Normal range		Concentration in water	Increment in sewage	Min	Max	Avg	Avg	Range	Avg	Range
COD				113	<10	143	22	159	87			143	84–163
COD filtered												95	70–120
COD filtered corrected for Cl⁻												82	46–112
BOD				29	<3	100–300	8	27	16	32	11–120	7.4	4.6–10.1
ABS				6	<0.1	7.4							
Total N	12	42	20–40	21			0	36	15	24	22–28	22.0	14.5–28.4
NH₃-N				19	<0.2	16.1						16.1	6.8–22.1
NO₂-N												0.30	0.04–0.58
NO₃-N	0	18	0–18[a]	1			−5	26	10			3.5	1.0[d]–7.7
Ortho PO₄				19									
Total PO₄	2[b]	50	20–40	25	<0.1	24.3	7	50	24	24[c]	24[c]	22.8	15.6–33.8
Complex PO₄				7								24.3	18.5–34.6
TDS	Trace	1,200	100–300	172	10–1,580	290	128	541	320			122	40–217
Total alkalinity as CaCO₃		230	100–150	96									
SO₄=	0	75	15–30	21	0–572	10–191	12	57	28			33	12–52
Cl⁻	20	550	20–50	24	0.5–510	22–162	6	200	74	144	96–494	56	40–102
Ca++ as CaCO₃		250	15–40		0–145	7–109	2.5	125	45			23[e]	1–43[e]
Mg++ as CaCO₃		118	15–40		0–120	1–90	0	65	26			7[f]	1–11[f]
Na+	30	290	40–70		0–198	57	8	101	66			57	42–98
K+		22	7–15		0–30	9.3						9.3	7.2–11.7
Bicarbonate, as CaCO₃					5–380	122	−44	265	100				
Silica					—	4–18	9	22	15				
Boron			0.1–0.4		—	0–0.4							
Suspended solids										40	13–74		
Total solids										645	362–1,619	291	128–457
Hydroxylated aromatic (tannic acid)												1.6	0.5–2.9
Carbohydrates, glucose												2.4	0.9–5.0
Organic nitrogen, N												2.2	1.0–4.0
Loss on ignition												69	21–145
Dissolved solids			100–300										

3.2-18 INCREMENTS OF COMMON IONS IN DOMESTIC WATER AFTER ONE CYCLE OF WATER USE (*Continued*)

[a] Results from biological action in organic nitrogen.
[b] Very low concentrations in natural waters.
[c] Reported as P.
[d] Loss.
[e] As Ca^{++}.
[f] As Mg^{++}.
[g] Summary of mineral constituents between source and disposal for 15 California cities.
[h] New England Interstate Water Pollution Control Commission.

REFERENCES

1. Data from *Studies of Waste Water Reclamation and Utilization,* State Water Pollution Control Board, Sacramento, Calif., quoted in *Summary Report – The Advanced Waste Treatment Research Program, June 1960–December 1961,* Public Health Service, U.S. Department of Health, Education, and Welfare, May 1962.
2. G.E. McCallum, *J. Water Pollut. Control Fed., 35*(1):1, 1963.
3. J.W. Masselli, N.W. Masselli, and M.G. Burford, *Controlling the Effects of Industrial Wastes on Sewage Treatment,* Industrial Waste Laboratory, Wesleyan University, June 1970.
4. R.L. Sanks and W.J. Kaufman, Taken from Neal, Robert A. Taft Sanitary Engineering Center, Cincinnati, Ohio.
5. D.M. Wells and E.F. Gloyna, Water Reuse in Texas. Presented at Joint Program of the Rocky Mountain Section AWWA and the WPCA, Albuquerque, October 4, 1965.
6. R.L. Bunch *et al., J. Water Pollut. Control Fed., 36*(11):1411, 1964.

3.2-19 DISTRIBUTION OF INORGANIC CONSTITUENTS IN A SAMPLE OF DOMESTIC SEWAGE

Constituents	Sample of sewage	
	Albertson[1]	ORF[2]
Ammonia (as N), mg/l	51.4	35.0
Nitrate (as N), mg/l	3.0	6.8
Nitrite (as N), mg/l	0.07	–
Borate (as B), mg/l	2.8	–
Chloride, mg/l	84.0	84.0
Sulfate, mg/l	69.0	57.0
Soluble phosphate (as P), mg/l	10.1	48.8
Bicarbonate, mg/l	690.0	–
Total hardness, mg/l	306.0	153
Sodium, mg/l	81.0	–
Potassium, mg/l	23.0	–
Calcium, mg/l	60.0	50.0
Total solids, mg/l	657	792.0
Conductivity, μmho/cm	1,380	900.0
pH	7.4	7.8

Source: F. Besik, *Water Sewage Works*, July 1971. Courtesy of Scranton Gillette Communications, Inc.

REFERENCES

1. O.E. Albertson and R.H. Sherwood, paper presented to Water Pollut. Control Fed., Washington, D.C., October 25–27, 1967.
2. Ontario Research Foundation sponsored by Central Mortgage and Housing Corp., Canada.

3.2-20 CONCENTRATION OF SOME INORGANIC CONSTITUENTS OF DOMESTIC SEWAGE

Constituent[a]	Concentration, mg/l	
	Whole sewage, U.S.[1] (soft water area)	Settled sewage, U.K.[2] (hard water area)
Cl	20.1	68
Si	3.9	–
Fe	0.8	0.8
Al	0.13	–
Ca	9.8	109
Mg	10.3	6.5
K	5.9	20
Na	23	100
Mn	0.47	0.05
Cu	1.56	0.2
Zn	0.36	0.65
Pb	0.48	0.08
S (all forms)	10.3	22
Phosphate (as P)	6.6	22

Source: Reprinted from H.A. Painter, in *Water and Water Pollution Handbook*. Vol. 1, L.E. Ciaccio, Ed. Marcel Dekker. New York, 1971, by courtesy of Marcel Dekker. Inc.

REFERENCES

1. H. Heukelekian and J.L. Balmat, *J. Water Pollut. Control Fed.*, 31:413, 1959.
2. H.A. Painter, *Water Waste Treat. J.*, 6:496, 1958.

3.2-21　CONCENTRATION OF MINERAL CONSTITUENTS
OF SEWAGE FRACTIONS

	Sewage fractions, ppm in raw sewage				
Element	Settleable (>100 μm)	Supracolloidal (1–100 μm)	Colloidal (1 nm–1 μm)	Dissolved	Sum of fractions
Cl	–	–	–	20	20
Si	3.3	0.42	0.11	0.07	3.9
Fe	0.35	0.10	0.17	0.15	0.77
Al	0.075	0.035	0.015	0.007	0.13
Ca	0.51	0.30	0.86	8.2	9.8
Mg	0.42	0.24	0.19	9.5	10
K	0.20	0.19	0.10	5.4	5.9
Na	0.18	0.10	0.16	23	23
Mn	0.12	0.007	0.015	0.33	0.47
Cu	0.11	0.037	0.082	1.3	1.6
Zn	0.086	0.059	0.21	0.04	0.36
Pb	0.30	0.067	0.11	0.05	0.48
S	0.046	0.12	0.18	10	10
P	0.55	0.99	0.57	4.5	6.6
Sum of elements	6.3	2.7	2.8	82	94
Mineral weight[a]	6.6	2.8	2.8	83	95
Recovery, %[b]	95	96	100	99	–

[a] Fixed solids by a separate determination and calculated in terms of the total elemental content.

[b] Calculated from the sum of individual determinations and separate total fixed solids determinations.

Source:　B.L. Goodman, *Design Handbook of Wastewater Systems*, Technomic Publishing, Lancaster, Pa., 1971. With permission.

3.2-22 METAL ANALYSES OF RAW AND EFFLUENT WASTEWATER

San Antonio Rilling Plant

Metal	No. analyses	Concentration in raw wastewater		No. analyses	Concentration in final effluent		Removal,[a] %
		Average,[a] μg/l	Range, μg/l		Average,[a] μg/l	Range, μg/l	
Phosphorus	6	–	<1000 — <3000	3	1,300	1,200–1,500	–
Zinc	6	380	190–620	3	132	85–190	65.2
Aluminum	6	2,295	920–4,000	3	520	460–550	77.5
Iron	6	710	580–900	3	210	160–290	70.5
Manganese	6	56	25–82	3	61	49–75	–
Barium	6	98	80–140	3	47	40–53	52.0
Chromium	6	237	160–400	3	57	36–70	76.0
Copper	6	86	47–120	3	34	31–38	60.5
Lead	6	55	40–60	3	<20	<20	–
Silver	6	13.3	8.5–17	3	11	11–12	17.3
Cadmium	6	<20	<20	3	<20	<20	–
Molybdenum	6	<20	<20	3	<20	<20	–
Nickel	6	<10	<10	3	<10	<10	–
Cobalt	6	<10	<10	3	<10	<10	–
Vanadium	6	<20	<20	3	<20	<20	–
Beryllium	6	<0.1	<0.1	3	<0.1	<0.1	–
Strontium	6	397	280–500	3	390	280–450	17.6
Arsenic	6	<100	<100	3	<100	<100	–
Boron	6	285	150–400	3	247	240–260	13.3

[a] Averages and percent removals computed by the editors.

Source: D. Vacker *et al., J. Water Pollut. Control Fed., 39* (5):763,1967. With permission.

REFERENCES

1. C.P. Priesing, Private communication, 1965.
2. Robert S. Kerr Water Research Center, Federal Water Pollution Control Administration, Ada, Oklahoma.

3.2.3   *INDUSTRIAL AND COMMERCIAL*

3.2-23   HEAVY METALS FOUND IN MAJOR INDUSTRIES

	Al	Ag	As	Cd	Cr	Cu	F	Fe	Hg	Mn	Pb	Ni	Sb	Sn	Zn
Pulp, paper mills, paperboard, building paper, board mills					x	x			x		x	x			x
Organic chemicals, petrochemicals	x		x	x	x		x	x	x		x			x	x
Alkalies, chlorine, inorganic chemicals	x		x	x	x		x	x	x		x			x	x
Fertilizers	x		x	x	x	x	x	x	x	x	x	x			x
Petroleum refining	x		x	x	x	x	x	x			x	x			x
Basic steel works, foundries			x	x	x	x	x	x	x		x	x	x	x	x
Basic nonferrous metal works, foundries	x	x	x	x	x	x	x		x		x		x		x
Motor vehicles, aircraft plating, finishing	x	x		x	x	x			x			x			
Flat glass, cement, asbestos products, etc					x										
Textile mill products					x										
Leather tanning, finishing					x										
Steam generation power plants					x										x

Note: Plastic materials-synthetics, meat products, dairy products, fruits and vegetables, grain milling, beet sugar, beverages, and livestock feedlot industries have no heavy metal discharges.

Source: J.G. Dean *et al., Environ. Sci. Technol.,* 6(6):518, 1972. Copyright 1972 American Chemical Society. With permission.

3.2-24 INDUSTRIAL WASTE: ORIGIN, CHARACTER, AND TREATMENT

Industries producing wastes	Origin of major wastes	Major characteristics	Major treatment and disposal methods
Food and Drug Industries			
Canned goods	Trimming, culling, juicing, and blanching of fruits and vegetables	High in suspended solids, colloidal and dissolved organic matter	Screening, lagooning, soil absorption or spray irrigation
Dairy products	Dilutions of whole milk, separated milk, buttermilk, and whey	High in dissolved organic matter, mainly protein, fat, and lactose	Biological treatment, aeration, trickling filtration, activated sludge
Brewed and distilled beverages	Steeping and pressing of grain, residue from distillation of alcohol, condensate from stillage evaporation	High in dissolved organic solids, containing nitrogen and fermented starches or their products	Recovery, concentration by centrifugation and evaporation, trickling filtration; use in feeds
Meat and poultry products	Stockyards, slaughtering of animals, rendering of bones and fats, residues in condensates, grease and wash water, picking of chickens	High in dissolved and suspended organic matter, blood, other proteins, and fats	Screening, settling and/or flotation, trickling filtration
Beet sugar	Transfer, screening and juicing waters, draining from lime sludge, condensates after evaporator, juice, extracted sugar	High in dissolved and suspended organic matter, containing sugar and protein	Reuse of wastes, coagulation, and lagooning
Pharmaceutical products	Mycelium, spent filtrate, and wash waters	High in suspended and dissolved organic matter, including vitamins	Evaporation and drying, feeds
Yeast	Residue from yeast filtration	High in solids (mainly organic) and BOD	Anaerobic digestion, trickling filtration
Pickles	Lime water, brine, alum and turmeric, syrup, seeds and pieces of cucumber	Variable pH, high suspended solids, color, and organic matter	Good housekeeping, screening, equalization
Coffee	Pulping and fermenting of coffee bean	High BOD and suspended solids	Screening, settling, and trickling filtration

Industries producing wastes	Origin of major wastes	Major characteristics	Major treatment
Fish	Rejects from centrifuge, pressed fish, evaporator and other wash water wastes	Very high BOD, total organic solids, and odor	Evaporation of total waste, barge remainder to sea
Rice	Soaking, cooking, and washing of rice	High in BOD, total and suspended solids (mainly starch)	Lime coagulation, digestion
Soft drinks	Bottle washing, floor and equipment cleaning, syrup storage tank drains	High pH, suspended solids and BOD	Screening, plus discharge to municipal sewer
Apparel Industry			
Textiles	Cooking of fibers, desizing of fabric	Highly alkaline, colored, high BOD and temperature, high suspended solids	Neutralization, chemical precipitation, biological treatment, aeration and/or trickling filtration
Leather goods	Unhairing, soaking, deliming and bating of hides	High total solids, hardness, salt, sulfides, chromium, pH precipitated lime and BOD	Equalization, sedimentation, and biological treatment
Laundry trades	Washing of fabrics	High turbidity, alkalinity, and organic solids	Screening, chemical precipitation, flotation, and adsorption
Chemical Industry			
Acids	Dilute wash waters; many varied dilute acids	Low pH, low organic content	Upflow or straight neutralization, burning when some organic matter is present
Detergents	Washing and purifying soaps and detergents	High in BOD and saponified soaps	Flotation and skimming precipitation with $CaCl_2$
Cornstarch	Evaporator condensate, syrup from final washes, wastes from bottling up process	High BOD and dissolved organic matter; mainly starch and related material	Equalization, biological filtration
Explosives	Washing TNT and guncotton for purification, washing and pickling of cartridges	TNT, colored, acid, odorous, and contains organic acids and alcohol from powder and cotton, metal, acid, oils, and soaps	Flotation, chemical precipitation, biological treatment, aeration, chlorination of TNT, neutralization

3.2-24 INDUSTRIAL WASTE: ORIGIN, CHARACTER, AND TREATMENT (continued)

Industries producing wastes	Origin of major wastes	Major characteristics	Major treatment and disposal methods
Insecticides	Washing and purification products such as 2,4-D and DDT	High organic matter, benzene ring structure, toxic to bacteria and fish, acid	Dilution, storage, activated carbon adsorption, alkaline chlorination
Phosphate and phosphorus	Washing, screening, floating rock, condenser bleed-off from phosphate reduction plant	Clays, slimes and tall oils, low pH, high suspended solids, phosphorus, silica, and fluoride	Lagooning, mechanical clarification, coagulation and settling of refined waste
Formaldehyde	Residues from manufacturing synthetic resins, and from dyeing synthetic fibers	Normally has high BOD and HCHO, toxic to bacteria in high concentrations	Trickling filtration, adsorption on activated charcoal
Material Industry			
Pulp and paper	Cooking, refining, washing of fibers, screening of paper pulp	High or low pH; colored; high suspended, colloidal, and dissolved solids; inorganic fillers	Settling, lagooning, biological treatment, aeration, recovery of by-products
Photographic products	Spent solutions of developer and fixer	Alkaline, contains various organic and inorganic reducing agents	Recovery of silver, plus discharge of wastes into municipal sewer
Steel	Coking of coal, washing of blast furnace flue gases, and pickling of steel	Low pH, acids, cyanogen, phenol, ore, coke, limestone, alkali, oils, mill scale, and fine suspended solids	Neutralization, recovery and reuse, chemical coagulation
Metal plated products	Stripping of oxides, cleaning and plating of metals	Acid, metals, toxic, low volume, mainly mineral matter	Alkaline chlorination of cyanide, reduction and precipitation of chromium, and lime precipitation of other metals
Iron foundry products	Wasting of used sand by hydraulic discharge	High suspended solids, mainly sand; some clay and coal	Selective screening, drying of reclaimed sand
Oil	Drilling muds, salt, oil, and some natural gas, acid sludges and miscellaneous oils from refining	High dissolved salts from field, high BOD, odor, phenol, and sulfur compounds from refinery	Diversion, recovery, injection of salts; acidification and burning of alkaline sludges

Industry	Origin of waste	Character of waste	Major treatment
Rubber	Washing of latex, coagulated rubber, exuded impurities from crude rubber	High BOD and odor, high suspended solids, variable pH, high chlorides	Aeration, chlorination, sulfonation, biological treatment
Glass	Polishing and cleaning of glass	Red color, alkaline nonsettleable suspended solids	Calcium chloride precipitation
Naval stores	Washing of stumps, drop solution, solvent recovery, and oil recovery water	Acid, high BOD	By-product recovery, equalization, recirculation and reuse, trickling filtration
Energy Industry			
Steam power	Cooling water, boiler blow-down, coal drainage	Hot, high volume, high inorganic and dissolved solids	Cooling by aeration, storage of ashes, neutralization of excess acid wastes
Coal processing	Cleaning and classification of coal, leaching of sulfur strata with water	High suspended solids, mainly coal; low pH, high H_2SO_4 and $FeSO_4$	Settling, froth flotation drainage control, and scaling of mines
Nuclear power and radioactive materials	Processing ores, laundering of contaminated clothes, research lab wastes, processing of fuel, power plant cooling waters	Radioactive elements; can be very acid and "hot"	Concentration and containing, or dilution and dispersion

Source: N.L. Nemerow, *Theories and Practices of Industrial Waste Treatment*, Addison-Wesley, Reading, Mass., 1963. With permission of N.L. Nemerow.

3.2-25   ORGANIC ACIDS FOUND IN INDUSTRIAL WASTEWATERS

Compound	Source	Ref.
Formic	Wood pulp production	1, 2
	Flax retting	3
Acetic	Wood pulp production	1, 2
	Flax retting	3
	Sugar refining	4
	Silage (grass)	5
Lactic	Silage (grass)	5
	Olive oil refining	6
Butyric	Wood pulp production	1
Oxalic	Wood pulp production	1
	Sugar refining	4
	Olive oil refining	4
Tartaric	Wine making	7
	Olive oil refining	6
Abietic	Wood pulp production	8
Dehydroabietic	Wood pulp production	8
p-Hydroxybenzoic	Wood pulp production	1
Pyruvic	Sugar refining	4
Adipic	Sugar refining	4
Malonic	Sugar refining	4
	Olive oil refining	6
Vanillic	Wood pulp production	1
Ferulic	Wood pulp production	1
Maleic	Olive oil refining	6
Fumaric	Olive oil refining	6

Source: Reprinted from S.D. Faust and J.V. Hunter, *Organic Compounds in Aquatic Environments,* Marcel Dekker, New York, 1971, by courtesy of Marcel Dekker, Inc.

REFERENCES
1. E. Eldridge, *Trans. 2nd Seminar Biological Problems on Water Pollution,* R.A. Taft Sanitary Engineering Center, Cincinnati Tech. Rept. W60-3, 255, 1959.
2. L. Ruus, *Sven. Papperstidn., 67:*221, 1964.
3. K. Menzel and I. Thomas, *Faserforsch. u. Textiltechnol., 8:*138, 1957. *Chem. Zbl., 128:*13846, 1957.
4. E. Leclerc and F. Edeline, *Centre Belge'etude Document. Eaux, 114-115:* 201, 1960.
5. W. Moore *et al., Water Waste Treat. J., 8:*226, 1961.
6. J. Ursinos, *Grasas Aceites, 10:*30, 1959.
7. N. Rizaev and K. Merenkov, *Teoriga i Prakt., donnoga Obmena. Akad. Nouk. Kaz. - USSR, Tr. Resp. Šoveskch,* 171, 1962.
8. R. Maenpaa, P. Hyminen, and J. Tikkai, *Paperi Puu, 50:*143, 1965.

3.2-26   CARBOHYDRATES FOUND IN INDUSTRIAL WASTEWATERS

Compound	Source	Ref
Sucrose	Sugar refining	1
	Pineapple cannery	2
	Sugar beet processing	3
Raffinose	Sugar beet processing	3
Glucose	Silage (grass)	4
	Wood pulp production	5
Galactose	Silage (grass)	4
	Wood pulp production	5
Fructose	Wood pulp production	5
	Silage (grass)	1
Xylose	Silage (grass)	1
	Wood pulp production	5
Arabinose	Wood pulp production	5
	Silage (grass)	1
Rhamnose	Wood pulp production	5
Cellobiose	Wood pulp production	5
Glacturonic acid	Wood pulp production	5

Source: Reprinted from S.D. Faust and J.V. Hunter, *Organic Compounds in Aquatic Environments,* Marcel Dekker, New York, 1971, by courtesy of Marcel Dekker, Inc.

REFERENCES
1. E. Leclerc and F. Edeline, *Centre Belgé 'etude Document. Eaux, 114-115:*201, 1960.
2. N. Burbank and J. Kumagi, *Proc. 20th Ind. Wastes Conf.,* Purdue Univ., Lafayette, Ind., 1965, p.365.
3. J. Laughlin, *Proc. 4th Ind. Wastes Conf.,* Texas Water Pollut. Control Assoc., 1964, p. 2.
4. W. Moore *et al., Water Waste Treat. J., 8:*226, 1961.
5. T. Maloney and E. Robinson, *TAPPI (Tech. Assoc. Pulp Pap. Ind.), 44:* 137, 1961.

3.2-27 ORGANIC BASES FOUND IN INDUSTRIAL WASTEWATERS

Compound	Source	Ref.
Choline	Brewery	1
Melamine	Resin manufacture	2
Methylamine	Sugar refining	3
Ethanolamine	Sugar refining	3
Allylamine	Sugar refining	3
Amylamine	Sugar refining	3
Benzidine	Dye manufacture	4
Naphthylamine	Dye manufacture	4
β-Naphthylamine	Dye manufacture	4
Pyridine	Ammoniacal liquor (coal gas)	5, 6
	Coke plants	7
Methylpyridenes	Ammoniacal liquor (coal gas)	6
	Coke plants	7
Dimethylpyridines	Ammoniacal liquor (coal gas)	6
Methylethylpyridines	Ammoniacal liquor (coal gas)	6
Diethylpyridines	Ammoniacal liquor (coal gas)	6
Trimethylpyridines	Ammoniacal liquor (coal gas)	6
Aniline	Ammoniacal liquor (coal gas)	6
Methylaniline	Ammoniacal liquor (coal gas)	6
Dimethylaniline	Ammoniacal liquor (coal gas)	6
Quinoline	Ammoniacal liquor (coal gas)	6
Isoquinoline	Ammoniacal liquor (coal gas)	6

Source: Reprinted from S.D. Faust and J.V. Hunter, *Organic Compounds in Aquatic Environments,* Marcel Dekker, New York, 1971, by courtesy of Marcel Dekker, Inc.

REFERENCES

1. F. Knorr, *Brauwiss., 18*:191, 1965.
2. A. Koganovskii *et al., Bum. Prom.,* 7, 1968.
3. E. Leclerc and F. Edeline, *Centre Belge'etude Document. Eaux,* 114-115, 201, 1960.
4. N. Takemura *et al., Int. J. Air-Water Pollut.,* 9:665, 1965.
5. A. Elliot and A. Lafreniere, *Water Sewage Works, 111*:R325, 1964.
6. M. Hughes, *J. Appl. Chem.,* 450, 1962.
7. M. Ettinger *et al., Ind. Eng. Chem., 46*:791, 1954.

3.2-28   PHENOLS FOUND IN INDUSTRIAL WASTEWATER

Compound	Source	Ref.
Dinitro-*o*-cresol	Pesticide manufacture	1
2,4-Dichlorophenol	Pesticide manufacture	2
p-Nitrophenol	Dye manufacture	3
p-Aminophenol	Dye manufacture	3
Phenol	Synthetic resin manufacture	4
	Coke manufacture	5
	Oil refining	6
o-Cresol	Coke manufacture	5
	Ammoniacal liquor (coal gas)	7
m-Cresol	Coke manufacture	5
	Ammoniacal liquor (coal gas)	7
p-Cresol	Ammoniacal liquor (coal gas)	7
	Coke manufacture	5
Xylenol	Ammoniacal liquor (coal gas)	7
Ethylphenol	Ammoniacal liquor (coal gas)	7
Guaiacol	Wood pulp production	8
Vanillin	Wood pulp production	8, 9
Vanillic acid	Wood pulp production	8

Source: Reprinted from S.D. Faust and J.V. Hunter, *Organic Compounds in Aquatic Environments*, Marcel Dekker, New York, 1971, by courtesy of Marcel Dekker, Inc.

REFERENCES

1. S. Jenkins and H. Hawkes, *Int. J. Air-Water Pollut.*, 5:407, 1961.
2. I. Oshina and N. Tyurina, *Khimiz, Sel'sk. – Khaz. Bashkirii, Ufa*, 5b:19, 1964.
3. K. Papov, *Khim. Ind.*, 37:203, 1965.
4. K. Singleton, *Purdue Univ. Engineering Bull. Extension Series, No. 121*, 62, 1967.
5. L. Semenchenko *et al.*, *Koko Khim.*, 42, 1967.
6. Y. Karelin *et al.*, *Khim. i. Teknol., Topl. i. Masel.*, 9:29, 1964.
7. A. Elliot and A. Lafreniere, *Water Sewage Works*, 111:R325, 1964.
8. T. Maloney and E. Robinson, *TAPPI (Tech. Assoc. Pulp Pap. Ind.)*, 44:137, 1961.
9. K. Christofferson, *Anal. Chim. Acta*, 31:233, 1969.

3.2-29 POLYNUCLEAR HYDROCARBONS FOUND IN INDUSTRIAL WASTEWATER

Compound	Source	Ref.
3,4-Benzpyrene	Shale oil	1
	Coke production	2, 3
	Oil refining	4
2,6-Dimethylnaphthalene	Oil refining	5
2,6-Dimethylanthracene	Oil refining	5
Benzpyrenes	Acetylene production	6
Benzperylenes	Acetylene production	6
Pyrene	Acetylene production	6
Fluorene	Acetylene production	6
Anthracene	Acetylene production	6
Perylene	Acetylene production	6
Acenaphthylene	Acetylene production	6
Naphthalene	Acetylene production	6
Fluoranthene	Acetylene production	6
Coronene	Acetylene production	6
9,10-Dibenzpyrene	Acetylene production	6

Source: Reprinted from S.D. Faust and J.V. Hunter, *Organic Compounds in Aquatic Environments*, Marcel Dekker, New York, 1971, by courtesy of Marcel Dekker, Inc.

REFERENCES

1. I. Veldre, L. Lake, and I. Arro, *Gig. Sanit., 30*:104, 1965.
2. Z. Fedorenko, *Vapr. Gigiena Naselen. Mest. Kiev, Sb., 5*:101, 1964.
3. Z. Fedorenko, *Gig. Sanit., 29*:17, 1964.
4. K. Ershova, *Gig. Sanit., 33*:102, 1968.
5. S. Brady, *Proc. Div. Refining, Amer. Petrol. Inst., 48*:556, 1968.
6. V. Livke and R. Vodyanik, *Khim. Prom., 30*, 1969.

3.2-30 MISCELLANEOUS AROMATICS FOUND IN INDUSTRIAL WASTEWATER

Compound	Source	Ref.
Divinylbenzene	Plastic manufacture	1
2,4-Dinitrobenzene	Dye manufacture	2
Nitrotoluene	TNT manufacture	3
Dinitrotoluene	TNT manufacture	3
2,4-Dinitrochlorobenzene	Dye manufacture	4
Nitrobenzene	Oil refining	5
Diphenyldisulfide	Oil refining	6

Source: Reprinted from S.D. Faust and J.V. Hunter, *Organic Compounds in Aquatic Environments*, Marcel Dekker, New York, 1971, by courtesy of Marcel Dekker, Inc.

REFERENCES

1. N. Progressov *et al., Vodosnabzh. Sanit. Tekh., 8*, 1968.
2. K. Papov, *Khim. Ind., 37*:203, 1965.
3. H. Kurmeier, *Wasser Luft., 8*:727, 1964.
4. K. Papov, *Khim. Ind., 37*:164, 1965.
5. Y. Karelin *et al., Khim i. Teknol., Topl. i. Masel., 9*:29, 1964.
6. S. Brady, *Proc. Div. Refining, Amer. Petrol. Inst., 48*:556, 1968.

3.2-31 MISCELLANEOUS ALIPHATICS FOUND IN INDUSTRIAL WASTEWATER

Compound	Source	Ref.
Dimethyldisulfide	Wood pulp production	1
Methylmercaptan	Wood pulp production	1
Methanol	Wood pulp production	1
Methanol	Synthetic resin manufacture	2
Formaldehyde	Wood pulp production	3
Formaldehyde	Synthetic resin manufacture	2
Furfural	Wood pulp production	2, 4, 5
Betanine	Sugar beet processing	6
Methyl mercury chloride	Plastic manufacture	7
Isoprene	Synthetic rubber production	8
Dimerized isoprene	Synthetic rubber production	8
Dimethyldioxane	Synthetic rubber production	8
Dimethylformamide	Synthetic fiber production	9
Caprolactam	Synthetic fiber production	10
Aminocaprolactam	Synthetic fiber production	11

Source: Reprinted from S.D. Faust and J.V. Hunter, *Organic Compounds in Aquatic Environments*, Marcel Dekker, New York, 1971, by courtesy of Marcel Dekker, Inc.

REFERENCES

1.   N. Kardos, *Khim. Ind., 9*:4, 1960.
2.   K. Singleton, *Purdue Univ. Engineering Extension Series, No. 121,* 62, 1967.
3.   K. Christofferson, *Anal. Chim. Acta, 31*:233, 1969.
4.   T. Maloney and E. Robinson, *TAPPI (Tech. Assoc. Pulp Pap. Ind.), 44*:137, 1961.
5.   Y. Tsirlin *et al., Gidroliz. Lesokhim. Prom., 19*:12, 1966.
6.   R. Pailthrop, *J. Water Pollut. Control Fed., 32*:1201, 1960.
7.   K. Irukayaina, *Proc. 3rd Int. Conf. Water Pollut. Res., 3*:153, 1967.
8.   V. Ivanov, *Okhr. Priro, Tsentr. — Chernozemn. Polsy,* 137, 1962.
9.   M. Thonke and D. Dittmann, *Fortschr. Wasserchem. Ihrer Grenzgeb, 4*:272, 1966.
10. V. Livke *et al., Vestn. Teklin. i Ekon. Inform. Nauchn. Issled. Inst. Teklin — Ekon. Issled Gas. Kom. Khim. i Neft. Prom. pri Gosplane,* USSR, 25, 1963.
11. J. Kaeding, *Fortsch. Wasserchem. Ihrer Grenzgeb, 5*:258, 1967.

3.2-32   COMPARATIVE STRENGTHS OF WASTEWATERS FROM INDUSTRY

Type of waste	5 day BOD, mg/l	COD, mg/l	Suspended solids, mg/l	pH value
Apparel				
Cotton	200–1,000	400–1,800	200	8–12
Wool scouring	2,000–5,000	2,000–5,000[a]	3,000–30,000	9–11
Wool composite	1,000	–	100	9–10
Tannery	1,000–2,000	2,000–4,000	2,000–3,000	11–12
Laundry	1,600	2,700	250–500	8–9
Food				
Brewery	850	1,700	90	4–6
Distillery	7,000	10,000	Low	–
Dairy	600–1,000	150–250[a]	200–400	Acid
Cannery				
citrus	2,000	–	7,000	Acid
pea	570	–	130	Acid
Slaughterhouse	1,500–2,500	200–400[a]	800	7
Potato processing	2,000	3,500	2,500	11–13
Sugar beet	450–2,000	600–3,000	800–1,500	7–8
Grass silo[b]	50,000	12,500[a]	Low	Acid
Farm	1,000–2,000	500–1,000[a]	1,500–3,000	7.5–8.5
Poultry	500–800	600–1,050	450–800	6.5–9
Materials				
Pulp				
sulfite	1,400–1,700	84–10,000	Variable	
Kraft	100–350	170–600	75–300	7–9.5
Paperboard	100–450	300–1,400	40–100	
Strawboard	950	850[a]	1,350	
Coke oven	780	1,650[a]	70	7–11
Oil refinery	100–500	150–800	130–600	2–6

[a] Permanganate value.
[b] Undiluted with washings.

Source: Reprinted from H.E. Painter, in *Water and Water Pollution Handbook,* Vol. 1, L.E. Ciaccio, Ed., Marcel Dekker, New York, 1971, p. 350, by courtesy of Marcel Dekker, Inc.

3.2-33   DAILY WASTE PER EMPLOYEE DAY FOR INDUSTRY

Industry	Total solids, lb	Sus- pended solids, lb	Settleable solids		Soluble solids, lb	Colloidal solids, lb	O₂ con- sumed lb
			lb	Sludge, %			
Tannery	15.81	1.92	0.51	8.50	13.89	1.41	1.73
Chemical manufacturing	27.73	1.18	0.33	10.07	26.55	0.85	1.53
Organic	2.95	0.39	0.10	8.54	2.56	0.29	0.20
Steel pickling	28.80	0.62	0.15	11.78	28.18	0.47	0.25
Dye	12.48	0.87	0.05	4.50	11.61	0.82	3.00
Distillery	92.3	29.16	4.32	10.96	63.14	24.84	31.89
Dairy	13.05	0.92	0.03	1.38	12.13	0.89	3.03
Laundry	12.28	3.61	0.05	3.75	8.90	3.56	1.55
Averages	16.48	1.61	0.17	7.44	14.60	1.34	1.67
Average without distillery	13.29	1.07	0.11	6.93	11.87	0.89	1.09
Average without distillery or dairy	13.33	1.10	0.14	—	11.84	0.88	0.92

Source: W. Rudolfs and L.R. Setter, *Industrial Wastes in New Jersey*, New Jersey Exp. Sta. Bull. 610, New Brunswick, N.J., 1936. With permission.

3.2-34   SUMMARY OF WASTE DISCHARGES FOR VARIOUS INDUSTRIAL WASTES

Industry	Unit of daily prod.	Employees/unit	Wastes, gal/unit	Typical analyses, mg/l		Sewered pop. equiv.		Remarks
				BOD	Suspended solids	BOD	Suspended solids	
Brewing	1 barrel beer (31 gal)	0.25	470	1,200	650	19	9	Spent grain dewatered
		0.25	470	800	450	12	6	Spent grain sold wet
Canning								
Apricots	100 cases #2 cans	—	8,000	1,020	—	410	—	
Asparagus	100 cases #2 cans	—	7,000	100	30	35	9	
Beans, green	100 cases #2 cans	—	3,500	200	60	35	9	
Beans, lima	100 cases #2 cans	—	25,000	190	420	240	440	
Beans, pork and	100 cases #2 cans	—	3,500	920	225	160	33	
Beets	100 cases #2 cans	45	3,700	2,600	1,530	480	240	
Corn, cream style	100 cases #2 cans	—	2,500	620	300	75	30	
Corn, whole kernel	100 cases #2 cans	—	2,500	2,000	1,250	250	130	
Grapefruit, juice	100 cases #2 cans	—	500	310[b]	170[b]	8[b]	3[b]	
Grapefruit, sections	100 cases #2 cans	—	5,600	1,850	270	520	63	Size of can unknown
Peaches-pears	100 cases cans	—	6,500	1,340	—	440	—	
Peas	100 cases #2 cans	1	2,500	1,700	400	210	40	
Pumpkin (squash)	100 cases #2½ cans	—	2,500	6,400	1,850	800	190	
Sauerkraut	100 cases #2 cans	—	300	6,300	630	100	8	
Spinach	100 cases #2 cans	—	16,000	620	—	490	—	
Succotash	100 cases #2 cans	—	12,500	520	250	330	130	
Tomatoes, products	100 cases #2 cans	—	7,000	1,000	500	350	150	
Tomatoes, whole	100 cases #2 cans	6.5	750	4,000	2,000	150	60	
Coal washery	1,000 tons coal washed	6	—	15	115,000	—	—	Wastes cause tastes and odors
Coke	100 tons coal carbonized	8	360,000	85	—	1,500	—	
Distilling, grain								
Combined wastes	1,000 bu grain mashed	40	600,000	230	360	3,500	2,300	Excluding intentionally discharged slop
Thin slop	1,000 bu grain mashed	—	—	34,000	—	55,000	—	
Tailings	1,000 bu grain mashed	—	—	740	—	50	—	
Evaporator condensate	1,000 bu grain mashed	—	—	1,200	—	1,500	—	

3.2-34   SUMMARY OF WASTE DISCHARGES FOR VARIOUS INDUSTRIAL WASTES (continued)

Industry	Unit of daily prod.	Employees/ unit	Wastes gal/ unit	Typical analyses, ppm		Sewered pop.[a] equiv.		Remarks
				BOD	Suspended solids	BOD	Suspended solids	
Distilling, molasses								
Distilling, molasses	1,000 gal 100 proof	8	8,400	33,000	3,270	12,000	1,000	Molasses slop
Cooling water	1,000 gal 100 proof	—	120,000	—	—	—	—	
Meat								
Packing house	100 hog units of kill	30	550	—	—	77[c]	25[c]	1 cow = 2½ hog units =
Packing house	100 hog units of kill	30	550	900	650	24	14	2½ calves = 2½ sheep
Slaughterhouse	100 hog units of kill	20	160	2,200	930	18	6	
Stockyards	1 acre	—	25,000	65	175	80	180	
Poultry	1,000 lb live weight	6	2,200	—	—	300	160	Avg weight = 4.5 lb/animal
Milk								
Receiving station	1,000 lb raw milk and cream	0.15	180	500	—	4	2	
Bottling works	1,000 lb raw milk and cream	0.89	250	—	—	6	3	
Cheese factory	1,000 lb raw milk and cream	0.38	200	100	750	16	9	
Creamery	1,000 lb raw milk and cream	0.16	110	1,250	660	6	3	
Condensery	1,000 lb raw milk and cream	0.47	150[d]	1,300	750	7	4	
Dry milk	1,000 lb raw milk and cream	0.39	150	480	—	6	3	
General dairy	1,000 lb raw milk and cream	1.09	340	570	540	10	5	
Oil field								
Oil field	100 bbl crude oil	1.3	18,000	—	—	—	—	
Oil refining	100 bbl crude oil	3	77,000	20	50	60	120	1 bbl = 42 gal
Paper								
Paper mill	1 ton paper	4.4	39,000	19	452	26	520	No bleaching
Paper mill	1 ton paper	4.6	47,000	24	156	40	220	With bleaching
Pasteboard	1 ton paper	2.1	14,000	121	660	97	445	
Strawboard	1 ton paper	1.4	26,000	965	1,790	1,230	1,920	
Deinking	1 ton paper	—	83,000	300	—	1,250	—	Old paper stock
Paper pulp								
Groundwood	1 ton dry pulp	2.5	5,000	645	—	16	—	
Soda	1 ton dry pulp	3.0	85,000	110	1,720	460	6,100	
Sulfate (kraft)	1 ton dry pulp	—	64,000	123	—	390	—	
Sulfite	1 ton dry pulp	3.1	60,000	443	—	1,330	—	

Process	Unit	a						Notes
Tanning, vegetable	100 lb raw hides	7	800	1,200	2,400	48	80	
Tanning, chrome	100 lb raw hides	—	—	—	—	24	40	
Textile								
Cotton								
Sizing	1,000 lb goods processed	—	60	820	—	2	—	
Desizing	1,000 lb goods processed	—	1,100	1,750	—	96	—	
Kiering	1,000 lb goods processed	—	1,700	1,240	—	108	—	
Bleaching	1,000 lb goods processed	—	1,200	300	—	17	—	
Scouring	1,000 lb goods processed	—	3,400	72	—	12	—	
Mercerizing	1,000 lb goods processed	—	30,000	55	—	83	—	
Dyeing								
Basic	1,000 lb goods processed	—	18,000	100	—	100	—	
Direct	1,000 lb goods processed	—	6,400	220	—	71	—	
Vat	1,000 lb goods processed	—	19,000	140	—	130	—	
Sulfur	1,000 lb goods processed	—	5,400	1,300	—	360	—	
Developed	1,000 lb goods processed	—	14,400	170	—	120	—	
Naphthol	1,000 lb goods processed	—	4,800	250	—	59	—	
Aniline black	1,000 lb goods processed	—	15,600	55	—	41	—	
Print works	1,000 lb goods processed	—	4,500	95	—	15	—	
Finishing	1,000 lb goods processed	—	6	1,250	—	0.4	—	
Rayon manufacture	1 cord wood distilled	35	680,000	30	—	1,000	—	Wood distillation process
Rayon manufacture	1,000 rayon produced	64	160	4.4	19	35	130	Cupra-ammonia process
Rayon manufacture	1,000 rayon produced	50	140	110	96	800	580	Viscose process
Rayon hosiery	1,000 hose produced	—	9,000	330	—	150	—	Boil-off and dye wastes
Silk hosiery	1,000 hose produced	—	13,700	1,720	—	1,180	—	Boil-off, dye and finish wastes
Woolen mill	1,000 finished goods	—	70,000	114	—	400	—	Scouring and dyeing—no grease wool
Woolen mill	1,000 finished goods	—	240,000	125	—	1,500	—	Scouring and dyeing—100% grease wool

a Persons/unit of daily production.
b Excluding peel bin wastes.
c Paunch manure to sewer.
d Excluding vacuum pan water.

Source: *The Cost of Clean Water*, U.S. Department of the Interior, Federal Water Pollution Control Administration, 1968.

3.2-35　CHARACTERISTICS OF MISCELLANEOUS INDUSTRIAL WASTES

Waste	Prod. unit	Flow, gal/ prod. unit	BOD, mg/l	BOD, lb/ prod. unit	Suspended solids, mg/l
Beet sugar					
Flume water	Ton of raw beets	2,200	200		800
Process water	Ton of raw beets	660	1,230		1,100
Lime cake drainage	Ton of raw beets	75	1,420		450
Steffans waste	Ton of raw beets	120	10,000		7,700
Beet sugar with Steffans waste	Ton of raw beets	3,000–4,000	450 700–2,000		800
Coal washing	1 ton of coal	600–2,400	–		3,000–150,000
Chemical manufacuring					2,680
Chemical processing					27–576
Coke	100 lb	180		4.3	
Gas and coke	1 ton of coal	200–400	1,000–6,000		200–3,000
Corn products	100 bushels				
Malthouse	100 bushels, barley	800	400		
Malthouse			390		72
Glue making	1 employee				
Soap making	1 employee				
Starch	1 bushel	146			
Roofing factory			915		820
Metal pickling					2,065

Waste	Pop. equiv./prod. unit BOD	Suspended solids	Total solids, mg/l	Oxygen consumed, mg/l	pH	Chlorides, mg/l	Ref.
Beet sugar							
Flume water	22	74					1
Process water	40	29					
Lime cake drainage	5	1					
Steffans waste	60	37					
Beet sugar with Steffans waste							3
Coal washing							3
Chemical manufacturing			40,840	95,100	7.3	2,563	4
Chemical processing							2
Coke	25						4
Gas and coke							3
Corn products	500						4
Malthouse							3
Malthouse			935	291			4
Glue making	102						4
Soap making	33						4
Starch	12.7						4
Roofing factory			3,196	326			4
Metal pickling			69,900	13,590	5.3	3,091	4

REFERENCES
1. Operations of wastewater treatment plants, *Manual of Practice No. 10,* Water Pollut. Control Fed., Washington, D.C., 1970 and *Ohio River Pollution Survey Report,* Suppl. D, U.S. Public Health Service, 1942. Also identified as *U.S. Public Health Service Industrial Waste Guide,* Ohio River Pollution Control, House Document 266, 78th Congress, 1st session.
2. W.L. Lencht *et al., J. Water Pollut. Control Fed.,* 34(10):999, 1962.
3. G.M. Fair and J.C. Geyer, *Water Supply and Waste-water Disposal,* John Wiley & Sons, New York, 1954, p.869.
4. C.E. Keefer, *Sewage-Treatment Works,* McGraw-Hill, New York, 1940, p.59.

3.2-36   ODOR THRESHOLD CONCENTRATIONS FOR VARIOUS CHEMICALS

Chemical	No. panelists	No. observations	Threshold odor level,[a] ppm	
			Avg	Range
Acetic acid	9	9	24.3	5.07–81.2
Acetone	12	17	40.9	1.29–330
Acetophenone	17	154	0.17	0.0039–2.02
Acrylonitrile	16	104	18.6	0.0031–50.4
Allyl chloride[b]	10	10	14,700	3,660–29,300
n-Amyl acetate	18	139	0.08	0.0017–0.86
Aniline	8	8	70.1	2.0–128
Benzene[c]	13	18	31.3	0.84–53.6
n-Butanol	32	167	2.5	0.012–25.3
p-Chlorophenol	16	24	1.24	0.02–20.4
o-Cresol	13	21	0.65	0.016–4.1
m-Cresol	29	147	0.68	0.016–4.0
Dichloroisopropylether	8	8	0.32	0.017–1.1
2,-4-Dichlorophenol	10	94	0.21	0.02–1.35
Dimethylamine	12	29	23.2	0.01–42.5
Ethylacrylate	9	9	0.0067	0.0018–0.0141
Formaldehyde	10	11	49.9	0.8–102
2-Mercaptoethanol	9	9	0.64	0.07–1.1
Mesitylene[c]	13	19	0.027	0.00024–0.062
Methylamine	10	10	3.33	0.65–5.23
Methyl ethyl pyridine	16	20	0.05	0.0017–0.225
Methyl vinyl pyridine	8	8	0.04	0.015–0.12
β-Naphthol[c]	14	20	1.29	0.01–11.4
Octyl alcohol[c]	10	10	0.13	0.0087–0.56
Phenol	12	20	5.9	0.016–16.7
Pyridine	13	130	0.82	0.007–7.7
Quinoline	11	17	0.71	0.016–4.3
Styrene[c]	16	23	0.73	0.02–2.6
Thiophenol[b]	10	10	13.5	2.05–32.8
Trimethylamine	10	10	1.7	0.04–5.17
Xylene[c]	16	21	2.21	0.26–4.13
n-Butyl mercaptan	8	94	0.006	0.001–0.06

[a] Threshold values based upon pure substances.
[b] Threshold of a saturated aqueous solution. Solubility data not available.
[c] Dilutions started with saturated aqueous solution at room temperature; solubility data obtained from literature for correction back to pure substances.

Source: Reprinted, with publisher's permission, from *J. Am. Water Works Assn., 55*(7) © 1963, by the American Water Works Association, Inc.

3.2.4   *AGRICULTURAL*

3.2-37   BACTERIAL DENSITIES AND FECAL STREPTOCOCCUS DISTRIBUTIONS IN WARM-BLOODED ANIMAL FECES

Fecal source	No. of samples	Densities per gram[a]		Ratio FC/FS	Total strains examined	Occurrence, %			
		Fecal coliforms	Fecal streptococci			Enterococci	S. bovis S. equinus	Atypical S. faecalis	S. faecalis liquifaciens
Human	43	13,000,000	3,000,000	4.4	1,067	73.8	None	None	26.2
Animal pets									
Cat	19	7,900,000	27,000,000	0.3	268	89.9	1.5	2.2	6.3
Dog	24	23,000,000	980,000,000	0.02	585	44.1	32.0	14.4	9.6
Rodents	24	160,000	4,600,000	0.04	539	47.3	17.1	0.4	35.3
Rabbit				0.0004[1]					
Chipmunk				0.03[1]					
Livestock									
Cow	11	230,000	1,300,000	0.2	438	29.7	66.2	None	4.1
Pig	11	3,300,000	84,000,000	0.04	296	78.7	18.9	None	2.4
Sheep	10	16,000,000	38,000,000	0.4	321	38.9	42.1	None	19.0
Poultry									
Duck	8	33,000 000	54,000,000	0.6	328	51.2	48.8	None	None
Chicken	10	1,300,000	3,400,000	0.4	275	77.1	1.1	None	21.8
Turkey	10	290,000	2,800,000	0.1	317	76.7	1.6	None	21.8

Note: Data from reference 1 added by editors of this book.

[a] Median values.

Source: E.E. Geldreich and B.A. Kenner, *J. Water Pollut. Control Fed., 41*(8)(Part 2):R336, 1969. With permission.

REFERENCE

1. E.E. Geldreich *et al.*, *J. Water Pollut. Control Fed., 40*(11):1861, 1968.

3.2-38   ANIMAL WASTES AND THEIR BOD

Animal	Daily solid wastes, lb	BOD, lb/ton of animal
Jaguar	0.264	0.38
Bengal tiger	1.32	0.42
Bengal tiger	0.743	0.48
Leopard	0.283	0.40
Cheetah	0.180	0.38
Lion	1.41	0.28
Kodiak bear	3.59	1.68
Sun bear	2.01	1.36
Monkey	0.0375	0.58
Dog faced baboon	0.182	0.50
Agouti	0.0124	0.118
Tasmanian devil	0.0158	0.42
Timber wolf	0.508	1.20
Collared peccary	0.273	0.78
Porcupine	0.163	0.134
Skunk	0.043	0.50
Coati mundi	0.299	0.70
Black bear	3.86	0.48

Source: F.W. MacDonald and H.R. Davis, *Water and Sewage Works, 113*(2):64, 1966. Courtesy of Scranton Gillette Communications, Inc.

3.2-39   ANIMAL WASTE CHARACTERISTICS

Item	Chickens	Hogs	Cattle
Animal size, lb	4−5	100	1,000
Wet manure, lb/day	0.12−0.39	2.8−9.5	38.5−74.0
Total solids, % wet basis	25−48	12−28	13−27
Total solids, lb/day	0.05−0.10	0.8−1.6	9.5−11.4
Volatile solids, % dry basis	74−79	83−87	−
Nitrogen,[b] lb/day	0.006[a]−0.0057	0.042−0.07[a]	0.35−0.45[a]
P_2O_5, lb/day	0.0010−0.0045	0.029−0.032	0.11−0.12
K_2O, lb/day	0.0005−0.0019	0.034−0.062	0.27−0.34

[a] Reference 2.
[b] Human 0.025 lb/day unit.[2]

Source: R.C. Loehr, *CRC Crit. Rev. Env. Contr. 1*(1):69, 1970.

REFERENCES

1. R.C. Loehr, Animal wastes—a national problem, *J. Sanit. Eng. Div. Proc. Am. Soc. Civ. Eng., SA2*:189, 1969.
2. *J. Am. Water Works Assn., 63*(5): © 1971 by Am. Water Works Assn., Inc., 2 Park Ave., New York, New York, 10016.

3.2-40 POLLUTIONAL CHARACTERISTICS OF ANIMAL WASTES WEIGHT UNITS

Total solids,[a] lb/day animal	Volatile solids,[a] lb/day	BOD lb/day	BOD lb/lb volatile solids	COD lb/day	COD lb/lb volatile solids	BOD to COD	Nitrogen Total, lb/day	Nitrogen Ammonia, lb/day	P_2O_5, lb/day	Ref.
Chickens (4–5 lb)										
0.063	0.035			0.033	0.93					1
0.066	0.044	0.015	0.34	0.050	1.14	0.30	0.0030	0.0006	0.0026	2
	0.051	0.015	0.29	0.057	1.11	0.26	0.0036	0.0023		3
0.057–0.084		0.015–0.032								4
		0.006–0.010								5
Swine (100 lb)										
0.80	0.62	0.20	0.32	0.75	1.20	0.27	0.032	0.024	0.025	3
0.97	0.80	0.43	0.54	0.96	1.20	0.45	0.064			6
0.50	0.35		0.30	0.47	1.56	0.19				7
		0.25								8
0.71	0.34	0.22		0.5			0.07			4
		0.56								9
	0.5	0.17		0.89			0.03			5
		0.38					0.045			10
Dairy cattle (1,000 lb)										
10.4	8.3	1.53	0.13	8.4	1.57	0.08	0.38		0.12	8
6.8	5.7	1.53	0.18	5.8	1.00	0.18	0.37			3
7.5		1.32	0.23		1.02	0.23	0.49	0.23		11
		0.44								5
Beef cattle (1,000 lb)										
3.62	3.17	1.02	0.32	3.26	1.03	0.31	0.26	0.11		11
9.00			0.28		1.15	0.40	0.26			12
Ducks										
0.16		0.02–0.04					0.8		0.006–0.01[b]	13

a Dry solids.
b As PO_4.

Source: R.C. Loehr, *J. Sanit. Eng. Div. Proc. Am. Soc. Civ. Eng.*, *95*(SA 2):189, 1969. With permission.

3.2-40   POLLUTIONAL CHARACTERISTICS OF ANIMAL WASTES WEIGHT UNITS (continued)

REFERENCES

.   E.P. Taigandies, and T.E. Hazen, *Trans. Am. Soc. Agr. Eng., 9*:374, 1966.
.   J.B. Dornbush and J.R. Anderson, *Proc. 19th Ind. Waste Conf.,* Purdue University, Lafayette, Ind., 1964, p. 317.
.   S.A. Hart and M.E. Turner, *J. Water Poilut. Control Fed., 37*:1578, 1965.
.   F.J. Little, *J. Proc. Inst. Sew. Purif.,* 1966, p. 452.
.   *Water Pollution Research–1963,* Department of Scientific and Industrial Research, Her Majesty's Stationery Office, London, 1964, p. 73.
.   E.P. Taiganides *et al., Trans. Am. Soc. Agr. Eng., 7*:123, 1964.
.   C.E. Clark, *J. Sanit. Eng. Div., Am. Soc. Civ. Eng., 91*(SA6):4567, 1965.
.   E.A. Jeffrey *et al.,* Aerobic and anaerobic digestion characteristics of livestock wastes, Eng. Series Bull., No. 57, University of Missouri, Columbus, 1963.
.   H.R. Poelma, The biological breakdown of pig urine, Bull. No. 28, Inst. Farm Buildings, Wageningen, Netherlands, 1966.
0.   R.J. Smith *et al.,* Piggery cleaning using renovated wastes, Dept. Agr. Eng., Iowa State Univ., Ames, 1970, unpublished data.
1.   S.A. Witzel *et al., Proc. Nat. Symp. Animal Wastes Management, Am. Soc. Agr. Eng.,* No. SP-0366, 1966, p. 10.
2.   R.C. Loehr and R.W. Agnew, *J. Sanit. Eng. Div., Am. Soc. Civ. Eng., 93*(SA4):72, 1967.
3.   Report on pollution of the navigable water of Moriches Bay and eastern section of Great South Bay, Long Island, New York, Fed. Water Pollut. Control Admin., Metuchen, New Jersey, September 1966.

3.2-41   THRESHOLD ODOR CONCENTRATION OF PESTICIDES AND SOLVENTS IN WATER

Compound	Threshold odor concentration,[a] ppm
Pesticides	
Parathion (technical grade)	0.003
Parathion (pure)	0.036
Endrin	0.009
Lindane	0.33
Formulation components	
Sulfoxide (synergist)	0.091
Aerosol OT (emulsifier)	14.6
Commercial solvents	
Deodorized kerosene	0.082
Solvent 1	0.016
Solvent 2	13.9
Solvent 3	0.090

[a] At room temperature by a panel of eight people.

Source: Reprinted, with publisher's permission, from *J. Am. Water Works Assn.*, *57*(2), © 1965, by the American Water Works Association, Inc.

REFERENCES

1. G.A. Malov, *Vodosnabzh. i Sanit. Tekh.*, *5*:31, 1957. *C.A. 52*:7582, 1958.
2. Unpublished data, Engineering Section, Robert A. Taft Sanitary Engineering Center, Cincinnati, Ohio.
3. J.M. Cohen *et al.*, *J. Am. Water Works Assn.*, *53*(1):49, 1961.

3.2-42   ODOR IMPARTED TO ODORFREE WATER BY INSECTICIDES AND HERBICIDES

Substance[a]	Concentration, ppm	Threshold odor number
Toxaphene	0.84	6
2,4-D (isooctyl)	2.0	17
2,2-D	92.5	4
D-D	0.0235	17
Rothane	50.0	nil
Chlordane	0.07	140
BHC	0.0175	8

[a] At room temperature by a panel of eight people.

Source: Reprinted, with publisher's permission, from *J. Am. Water Works Assn.*, *64*(5), © 1972, by the American Water Works Association, Inc.

3.2-43   ODOR AND TOXICITY DATA ON HERBICIDES AND PESTICIDES

Herbicide or pesticide	Characteristic odor	Concentration at which odor is just detectable, ppm	Toxicity[a] Fish, ppm	Toxicity[a] Rats, mg/kg
Benzene hexachloride delta isomer	Earthy to musty	0.00013	(0.79)	(30)
Guthion®	Hydrocarbon	0.0002	0.0052	20
Benzene hexachloride beta isomer	Musty	0.00032	(0.79)	(30)
Chlordane (5% granular)	Musty	0.0005	0.022	500
Benzene hexachloride, 11.7% gamma and 13.6% other isomers	Musty	0.00125	0.056	125
D-D	Horseradish	0.0014		
Chlordane (40% wettable powder)	Musty and chlorinous	0.0025	0.022	500
Isopropyl ester of 2,4-D	Chlorophenol or iodoform	0.0031		375
Pro-noxfish®	Camphor to kerosene	0.004		
Methyl parathion	Hydrocarbon to skunky	0.0123	1.9	9–25
Aldrin	Musty to moldy	0.017	0.013	50
Endrin	Musty and chlorinous	0.018	0.0006	5–45
EPN-300	Skunky to rotten cabbage	0.018	0.10	13.6
Heptachlor	Musty and chlorinous	0.02	0.019	90
Co-ral®	Chlorinous	0.02		15.5
Parathion (commercial formulation)	Rotten onion	0.04	0.095	6
Dieldrin	Musty to camphor	0.041	0.0079	60
Isopropyl 2,4-dichlorophenoxyacetate	Chlorophenol or iodoform	0.055		(375)
Delnav	Kerosene to skunky	0.06	0.034	
Malagran 5 (5% malathion)	Garlic	0.081	0.09	1,200
Benzene hexachloride alpha isomer	Musty	0.088	(0.079)	(30)
Isooctyl 2,4-dichlorophenoxyacetate	Chlorophenol or iodoform	0.12		(375)
Toxaphene	Musty to moldy	0.14	0.0035	40
Dichlorodiphenyltrichloroethane	Chlorophenol and iodoform	0.35	0.016	250
Rotenone	Musty and chlorinous	0.36	0.011	high
2-(2,4,5-Trichlorophenoxypropionic acid	Iodoform	0.78	1.23	500
Malathion (commercial formulation)	Onion	1.0	0.09	1,156
2,4,5-Trichlorophenoxyacetic acid	Iodoform to musty	2.92	2.3	500
2,4-Dichlorophenoxyacetic acid	Chlorophenol to musty	3.13		375
Methoxychlor	Musty to chlorinous	4.7	0.035	5,000
Benzene hexachloride gamma isomer (lindane)	Chlorinous	12.0	0.056	125
2,2-Dichloropropionic acid	Acetic acid	23.2	340	2.86

[a]   Toxicity data are the lowest reported in the available literature. The figures in parentheses are the lowest values reported for the specific class involved.

Source: Reprinted, with publisher's permission, from *J. Am. Water Works Assn.*, *57*(8) © 1965, by the American Water Works Association, Inc.

3.2.5   *MARINE*

3.2-44   MAJOR CATEGORIES OF MARINE POLLUTION

Symbols: ++   important
          +   significant
          (+) slight
          ?   uncertain
          –   negligible

Category	Harm to living resources	Hazards to human health	Hindrance to maritime activities	Reduction of amenities
Domestic sewage (including food processing wastes)	+ +	+ +	(+)	+ +
Pesticides	+ +	+	–	–
Inorganic wastes (including heavy metals)	+ +	+ +	–	(+)
Radioactive materials	?	+	(+)	–
Oil and oil dispersants	+	?	+	+ +
Petrochemicals and organic chemicals	+	?	(+)	(+)
Organic wastes (including pulp and paper wastes)	+ +	?	(+)	+
Military wastes	+	?	+	?
Heat	+	–	+	–
Detergents	(+)	–	–	(+)
Solid objects	+	–	+	+ +
Dredging spoil and inert wastes	+	–	+	+

Source: World Health Organization, March 1970.

3.3   COLLECTION

3.3-1   FLOW FORMULAE

Hazen-Williams[1]

$V = 1.318 \, CR^{0.63} \, S^{0.54}$ (used primarily for pressure conduits)

where:

V = velocity of flow, ft/sec
C = constant depending on the material, age and roughness of pipe (See values given in 1.1–43.)
R = hydraulic radius of pipe, ft (flow area/wetted perimeter)
S = hydraulic gradient, feet per foot of length.

For circular conduits flowing full, the equation may be restated as

$Q = 0.279 \, CD^{2.63} \, S^{0.54}$

where:

Q = flow, ft³/sec or cfs
D = diameter of pipe, ft

The nomographic chart (1.1–46) is used to calculate the head loss and is based on a value of C = 100 applied to pipes in use 15–20 years with normal water. To determine the loss of head for any value of C, the loss of head obtained by the nomogram is multiplied by a value k, as given below:

C	40	60	80	90	100	110	120	130	140
k	5.46	2.58	1.51	1.22	1.00	0.838	0.713	0.615	0.536

Manning[2]

$V = \dfrac{1.49}{n} R^{2/3} S^{1/2}$ (applied primarily to open channel problems)

where:

V = velocity of flow, ft/sec
n = coefficient of roughness
R = hydraulic radius, ft
S = slope of the energy gradient, feet per foot of length.

A nomographic chart (1.1–45) is used to calculate velocity and quantity of flow on the basis of Manning's $n = 0.013$.

REFERENCES

1. G.S. Williams and A. Hazen, *Hydraulic Tables*, 1st ed., John Wiley and Sons, New York, 1905.
2. R. Manning, *Trans. Inst. Civil Eng. Ireland*, Volume 20, 1890.

3.3-2   VALUES OF EFFECTIVE ABSOLUTE ROUGHNESS AND FRICTION FORMULA COEFFICIENTS

Conduit material	Effective absolute roughness, Darcy-Weisbach, k, ft	Manning, n, $ft^{1/6}$	Hazen-Williams, C[a]
Closed conduits			
Asbestos-cement pipe	0.001–0.01	0.011–0.015	100–140
Brick	0.005–0.02	0.013–0.017	–
Cast iron pipe		0.015	–
Uncoated (new)	0.00085	–	130
5 yr old	–	–	120
10 yr old	–	–	107–113[b]
20 yr old	–	–	90–100
30 yr old	–	–	75–90
Asphalt dipped (new)	0.0004	–	–
Cement lined and seal coated	0.001–0.01	0.011–0.015	100–140(151)[3]
New			146 (134–151)[3]
0–5 yr old			148 (145–151)[3]
5–10 yr old			141[3]
10–20 yr old			142 (132–151)[2]
20–30 yr old			139 (132–146)[2]
30–40 yr old			135 (125–146)[3]
Concrete (monolithic)			
Smooth forms	0.001–0.005	0.012–0.014	–
Rough forms	0.005–0.02	0.015–0.017	–
Concrete pipe	0.001–0.01	0.011–0.015	100–140
Bituminous concrete	–	0.015	–
Corrugated metal pipe			
(1/2 in. x 2 2/3 in. corrugations)	–	0.022	
Plain	0.1–0.2	0.022–0.026	–
Paved invert	0.03–0.1	0.018–0.022	–
Spun asphalt lined	0.001–0.01	0.011–0.015	100–140
Plastic pipe (smooth)	0.01	0.011–0.015	100–140
Steel pipe			
Riveted, as for cast iron, 20 yr old	–	–	107–113
Welded, as for cast iron, 5 yr old	–	–	120
Vitrified clay			
Pipes	0.001–0.01	0.011–0.015	100–140
Liner plates	0.005–0.01	0.013–0.017	–
Wood stave	–	–	120
Open channels			
Lined channels			
Asphalt	–	0.013–0.017	–
Brick	–	0.012–0.018	–
Concrete	0.001–0.03	0.011–0.020	–
Rubble or riprap	0.02	0.020–0.035	–
Vegetal	–	0.030–0.40	–

3.3-2 VALUES OF EFFECTIVE ABSOLUTE ROUGHNESS AND FRICTION FORMULA COEFFICIENTS (continued)

Conduit material	Effective absolute roughness, Darcy-Weisbach, k, ft	Manning, n, $ft^{1/6}$	Hazen-Williams, C^a
Excavated or dredged			
Earth, straight, and uniform	0.01	0.020–0.030	–
Earth, winding, fairly uniform	–	0.025–0.040	–
Rock	–	0.030–0.045	–
Unmaintained	–	0.050–0.14	–
Natural channels (minor streams, top width of flood stage <100 ft)	0.1–3.0	–	–
Fairly regular section	–	0.03–0.07	–
Irregular section with pools	–	0.04–0.10	–

[a] Assume dimensional units contained in 1.318 term in formula. Varies with velocity and depth.
[b] Larger values apply to larger sizes of pipes–24 in. diam or larger; smaller values to pipes 4 in. in diameter.

REFERENCES

1. *Manual of Practice No. 9*, Water Pollut. Control Fed., 1969.
2. J.W. Clark and W. Viessman, Jr., *Water Supply and Pollution Control*, International Textbook Co., Scranton, Pa., 1965.
3. W.T. Miller, *J. Am. Water Works Assn.*, 57(6):773, 1965.

3.3-3 MANNING'S ROUGHNESS COEFFICIENTS

Used for Various Materials

Conduit size and shape	Material	Manning's n
12–15 in. (30.5–38 cm) pipe	Vitrified clay	0.014
18–30 in. (45.7–76.2 cm) pipe	Vitrified clay	0.013
Pipes larger than 30 in. (76.2 cm)	Reinforced concrete	0.013
All sizes and shapes	Brick	0.016
Type "A" or "B" canals[a]	Reinforced concrete	0.013
Type "A" or "B" canals[a]	Brick or soil cement	0.016
Type "A" or "B" canals[a]	Sodded slopes	0.035

[a] "A" canal – conduits for unchlorinated stormwater or chlorinated mixed sewage and stormwater. "B" canal – conduits for unchlorinated mixed sewage and stormwater.

REFERENCE

1. D.R. Horsefield, *J. Water Pollut. Control Fed.*, 40:1455, 1968.

3.3-4 COEFFICIENTS FOR DIFFERENT PIPE MATERIALS

Pipe material	C^a	n^b
Asbestos cement	115 (35)[c]	0.0112 (34)
Vitrified clay	106 (34)	0.0125 (31)
Concrete	86 (13)	0.0151 (11)

[a] Estimated Hazen-Williams full pipe coefficient C_F.
[b] Full pipe coefficient, Manning's, n_F estimated.
[c] Values in parentheses are number of lines studied.

REFERENCE

1. R.D. Pomeroy, *J. Water Pollut. Control Fed.*, 39:1546, 1967.

3.3-5 MAXIMUM PERMISSIBLE CHANNEL VELOCITIES

Channel material	Maximum velocity (after aging), ft/sec		
	Clear water	Water with silts	Water with sand, gravel, or rock fragments
Fine sand (noncolloidal)	1.50	2.50	1.50
Sandy loam (noncolloidal)	1.75	2.50	2.00
Silt loam (noncolloidal)	2.00	3.00	2.00
Alluvial silt (noncolloidal)	2.00	3.50	2.00
Firm loam	2.50	3.50	2.25
Volcanic ash	2.50	3.50	2.00
Fine gravel	2.50	5.00	3.75
Stiff clay (very colloidal)	3.75	5.00	3.00
Graded loam to cobble (noncolloidal)	3.75	5.00	5.00
Alluvial silt (colloidal)	3.75	5.00	3.00
Graded silt to cobbles (colloidal)	4.00	5.50	5.00
Coarse gravel (noncolloidal)	4.00	6.00	6.50
Cobbles and shingles	5.00	5.50	6.50
Shales and hardpans	6.00	6.00	5.00

Source: S. Fortier and F.C. Scobey, *Proc. Am. Soc. Civil Eng., 51*(2):1397, 1926. With permission.

3.3-6 PERMISSIBLE VELOCITIES FOR CHANNELS LINED WITH VEGETATION[a]

Cover	Slope range,[b] %	Permissible velocity	
		Erosion resistant soils, ft/sec	Easily eroded soils, ft/sec
Bermudagrass	0–5	8	6
	5–10	7	5
	over 10	6	4
Buffalograss Kentucky bluegrass Smooth brome Blue grama	0–5	7	5
	5–10	6	4
	over 10	5	3
Grass mixture	0–5[b]	5	4
	5–10	4	3
Lespedeza sericea Weeping lovegrass Yellow bluestem Kudzu Alfalfa Crabgrass	0–5[c]	3.5	2.5
Common lespedeza[d] Sudangrass[d]	0–5[e]	3.5	2.5

Note: Values apply to average, uniform stands of each type of cover.

[a] Use velocities exceeding 5 ft/sec only where good covers and proper maintenance can be obtained.

[b] Do not use on slopes steeper than 10% except for side slopes in a combination channel.

[c] Do not use on slopes steeper than 5% except for side slopes in a combination channel.

[d] Annuals — used on mild slopes or as temporary protection until permanent covers are established.

[e] Use on slopes steeper than 5% is not recommended.

Source: D.K. Todd, Ed., *The Water Encyclopedia*, Water Information Center, Manhasset Isle, Port Washington, New York, 1970. **Original source:** U.S. Soil Conservation Service.

3.3-7   MAXIMUM LOADS FOR HORIZONTAL DRAINS

| | | Building drain or building sewer | | | |
| | | Slope | | | |
Diameter of drain, in.	Horizontal fixture branch, f.u.	1/16 in./ft, f.u.	1/8 in./ft, f.u.	1/4 in./ft, f.u.	1/2 in./ft, f.u.
1¼	1				
1½	3				
2	6			21	26
2½	12			24	31
3 (waste)	32[a]		36[a]	42[a]	50[a]
4	160		180	216	250
5	360		390	480	575
6	620		700	840	1,000
8	1,400	1,400	1,600	1,920	2,300
10	2,500	2,500	2,900	3,500	4,200
12	3,900	3,900	4,600	5,600	6,700
15	7,000	7,000	8,300	10,000	12,000

[a] Not more than two water closets or two bathroom groups.

Source: *Report of Public Health Service Committee on Plumbing Standards,* U.S. Dept. of Health, Educ., and Welfare, Public Health Service, Washington, D.C., 1963.

3.3-8   MAXIMUM LOADS FOR SOIL AND WASTE STACKS

One or Two Branch Intervals

Diam stack, in.	Max load on stack, f.u.	Diam stack, in.	Max load on stack, f.u.
1¼	2	5	540
1½	4	6	930
2	8	8	2,100
2½	20	10	3,750
3	48[a]	12	5,850
4	240	15	10,500

[a] Not more than two water closets or bathroom groups within each branch interval nor more than four water closets or bathroom groups on the stack.

Source: *Report of Public Health Service Committee on Plumbing Standards,* U.S. Dept. of Health, Educ., and Welfare, Public Health Service, Washington, D.C., 1963.

3.3-9   MAXIMUM LOADS FOR ANY ONE BRANCH INTERVAL ON MULTISTORY SOIL AND WASTE STACKS[a]

Diameter of stack	Number of branch intervals													Load limit for tall stacks, f.u.
	3 f.u.	4 f.u.	5 f.u.	6 f.u.	7 f.u.	8 f.u.	9 f.u.	10 f.u.	11 f.u.	12 f.u.	13 f.u.	14 f.u.	15 f.u.	
2	3													10
2½	8	7												28
3[b]	20	18	17	16										102
4	100	90	84	80	77	75	73	72	71	70	69	68	68	530
5	225	205	190	180	175	170	165	162	159	157	156	154	153	1,400
6	385	350	325	310	300	290	285	280	275	271	268	266	263	2,900
8	875	785	735	700	675	655	640	630	620	612	606	600	594	7,600
10	1,560	1,405	1,310	1,250	1,205	1,170	1,140	1,125	1,110	1,095	1,080	1,075	1,062	15,000
12	2,435	2,195	2,045	1,950	1,800	1,825	1,790	1,755	1,730	1,705	1,685	1,670	1,655	26,000
15	4,375	3,935	3,675	3,500	3,380	3,280	3,210	3,150	3,110	3,060	3,030	3,000	2,975	50,000

[a] These limits are applicable only when the maximum load within any one branch interval is not greater than $N(\frac{1}{2n} + \frac{1}{4})$, where N = permissible load on a stack of one or two branch intervals, and n = number of branch intervals on the stack under consideration.

[b] Not more than two water closets or bathroom groups within each branch interval nor more than six water closets or bathroom groups on the stack.

Source: *Report of Public Health Service Committee on Plumbing Standards*, U.S. Dept. of Health, Educ., and Welfare, Public Health Service, Washington, D.C., 1963.

3.3-10   SIZE OF HORIZONTAL STORM DRAINS

Diameter of drain, in.	Maximum projected roof area for drains of various slopes		
	1/8 in. slope, ft²	1/4 in. slope, ft²	1/2 in. slope, ft²
3	822	1,160	1,644
4	1,880	2,650	3,760
5	3,340	4,720	6,680
6	5,350	7,550	10,700
8	11,500	16,300	23,000
10	20,700	29,200	41,400
12	33,300	47,000	66,600
15	59,500	84,000	119,000

Note: This table is based upon a maximum rate of rainfall of 4 in./hr. If in any state, city, or other political subdivision, the maximum rate of rainfall is more or less than 4 in./hr, then the figures for the roof area must be adjusted proportionately by multiplying the figure by 4 and dividing by the maximum rate of rainfall in in./hr.

Source: *Report of Public Health Service Committee on Plumbing Standards,* U.S. Dept. of Health, Educ., and Welfare, Public Health Service, Washington, D.C., 1963.

3.3-11   SIZE OF VERTICAL LEADERS

Size of leader or conductor,[a] in.	Maximum projected roof area, ft²
2	720
2½	1,300
3	2,200
4	4,600
5	8,650
6	13,500
8	29,000

Note: This table is based upon a maximum rate of rainfall of 4 in./hr. If in any state, city or other political subdivision, the maximum rate of rainfall is more or less than 4 in./hr, then the figures for roof area must be adjusted proportionately by multiplying the figure by 4 and dividing by the maximum rate of rainfall in in./hr.

[a] The equivalent diameter of square or rectangular leader may be taken as the diameter of that circle which may be inscribed within the cross-sectional area of the leader.

Source: *Report of Public Health Service Committee on Plumbing Standards,* U.S. Dept. of Health, Educ., and Welfare, Public Health Service, Washington, D.C., 1963.

3.3-12　SIZE OF ROOF GUTTERS

**Maximum projected roof area for gutters
of various slopes**

Diameter of gutter,[a] in.	1/16 in. slope, ft²	1/8 in. slope, ft²	1/4 in. slope, ft²	1/2 in. slope, ft²
3	170	240	340	480
4	360	510	720	1,020
5	625	880	1,250	1,770
6	960	1,360	1,920	2,770
7	1,380	1,950	2,760	3,900
8	1,990	2,800	3,980	5,600
10	3,600	5,100	7,200	10,000

[a] Gutters other than semicircular may be used provided they have an equivalent cross-sectional area.

Source: *Report of Public Health Service Committee on Plumbing Standards,* U.S. Dept. of Health, Educ., and Welfare, Public Health Service, Washington, D.C., 1963.

3.3-13 DISCHARGE FROM TRIANGULAR NOTCH WEIRS WITH END CONTRACTIONS

	Flow, gal/min			Flow, gal/min			Flow, gal/min	
Head (H), in.	90° notch	60° notch	Head (H), in.	90° notch	60° notch	Head (H), in.	90° notch	60° notch
1	2.19	1.27	6¾	260	150	15	1,912	1,104
1¼	3.83	2.21	7	284	164	15½	2,073	1,197
1½	6.05	3.49	7¼	310	179	16	2,246	1,297
1¾	8.89	5.13	7½	338	195	16½	2,426	1,401
2	12.4	7.16	7¾	367	212	17	2,614	1,509
2¼	16.7	9.62	8	397	229	17½	2,810	1,623
2½	21.7	12.5	8¼	429	248	18	3,016	1,741
2¾	27.5	15.9	8½	462	267	18½	3,229	1,864
3	34.2	19.7	8¾	498	287	19	3,452	1,993
3¼	41.8	24.1	9	533	308	19½	3,684	2,127
3½	50.3	29.0	9¼	571	330	20	3,924	2,266
3¾	59.7	34.5	9½	610	352	20½	4,174	2,410
4	70.2	40.5	9¾	651	376	21	4,433	2,560
4¼	81.7	74.2	10	694	401	21½	4,702	2,715
4½	94.2	54.4	10½	784	452	22	4,980	2,875
4¾	108	62.3	11	880	508	22½	5,268	3,401
5	123	70.8	11½	984	568	23	4,565	3,213
5¼	139	80.0	12	1,094	632	23½	5,873	3,391
5½	156	89.9	12½	1,212	700	24	6,190	3,574
5¾	174	100	13	1,337	772	24½	6,518	3,763
6	193	112	13½	1,469	848	25	6,855	3,958
6¼	214	124	14	1,609	929			
6½	236	136	14½	1,756	1,014			

Note: Based on formula:

$$Q = (C)\,(4/15)\,(L)\,(H)\,\sqrt{2gH}$$

in which

    Q = flow of water in cfs
    L = width of notch in ft at H distance above apex
    H = head of water above apex of notch in ft
    C = constant varying with conditions, .57 used for this table
    a = should be not less then ¾ L

For 90° notch the formula becomes
    $Q = 2.536H^{5/2}$.

For 60° notch the formula becomes
    $Q = 1.4076H^{5/2}$.

Source: *Manual for Sewage Plant Operators,* Texas Water and Sewage Works Assn., Austin, Texas, 1955. With permission.

3.3-14 DISCHARGE FROM RECTANGULAR WEIR WITH END CONTRACTIONS

Head (H), in.	Length (L) of Weir, ft — 1, gal/min	3, gal/min	5, gal/min	Additional gal/min for each ft over 5 ft	Head (H), in.	Length (L) of Weir, ft — 3, gal/min	5, gal/min	Additional gal/min for each ft over 5 ft
1	35.4	107.5	179.8	36.05	8	2,338	3,956	814
1¼	49.5	150.4	250.4	50.4	8¼	2,442	4,140	850
1½	64.9	197	329.5	66.2	8½	2,540	4,312	890
1¾	81	248	415	83.5	8¾	2,656	4,511	929
2	98.5	302	506	102	9	2,765	4,699	970
2¼	117	361	605	122	9¼	2,876	4,899	1,011
2½	136.2	422	706	143	9½	2,985	5,098	1,051
2¾	157	485	815	165	9¾	3,101	5,288	1,091
3	177.8	552	926	187	10	3,216	5,490	1,136
3¼	199.8	624	1,047	211	10½	3,480	5,940	1,230
3½	222	695	1,167	236	11	3,716	6,355	1,320
3¾	245	769	1,292	261	11½	3,960	6,780	1,410
4	269	846	1,424	288	12	4,185	7,165	1,495
4¼	293.6	925	1,559	316	12½	4,430	7,595	1,575
4½	318	1,006	1,696	345	13	4,660	8,010	1,660
4¾	344	1,091	1,835	374	13½	4,950	8,150	1,780
5	370	1,175	1,985	405	14	5,215	8,980	1,885
5¼	395.5	1,262	2,130	434	14½	5,475	9,440	1,985
5½	421.6	1,352	2,282	465	15	5,740	9,920	2,090
5¾	449	1,442	2,440	495	15½	6,015	10,400	2,165
6	476.5	1,535	2,600	538	16	6,290	10,900	2,300
6¼		1,632	2,760	560	16½	6,565	11,380	2,410
6½		1,742	2,920	596	17	6,925	11,970	2,520
6¾		1,826	3,094	630	17½	7,140	12,410	2,640
7		1,928	3,260	668	18	7,410	12,900	2,745
7¼		2,029	3,436	701.5	18½	7,695	14,410	2,855
7½		2,130	3,609	736	19	7,980	13,940	2,970
7¾		2,238	3,785	774	19½	8,280	14,460	3,090

Note: This table is based on Francis formula:

$$Q = 3.33 (L - 0.2H) H^{1.5}$$

in which
 $Q =$ ft³ of water flowing per second
 $L =$ length of weir opening in feed (should be 4 to 8 times H)
 $H =$ head on weir in feet (to be measured at least 6 ft back of weir opening)
 $a =$ should be at least 3H.

REFERENCE

1. *Manual for Sewage Plant Operators,* Texas Water and Sewage Works Assn., Austin, Texas, 1955.

3.3-15   SEWER GRADIENTS

Size of pipe, in.	Fall, ft/100 ft sewer	Capacity, flowing full, mgd
6	0.60	0.27
8	0.40	0.46
10	0.29	0.72
12	0.22	1.00
15	0.16	1.60
18	0.12	2.30
21	0.10	3.20
24	0.08	4.00

Note: The gradients for sewers flowing half full are suggested as minimum grades for ordinary use, as with careful construction a theoretical velocity of approximately 2 ft/sec can be obtained.

Source: *Manual for Sewage Plant Operators,* Texas Water and Sewage Works Assn., Austin, Texas, 1955. With permission.

3.3-16   LOCATION OF COMPONENTS OF SEWAGE DISPOSAL SYSTEM

Type of system	Distance, ft						
	Well or suction line	Water supply line (pressure)	Stream	Dwelling	Property line	Disposal field	Seepage pits
Building sewer	50	10					
Septic tank	50			5			
Disposal field	100[a]		25	10	10		
Seepage pit	100		50	20	10	20	20
Dry well	50			10			
Cesspool[b]	150		50	20	15	15	15

[a] When approved by the authority having jurisdiction and, under special soil conditions, this distance may be reduced. However, a minimum separation of not less than 50 feet shall be maintained.

[b] Not recommended as a substitute for a septic tank. To be used only when found necessary and approved by the Administrative Authority.

Source: *Report of Public Health Service Committee on Plumbing Standards,* U.S. Dept. of Health, Educ., and Welfare, Public Health Service, Washington, D.C., 1963.

3.3-17   HERBICIDES TESTED IN ROOT CONTROL EXPERIMENTS

Common name	Chemical name	Trade name	Formulation
Bensulide	0,0-diisopropyl phosphorodithioate S-ester with N-(2-mercaptoethyl) = benzenesulfonamide	Prefar®	Emulsifiable concentrate
Dichlobenil	2,6-dichlorobenzonitrile	– Casoron W50®	1% emulsion 50% wettable powder
Dinoseb	2-sec-butyl-4,6-dinitrophenol	Premerge®	Alkanolamine salt
Endothal	7-oxabicyclo (2.2.1) = heptane-2,3-dicarboxylic acid	Dipotassium endothal Hydrothol 191®	Dipotassium salt Cocoamine salt
Metham (SMDC)	sodium methyl = dithiocarbamate	Vapam® Vaporooter®	Water soluble concentrate Vapam + surfactant
Paraquat	1,1'-dimethyl-4,4'-bipyridinium ion	Paraquat	Water soluble concentrate
Trifluralin	a,a,a-trifluoro-2,6-dinitro-N,N-dipropyl-p-toluidine	Treflan®	Emulsifiable concentrate
2,4-D	(2,4-dichlorophenoxy) acetic acid	Dow Formula 40 Weedone LV4®	Alkanolamine salts Butoxyethanol ester
2,4,5-T	(2,4,5-trichloro = phenoxy) acetic acid	Weedar 2,4,5-T®	Triethylamine salt
Copper sulfate	copper sulfate $5H_2O$	Bluestone	94.3% pentahydrate
Chlorthiamid	2,6-dichlorothiobenzamide	Prefix®	7.5% granules

Source: J.F. Ahrens *et al., J. Water Pollut. Control Fed., 42*(9):1643, 1970. With permission.

3.3-18 CHARACTERISTICS OF COMMON GASES ENCOUNTERED IN
SEWERS, WASTEWATER PUMPING STATIONS, AND TREATMENT PLANTS

Table A

Class	Gas	Chemical formula	Common properties[a]	Specific gravity or vapor density (Air = 1)	Physiological effect[a]
1	Carbon dioxide	CO_2	Asphyxiant. Colorless, odorless. When breathed in large quantities may cause acid taste. Nonflammable. Not generally present in dangerous amounts unless oxygen deficiency exists	1.53	Cannot be endured at 10% more then few min, even if subject is at rest and oxygen content normal. Acts on respiratory nerves
1b	Carbon monoxide	CO	Chemical asphyxiant. Colorless, odorless, tasteless, flammable. Poisonous	0.97	Combines with hemoglobin of blood. Unconsciousness in 30 min at 0.2–0.25%. Fatal in 4 hr at 0.1%. Headache in few hr at 0.02%
2	Chlorine	Cl_2	Irritant. Yellow green color. Choking odor detectable in very low conc. Nonflammable	2.49	Irritates respiratory tract. Kills most animals in very short time at 0.1%
3	Gasoline	$C_5H_{12} - C_9H_{20}$	Volatile solvent. Colorless. Odor noticeable at 0.03%. Flammable	3.0–4.0	Anesthetic effects when inhaled. Rapidly fatal at 2.4%. Dangerous for short exposure at 1.1–2.2%
1a	Hydrogen	H_2	Simple asphyxiant. Colorless, odorless, tasteless. Flammable	0.07	Acts mechanically to deprive tissues of oxygen. Does not support life
2 and 4	Hydrogen sulfide	H_2S	Irritant and poisonous volatile compound. Rotten egg odor in small concentration. Exposure for 2–15 min at 0.01% impairs sense of smell. Odor not evident at high conc. Colorless. Flammable	1.19	Impairs sense of smell rapidly as concentration increases. Death in few min at 0.2%. Exposure to 0.01–0.1% rapidly causes acute poisoning. Paralyzes respiratory center
1a	Methane	CH_4	Simple asphyxiant. Colorless, odorless, tasteless. Flammable	0.55	Acts mechanically to deprive tissues of oxygen. Does not support life
1a	Nitrogen	N_2	Simple asphyxiant. Colorless, tasteless. Nonflammable. Principal constituent of air (about 79%)	0.97	Physiologically inert

Table B

Class	Name	Formula	Sp. gr.	Properties	Max safe 60 min exposure, % by volume in air	Max safe 8 hr exposure, % by volume in air	Explosive range, % by volume in air – Lower limit	Explosive range, % by volume in air – Upper limit	Likely location of highest concentration	Most common sources
—	Oxygen (in air)	O_2	1.11	Colorless, odorless, tasteless. Supports combustion						Normal air contains 20.8% of O_2. Man can tolerate down to 12%. Min safe 8 hr exposure, 14–16%. Below 10% dangerous to life. Below 5–7% probably fatal
Mainly 1a	Sludge gas	c	Variable	Mostly a simple asphyxiant. May be practically odorless, colorless. Flammable						Will not support life
1					4.0–6.0	0.5	–	–	At bottom; when heated may stratify at points above bottom	Products of combustion, sewer gas, sludge. Also issues from carbonaceous strata
1b					0.04	0.005	12.5	74.0	Near top, especially if present with illuminating gas	Manufactured gas, flue gas, products of combustion, motor exhausts. Fires of almost any kind
2					0.0004	0.0001	–	–	At bottom	Chlorine cylinder and feed line leaks
3					0.4–0.7	varies	1.3	6.0	At bottom	Service stations, garages, storage tanks, and houses

3.3-18 CHARACTERISTICS OF COMMON GASES ENCOUNTERED IN SEWERS, WASTEWATER PUMPING STATIONS, AND TREATMENT PLANTS (continued)

Class	Max safe 60 min exposure, % by volume in air	Max safe 8 hr exposure, % by volume in air[b]	Explosive range, % by volume in air		Likely location of highest concentration	Most common sources
			Lower limit	Upper limit		
1a	–	–	4.0	74.0	At top	**Manufactured gas, sludge digestion tank gas, electrolysis of water. Rarely from rock strata**
2 and 4	0.02–0.03	0.001	4.3	46.0	Near bottom, but may be above bottom if air is heated and highly humid	Coal gas, petroleum, sewer gas. Fumes from blasting under some conditions. Sludge gas
1a	Probably no limit, provided O_2 % is sufficient for life	–	5.0	15.0	At top, increasing to certain depth	Natural gas, sludge gas, manufactured gas, sewer gas. Strata of sedimentary origin. In swamps or marshes
1a	–	–	–	–	Near top, but may be found near bottom	Sewer gas, sludge gas. Also issues from some rock strata
–	–	–	–	–	Variable at different levels	Oxygen depletion from poor ventilation and absorption, or chemical consumption of oxygen
Mainly 1a	No data. Would vary widely with composition	–	5.3	19.3	Near top of structure	From digestion of sludge

a Percentages shown represent volume of gas in air.
b Conforms to "Threshold Limit Values of Air-Borne Contaminents for 1968" adapted at the 30th Annual Meeting, American Conference of Governmental Industrial Hyigenists, St. Louis, May 13, 1968.
c Mostly methane and carbon dioxide, with small amounts of nitrogen sulfide, and oxygen; occasionally traces of carbon monoxide.

Source: *Manual of Practice No. 1,* Water Pollution Control Federation, Alexandria, Va., 1969. With permission.

3.4   TREATMENT

3.4-1   RELATIVE EFFICIENCIES OF SEWAGE TREATMENT PROCESSES

% Removals

Treatment operation or process	5 day 20°C BOD	Suspended solids	Bacteria	COD	P	N
Fine screening	5–10	2–20	10–20	5–10	–	–
Chlorination of raw settled sewage	15–30	–	90–95	–	–	–
Plain sedimentation	25–40	40–70	25–75	20–35	5	5
Chemical precipitation	50–85	70–90	40–80	40–70	–	–
Trickling filtration, preceded and followed by plain sedimentation	50–95	50–92	90–95	50–80	25–45	30–40
Activated sludge, preceded and followed by plain sedimentation	55–95	55–95	90–98	50–80	–	–
Stabilization ponds	90–95	85–95	95–98	70–80		
Chlorination of biologically treated sewage	–	–	98–99	–	–	–
Intermittent sand filtration	80–95	85–95+	95–98	–	–	–
Activated carbon	>95	>95	–	–	–	–
Chemical precipitation	–	–	–	–	90–95	–
Ammonia stripping	–	–	–	–	–	90
Electrodialysis	–	–	–	–	>95	>95

REFERENCES

1. G.M. Fair, J.C. Geyer, and D.A. Okun, *Water Purification and Wastewater Treatment and Disposal,* Vol. 2, John Wiley & Sons, New York, 1968, p. 212.
2. G.M. Fair and J.C. Geyer, *Water Supply and Waste Water Disposal,* John Wiley & Sons, New York, 1954.

3.4-2   WET-WEATHER TREATMENT PLANT PERFORMANCE DATA

Device	Control alternatives	Design loading rate (gpmft^{-2})	Removal efficiency (%) BOD$_5$	SS
Primary	Swirl concentrator	60	25—60	50
	Microstrainer	20	40—60	70
	High-rate filtration	24	60—80	90
	Dissolved air flotation	2.5	50—60	80
	Sedimentation	0.5	25—40	55
	Representative performance		40	60
Secondary	Contact stabilization		75—88	90
	Physical-chemical		85—95	95
	Representative performance		85	95

Source: R. Field, *CRC Crit. Rev. Environ. Control, 16*:147, 1986.

3.4-3   REMOVAL OF MICROORGANISMS[a]

Process	% Removal Bacterial indicators[a]	Viruses	Ref.
Primary treatment	5—10	10	2
Grit chambers		0—50	3
Plain sedimentation		0—?	3
		34	4
Septic tanks	25—75		
Secondary treatment			
Trickling filters	18—99	0—85	3
		83—95	5
		54	2
Activated sludge	25—99	75—99	3
		80	4
		94	2
Stabilization ponds	60—99	99	4
Tertiary treatment	93—99.99		

[a] Compiled from C.L. Haas, *CRC Crit. Rev. Environ. Control, 17*:1, 1986.

REFERENCES

1. E.E. Geldreich, Bacterial populations and indicator concepts in feces, sewage, stormwater, and solid wastes, in *Indicators of Viruses in Water and Food,* G. Berg, Ed., Ann Arbor Science, Ann Arbor, Mich., 1978, 51.
2. L.Y.C. Leong, *Water Sci. Technol., 15*:91, 1983.
3. R.S. Engelbrecht, Removal and inactivation of enteric viruses by wastewater and water treatment processes, paper presented at Sem. Advanced Wastewater Treatment, Japan Research Group of Water Pollution, Tokyo, October 11 and 12, 1976.
4. R.C. Cooper, *J. Environ. Health, 37*:342, 1975.
5. G. Berg and T.G. Metcalf, Indicators of viruses in waters, in *Indicators of Viruses in Water and Food,* G. Berg, Ed., Ann Arbor Science, Ann Arbor, Mich., 1978.

3.4-4 SUMMARY OF OPERATING DATA

Characteristic	Galesburg, Ill.[1]		DeKalb, Ill.[2]		Oshkosh, Wisc.[3]		Minneapolis–St. Paul, Minn.[4]		New Britain, Conn.[5]	Niles, Mich.[6]	Aurora Ill.[7]	Orange County, Calif.[8]	Belvidere, Ill.[8]	International Airport[8] Miami, Fla.
Year of record	1953	1954	1954	1955	1954	1955	1953	1954	1949–50	1954	1949	1950–51	1952	1951
Raw sewage														
BOD, mg/l	176	194	240	213	170	147	188	212	187	106	112	565	191	296
Suspended solids, mg/l	170	179	203	190	179	174	265	276	209	250	182	334	191	184
BOD, lb/cap day	0.22	0.26									0.106			
Suspended solids, lb/cap day	0.21	0.21									0.173		0.165	
Population served, 1,000's	33	34	14.8	15.0	43	43	914	932		14	56			
Sewage flow,														
Total, mgd	5.09	4.81	1.75	1.83	5.17	5.36	142.6	138.59	11.25	2.3	6.33	11.023	1.039	0.15
Max, mgd	5.66	5.41	2.62	3.15					14.9					
per capita, gal/day													103	
Grit removal,														
ft³/mil gal	1.79	1.3			1.03	0.95	4.9	5.7		0.16	2.11			
Volatile solids in grit, %							7.7	10.2				2.4		
Primary settling tank effluent														
BOD, mg/l	132	140	144	119	115	93	118	133	156		72	467	111	201
Suspended solids, mg/l	87	88	103	88	87	79	83	81	159		84	155	112	100
Trickling filter effluent														
BOD, mg/l	71	64	61	51							30			
Suspended solids, mg/l	83	78	79	71							56			
Secondary settling tank effluent														

3.4-4 SUMMARY OF OPERATING DATA (continued)

Characteristic	Galesburg, Ill.[1]		DeKalb, Ill.[2]		Oshkosh, Wisc.[3]		Minneapolis–St. Paul, Minn.[4]		New Britain, Conn.[5]	Niles, Mich.[6]	Aurora, Ill.[7]	Orange County, Calif.[8]	Belvidere, Ill.[8]	International Airport[8] Miami, Fla.
Year of record	1953	1954	1954	1955	1954	1955	1953	1954	1949–50	1954	1949	1950–51	1952	1951
Trickling filter loadings														
Population persons/ acre foot														
BOD, lb/ 1,000 ft³ day	5.5	10.8												
BOD removal, %														
BOD removal, lb/ 1,000 ft³ day														
Activated sludge plant														
Return sludge, %									9.7				20	
Aeration, hr									0.12				9.5	
Air used, ft³/gal														
Air used, ft³/lb														
BOD removed, % (final)[f]	77.8	82.2	77	79	35	35.3	37.2	37.5		52	81.2	19.2	97.5	95
BOD removed, lb/day														
BOD load, lb/1,000 lb M.L.S.[a]														
BOD load, lb/1,000 ft³ aeration tank volume														
Suspended solids removal, % (final)	79.5	83.8	68	70	50	47.7	68.8	70.8		65	82.4	53.8	97.4	96.2
Mixed liquor, DO mg/l														1.6
Mixed liquor, Suspended solids mg/l									878				553	800

Activated sludge (*cont.*)										
Power consumed, kWh/mil gal									0.633	
kWh/lb BOD removed										1.2
Gas production										
1,000 ft³/day							10.65	15.1	7.92	
ton/day	7.82	8.10	9.7	9.4						
ft³/lb VSᵇ added	0.69	0.62	0.758	0.753						1.61
ft³/cap day							0.7		0.721	
CH₄, %							64.0		64.2	
Digestion tanks										
Capacity 1,000 ft³										
ft³/cap										
Digestion period days										
Dry solids added ton/day										
1,000 ft³ day										
Raw sludge, solids concentration, %	7.1	7.5	5.8	7.9	7.25	6.65	5.2	3.9	5.70	4.8
volatile solids, % total solids	62.9	61.7	67.6	63.3	70.3	73.4	44.0	69.0	74.0	72.8
Sludge withdrawn, dry solids, ton/day										
Sludge withdrawn, solids concentration, %	8.8	8.8	10.3	11.3			12.0	10.9	4.37	
Sludge withdrawn, VS,ᵇ % total solids	44.4	43.5	49.8	47.2			54.0	44.4	64.0	
Tank temperature, °F										
Sludge conditioning and filtering										
Unelutriated sludge alkalinity, mg/l					945	1,000				

3.4-4 SUMMARY OF OPERATING DATA (continued)

Characteristic	Galesburg, Ill.[1]		DeKalb, Ill.[2]		Oshkosh, Wisc.[3]		Minneapolis–St. Paul, Minn.[4]		New Britain, Conn.[5]	Niles, Mich.[6]	Aurora, Ill.[7]	Orange County, Calif.[8]	Belvidere, Ill.[a]	International Airport[8] Miami, Fla.
Year of record	1953	1954	1954	1955	1954	1955	1953	1954	1949–50	1954	1949	1950–51	1952	1951
Filter cake, ton/day							10.4e	11.1d	6.01			1.88		
Hz/hr														
moisture %														
Screenings, ft³/mg	0.47	0.4	3.0	3.0	0.99	0.84	0.8	1.1						
pH, raw sewage	7.4	7.5			7.5	7.6								
pH, final effluent	7.4	7.5			7.7	7.6								

[a] MLS – mixed liquor solids.
[b] VS – volatile solids.
[d] Low value for use of FeCl₃, high for chlorinated coppers; change made from latter to former in 1952.
[e] Wet cake.
[f] Final efficiency.

REFERENCES

1. L.W. Hunt, *Sew. Ind. Wastes*, 27:1209, 1955.
2. D.E. Henn, *Sew. Ind. Wastes*, 28(9):1184, 1956.
3. R.W. Frazier, *Sew. Ind. Wastes*, 28(9):1181, 1956.
4. K.L. Mick, *Sew. Ind. Wastes*, 28(8):1052, 1956.
5. J.R. Szymanski, *Sew. Ind. Wastes*, 22(11):1489, 1950.
6. C.V. Clem, *Sew. Ind. Wastes*, 28(3):333, 1956.
7. W.A. Sperry, *Sew. Ind. Wastes*, 22(11):1490, 1950.
8. I.M. Launer, *Sew. Ind. Wastes*, 25(7):852, 1953.

3.4.1   PRIMARY TREATMENT

3.4-5   VELOCITIES AT WHICH PARTICLES OF SAND
AND SILT SUBSIDE IN STILL WATER

Diam of particle, mm	Order of magnitude[a]	Hydraulic subsiding value, mm/sec	Time required to settle 1 ft
10.0	Gravel	1,000	0.3 sec
1.0		100	3.0 sec
0.8		83	4.9 sec
0.5		53	6.0 sec
0.4	Coarse sand	42	7.3 sec
0.3		32	9.5 sec
0.2		21	14.5 sec
0.1		8	38.0 sec
0.08		6	51.0 sec
0.06		3.8	1.3 min
0.05		2.9	1.8 min
0.04	Fine sand	2.1	2.4 min
0.03		1.3	3.9 min
0.02		0.62	4.52 min
0.01		0.154	33.0 min
0.005		0.0385	2.2 hr
0.004	Silt	0.0247	3.5 hr
0.003		0.0138	6.2 hr
0.002		0.0062	13.6 hr
0.0015		0.0035	35.0 hr
0.001	Bacteria	0.00154	230.0 days
0.0001	Clay particles	0.0000154	63.0 years
0.00001	Colloidal particles	0.000000154	19,765 years

Note: Temperature, 10°C; specific gravity of sand and silt particles, 2.65; values for 10 mm–0.1 mm particles from Hazen's experiments; values for 0.02 mm–0.00001 mm particles from Wiley's formula; intermediate values interpreted from connecting curve.

[a] Order of magnitude given is merely a comparative expression; the particles throughout were computed on the basis of sand and silt of 2.65 specific gravity.

Source: Reprinted, with publisher's permission, from *J. Am. Water Works Assn.*, 47(8), © 1955, by the American Water Works Association, Inc.

3.4-6 TREATMENT PLANT PERFORMANCE FOR REMOVAL OF GREASE

Parameter	Primary treatment, %		Secondary treatment, %		Complete treatment, %	
	Average	Range	Average	Range	Average	Range
BOD$_5$, total	32	3–67	78	56–90	85	69–94
COD, total	32	0–52	66	32–80	76	50–87
Suspended solids	48	4–65	69	0–96	82	45–98
Grease	45	17–65	74	48–96	84	57–98

Source: R.C. Loehr *et al., J. Water Pollut. Control Fed., 41*(5): R142, 1969. With permission.

3.4-7 COMMINUTOR SIZE SELECTION

Drum diam, in.	Drum, RPM	Avg slot width, in.	Horse-power	Standard sizes Height ft	Standard sizes Height in.	Net weight, lb	Rates of flow, avg 12 hr day time, mgd	Maximum hourly rates of flow, mgd
4	56	1/4	1/4	2	3 1/4	175	–0.035	0.09
7	56	1/4	1/4	4	3	450	.03–0.113	0.24
7	56	1/4	1/4	4	3	450	.06–0.20	0.36
10	45	1/4	1/2	4	5	650	.17–0.72	1.08
15	37	1/4	3/4	4	11 1/2	1,100	.25–1.82	2.4
25	25	3/8	1 1/2 in.	5	9 1/2	2,100	.97–5.10	6.1
25	25	3/8	1 1/2 in.	6	11 1/2	3,500	1.00–9.40	11.1
36	15	3/8	2	9	4 1/2	8,500	1.30–20.00	24.0

Source: B.L. Goodman, *Design Handbook of Wastewater Systems,* Technomic Publishing, Lancaster, Pa., 1971. With permission.

3.4-8   SEPTIC TANK CAPACITIES

No. bedrooms	Individual dwellings		Camps and day schools		
	Max. no. persons served	Nominal liquid capacity of tank (m³)[a]	Maximum no. persons served		Nominal liquid capacity of tank (m³)
			Camps	Day schools	
2 or less	4	1.89	40	60	3.80
3	6	2.27	80	120	7.57
4	8	2.84	120	180	11.36
5	10	3.41	160	240	15.14
6	12	4.16	200	300	18.92
7	14	4.92	240	360	22.71
8	16	5.68	280	420	26.50
			320	480	30.28[b]

[a] Liquid capacity is based on the number of bedrooms in a dwelling. The total volume in m³ of a tank includes air space above liquid level.

[b] Tanks with capacities in excess of 30.28 m³ should be designed for the specific requirements involved; however, in such cases the necessity of a more complete type of treatment should receive consideration.

Adapted from: E.W. Steel, *Water Supply and Sewerage,* 4th ed., McGraw-Hill, N.Y., 1960.

3.4.2   SECONDARY TREATMENT

3.4-9   MINIMUM STANDARDS FOR ABSORPTION SYSTEM CONSTRUCTION

	Trenches	Seepage beds
Individual lines, max. length	100 ft	100 ft
Trench or bed bottom, min. depth	18 in.	24 in.
Trench bottom, min. width	12 in.	36 in.
Field tile, min. diameter	4 in.	4 in.
Field tile lines, max. slope	6 in. in 100 ft	Level grade
Trench or bed, min. separation	6 ft	6 ft
Effective absorption area, min. per dwelling unit	See 1.1–37	See 1.1–37

Source: *Report of Public Health Service Committee on Plumbing Standards,* U.S. Dept. of Health, Educ., and Welfare, Public Health Service, Washington, D.C., 1963.

3.4-10   ABSORPTION AREA REQUIREMENTS FOR INDIVIDUAL RESIDENCES[a]

Percolation rate (time required for water to fall 1 inch in minutes)	Required absorption area, in ft² /bedroom,[b] standard trench,[c] seepage beds,[c] and seepage pits[d]	Percolation rate (time required for water to fall 1 inch in minutes)	Required absorption area, in ft² /bedroom,[b] standard trench,[c] seepage beds,[c] and seepage pits[d]
1 or less	70	10	165
2	85	15	190
3	100	30[e]	250
4	115	45[e]	300
5	125	60[e,f]	330

Note: Provides for garbage grinder and automatic washing machine.

[a] Desirable to provide sufficient additional area for entire new absorption area if need in the future.
[b] In every case, sufficient area should be provided for at least two bedrooms.
[c] Absorption area is figured as trench bottom area.
[d] Absorption area for seepage pits is figured as effective sidewall area beneath the inlet.
[e] Unsuitable for seepage pits if over 30.
[f] Unsuitable for leaching systems if over 60.

Source: *Report of Public Health Service Committee on Plumbing Standards,* U.S. Dept. of Health, Educ., and Welfare Public Health Service, Washington, D.C., 1963.

3.4-11 SIZE AND SPACING FOR TRENCH DISPOSAL SYSTEMS

Width of trench at bottom, in.	Recommended depth of trench, in.	Spacing tile lines[a] center to center, ft	Effective absorption area per lineal foot of trench, ft²
18	18–30	6.0	1.5
24	18–30	6.0	2.0
30	18–36	7.6	2.5
36	24–36	9.0	3.0

[a] A greater spacing is desirable where available area permits.

Source: *Report of Public Health Service Committee on Plumbing Standards,* U.S. Dept. of Health, Educ., and Welfare, Public Health Service, Washington, D.C., 1963.

3.4-12 CLASSIFICATION OF TRICKLING FILTERS

Type of filter	Hydraulic load	Organic load, lb BOD/day 1,000 ft³	Application	Sloughing	BOD removal through filter, %
Standard rate	25–100	5–25	Intermittent	Largely periodic	80–85
High rate	200–1,000	25–300	Continuous	Continuous	65–80
Roughing filters	500	300	Continuous	Continuous	25–65

Source: Reprinted from E.J. Genetelli, in *Water and Water Pollution Handbook,* Vol. I, L.E. Ciaccio, Ed., Marcel Dekker, New York, 1971, p. 440, by courtesy of Marcel Dekker, Inc.

REFERENCE

1. Report of the National Research Council Subcommittee on Military Sewage Treatment, *Sewage Works J., 18*:794, 1946.

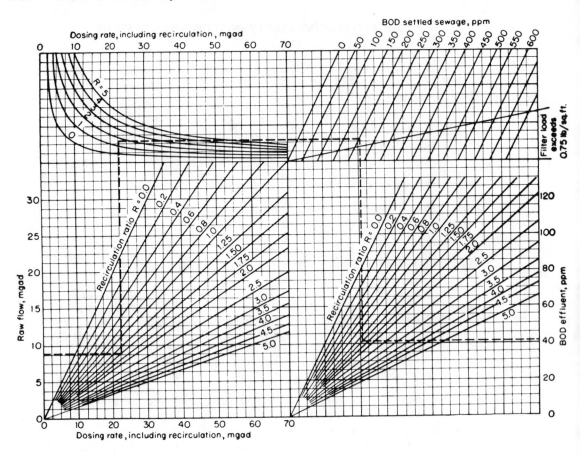

3.4-13   Expected performance of single stage, high rate filters based on Upper Mississippi Group Policy.

Courtesy of Dorr-Oliver Incorporated, Stamford, Conn.

3.4-14 COMPARATIVE COMPOSITION OF A DOMESTIC SEWAGE AND TRICKLING FILTER EFFLUENT[a]

Sewage in mg/l carbon

	Influent		Effluent	
Constituent	Particulate	Soluble	Particulate	Soluble
Fat acid	7.10	0	0.12	0
Fat ester	28.2	0	0.12	0
Protein	23.0	8.0	2.74	0.25
Amino acids	0	5.0	0	0.06
Carbohydrates	15.0	40.0	1.39	0.24
Soluble acids	4.0	17.0	0.13	1.65
Amides	1.5	0	–	–
MBAS	3.0	11.0	0.05	1.40
Creatinine	0	3.5	–	–
Amino sugars	1.8	0	0.38	0
Muramic acid	0.2	0	0.05	0
Sum	147	85	5.0	3.6
Total carbon	205	106	12.9	14.0

[a] Great Britain.

Source: S.D. Faust and J.V. Hunter, *Organic Compounds in Aquatic Environments,* Marcel Dekker, New York, 1971, by courtesy of Marcel Dekker, Inc.

3.4-15 RESULTS OF TRICKLING FILTER OPERATION

No. plants	Group designation	Range of BOD loading, lb/acre ft	BOD of effluent from filter and secondary settling, mg/l	Overall BOD reduction, %
4	No deep recirculation	253–429	86.9	92.2
4	With deep recirculation	245–1,170	85.7	90.8
8	2 stage	192–930	66.5	90.2
4	With shallow recirculation	792–2,320	81.8	88.1
4	With deep recirculation	1,100–1,950	78.7	87.5
3	No deep recirculation	720–2,130	75.7	82.8
3	With deep recirculation	2,530–5,750	70.3	78.3
4	With shallow recirculation	2,370–8,250	70.1	74.8

Source: F.W. Mohlmann, *Water Sewage Works,* June 1, 1955, R-261. Courtesy of Scranton Gillette Communications, Inc.

3.4-16   CONCENTRATIONS OF VARIOUS CHEMICALS HAVING RETARDANT OR INHIBITING EFFECT ON BIOLOGICAL FILTRATION

Waste or chemical	Concentration, mg/l
Boron	1
Chromium	3
Chromium, trivalent chromate	10
Copper (Cu)	1
Copper plating	5
Cyanide	1–2
Iron	5
Lead	0.1
Nickel	1–3

Source: *Manual of Practice No. 8*, Water Pollution Control Federation, Washington, D.C., 1959. With permission.

3.4-17   HYDRAULIC CHARACTERISTICS OF SELECTED FILTER MEDIA

Media	A_v[a] (ft^2/ft^3)	Exponent n[b]	Coefficient C[b]	Ref.
Polygrid	30	0 65	9.5	1
Glass spheres, 0.5 in. (2.8 cm) diam	85	0.82	22.5	2
Glass spheres, 0.75 in. (1.9 cm) diam	60.3	0.80	15.8	2
Glass spheres, 1.0 in. (2.5 cm) diam	41.6	0.75	12.0	2
Porcelain spheres, 3.0 in. (7.6 cm) diam	12.6	0.53	5.1	2
Rock, 2.5–4.0 in. (6.3–10.2 cm)[c]	–	0.408	4.15	3
Dowpac	25	0.50	4.84	4
Asbestos[d]	25	0.50	5.10	1
Mead-Cor[e]	30	0.70	5.6	1
Asbestos[f]	50	0.75	7.2	1
Asbestos[g]	85	0.80	8.0	1

[a] Specific surface. Multiply by 3.32 to obtain m^2/m^3.
[b] n and C obtained in formula $\dfrac{t}{D} = \dfrac{C}{Q^n}$

where
t = time, sec
D= depth, ft
C = coefficient
Q= flow rate, gal min/ft²
n = coefficient (exponent)

[c] Reported as model time.
[d] Corrugated asbestos packing.
[e] The Mead Corp.
[f] B.F. Goodrich Co.
[g] Johns-Manville Corp.

Source: W.W. Eckenfelder, Jr. and E.L. Barnhart, *J. Water Pollut. Control Fed.*, *35*(12):1535, 1963. With permission.

REFERENCES

1. W.W. Eckenfelder, Jr. and E.L. Barnhart, *J. Water Pollut. Control Fed.*, *35*(12):1535, 1963.
2. M.D. Sinkoff *et al., J. Sanit. Eng. Div. Proc. Am. Soc. Civ. Eng.*, *85*(SA6):51, 1959.
3. F.J. Burgess *et al., J. Water Pollut. Control Fed.*, *33*(8):787, 1961.
4. E.H. Bryan and D.H. Moeller, *Proc. 3rd Biol. Waste Treat. Conf.*, Manhattan College, New York, 1961.

3.4-18   RESIDENCE CHARACTERISTICS IN
THE PRESENCE OF FILTER SLIME

Media	Hydraulic loading,[a] gal/min ft^2	Residence time, min	Depth,[b] ft
Rock − ¾–1½ in. (1.9–3.8 cm)	0.093	35	6.0
Rock − 1½–3 in. (3.8–7.6 cm)	0.093	26	6.0
Polygrid	1.0	6	18.0
Asbestos	1.0	12.2	18.0
Asbestos	5.0	3.3	18.0
Polygrid[c]	2.0	3.8	18.0

[a] Multiply by 40.7 to obtain 1/sec/m.^{2}
[b] Multiply by 0.3 to obtain m.
[c] Thin slime cover.

Source: W.W. Eckenfelder, Jr. and E.L. Barnhart, *J. Water Pollut. Control Fed.*, *35*(12):1535, 1963. With permission.

3.4-19   OXYGEN TRANSFER
CHARACTERISTICS FOR SELECTED MEDIA

Filter media	Flow rate,[a] gal/min ft^2	Transfer[b] coefficient K, ft^{-1}	lb O$_2$/hr/ft^2/ft[c,d]
Polygrid	0.75	3.3	0.0111
	1.50	2.0	0.0135
	3.00	1.2	0.0162
	4.50	0.9	0.0182
	6.00	0.75	0.2000
Asbestos	0.75	2.6	0.0087
	1.50	1.8	0.0121
	3.00	1.2	0.0162
	4.50	0.95	0.0193
	6.00	0.80	0.0216

[a] Multiply by 40.7 to obtain 1/min m^2.
[b] From expression $-\dfrac{dC}{dD} = K(C_s-C)$ where (C_s-C) = oxygen deficit (same units as oxygen concentration). D = filter depth, ft and K = transfer coefficient, ft^{-1}.
[c] Multiply by 1,628 to obtain mg O$_2$/hr/cm^2/m.
[d] 20°C, 0 dissolved oxygen.

Source: W.W. Eckenfelder, Jr. and E.L. Barnhart, *J. Water Pollut. Control Fed.*, *35*(12):1535, 1963. With permission.

3.4-20 CHARACTERISTICS OF COMMERCIALLY AVAILABLE DIFFUSERS

Type diffuser	Description
Nonporous	Stainless steel nozzle type with check valve
Porous	Ceramic plates and tubes
Porous plastic	Saran wrapped media, metal core with integral end caps and control orifice
Porous plastic	Saran cloth flexible media, metal frame with integral end cap and orifice
Nonporous	Plastic nozzle type with integral control orifice
Porous	Ceramic tubes with cast iron end caps and control orifice
Nonporous	Plastic and metal valve type with integral control orifice
Nonporous	Nonporous metal shear type with control orifice
Nonporous	Tubular metal grid, nozzle type diffusion, 0.5 to 1 m below water surface
Nonporous	Plastic base with elastomer cover
Porous	Ceramic plates and ceramic tubes with and without integral end caps and control orifice
Nonporous	Metal nozzle type, adjustable, with 4–12 openings; ball check valve
Porous plastic	Porous plastic media, plastic pan-type holder
Nonporous	Plastic aeration tubing
Nonporous	Variable flow multiple orifice type, made of cast bronze or aluminum magnesium
Porous	Ceramic plates and tubes
Porous	Plastic media, various mountings
Nonporous	Nonporous metal valve type with plastic ball valve
Nonporous	Plastic nozzle type

Source: Aeration in wastewater treatment – Manual of Practice No. 5, *J. Water Pollut. Control Fed.*, *41*(11):1863, 1969. With permission.

3.4-21 CHARACTERISTICS OF COMMERCIAL MECHANICAL AERATORS

Type	Aerator characteristics
Updraft	Chicago — propeller-driven flow discharged against diffuser cone at top
Combination	D-O Aerator — induced by subsurface rotor and compressed air
Updraft	Simcar — induced by rotating impeller at top
Plate	Vortair — induced by horizontal, radially vaned impeller at top
	Aero-Accelator — induced by subsurface rotor and compressed air
Plate	American — induced by horizontal vaned impeller at top
Updraft	Lightnin — induced by rotating impeller at top with rotor below surface
Combination	Permaerator — induced by surface, subsurface, and compressed air
Downdraft	Produced by impeller in tube; with radial inlet troughs
Updraft	Aqua-Lator — either submersible or non-submersible pump discharge through vertical tube
Updraft	YeoCone — induced by spiral vaned revolving cone at top and draft tube
Updraft	Sigma — induced by scoop-type revolving blades at top

Source: Aeration in wastewater treatment — Manual of Practice No. 5, *J. Water Pollut. Control Fed., 41*(10):1863, 1969. With permission.

3.4-22 DIFFUSER EFFICIENCIES AT VARIOUS LOCATIONS IN TANK UNDER STANDARD CONDITIONS

Location	Type	Depth, ft	Air rate	Efficiency, %	Tank width, ft
Diffusers mounted on a header on the wall side	Porous	12.75	4 ft³/min ft		
			8 ft³/min dif	9.7	24
			12 ft³/min ft		
			8 ft³/min dif	11.5	
Diffusers mounted on both sides of a header near tank wall	Porous	12.4	8 ft³/min ft		
			8 ft³/min dif	12	24
			16 ft³/min ft		
			8 ft³/min dif	12.8	
			20 ft³/min ft		
			10 ft³/min dif	12.5	24
Diffusers mounted on both sides of a header located near both tank walls	Porous	12.4	16 ft³/min ft		
			8 ft³/min dif	14	24
Diffusers mounted on both sides of multiple headers	Porous	12.4	72 ft³/min ft		
			8 ft³/min dif	16	24

Source: Aeration in wastewater treatment — MOP 5, *J. Water Pollut. Control Fed., 41*(12):2026, 1969. With permission.

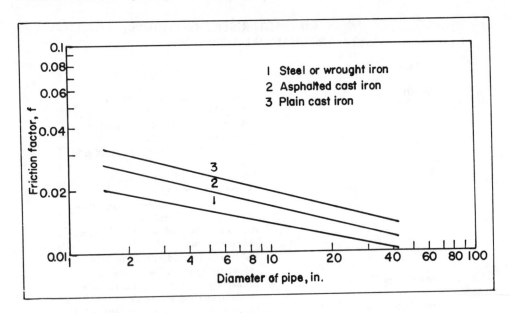

3.4-23   Values of friction factor f for flow of air in pipes.

Note: $\Delta P = \dfrac{f\,L\,T\,Q^2}{38{,}000\,P\,D^5}$

where ΔP = pressure drop due to friction, lb in^{-2}; f = friction factor; L = length of pipe, ft; T = absolute temperature, degrees Rankine; Q = volume of free air, ft^3 min^{-1}; P = compressed air pressure, lb in^{-2}; and D = diameter of pipe, in.

Adapted from: J.W. Clark and W. Viessman, Jr., *Water Supply and Pollution Control,* International Textbook Co., Ltd. (Blackie & Son, Ltd.), Glasgow, 1965, p. 414.

3.4-24   VOLATILE FATTY ACIDS IN SECONDARY EFFLUENTS

Average of 13 sewage effluents

Acid	Concentration, μg/l	Acid	Concentration, μg/l
Formic	91.0	Butyric	30.7
Acetic	130.0	Isovaleric	73.4
Propionic	13.7	Valeric	8.1
Isobutyric	26.5	Caproic	47.9

REFERENCE

1. J. Murtaugh and R. Bunch, *J. Water Pollut. Control Fed.,* *37*:410, 1965.

3.4-25   OPERATIONAL CHARACTERISTICS OF ACTIVATED SLUDGE PROCESSES[1-3]

Process modification	Hydraulic displacement	Aeration system	Application	BOD removal (%)
Conventional	Uniform	Diffused air, mechanical aerators	Low strength domestic wastes, susceptible to shock loads	85—95
Tapered aeration	Uniform	Diffused air mechanical aerators	Air supply tapered to match organic loading demand	85—95
Complete mix	Homogeneous	Diffused air, mechanical aerators	General application, resistant to shock loads, surface aerators	85—95
Step aeration	Increasing	Diffused air	General application to wide range of wastes	85—95
Modified aeration	Uniform	Diffused air	Intermediate degree of treatment where cell tissue in the effluent is not objectionable	60—75
Contact stabilization	Uniform	Diffused air, mechanical aerators	Expansion of existing systems, package plants, flexible	80—90
Two-state aeration	Subdivided	Diffused air, mechanical aerators	Provides opportunity to exploit quality of sludge in each stage of process	86—98[5]
Extended aeration	Extended	Diffused air, mechanical aerators	Small communities, package plants, flexible, surface aerators	75—95 0—96[4] 37—99[5]
Kraus process	Uniform	Diffused air	Low nitrogen, high strength wastes	85—95
High rate aeration	Reduced, uniform	Mechanical aerators	Use with turbine aerators to transfer oxygen and control the floc size, general application	75—90
Pure oxygen systems	Homogenous, reactors in series	Mechanical aerators	General application: use where limited volume is available: use near economical source of oxygen, turbine, or surface aerators	85—90

REFERENCES

1. Metcalf and Eddy, Inc., *Wastewater Engineering,* McGraw-Hill, N.Y., 1972.
2. Metcalf and Eddy, Inc., *Wastewater Engineering: Treatment, Disposal, Reuse,* 2nd ed., revised by G. Tchobanoglous, McGraw-Hill, N.Y., 1979.
3. G.M. Fair *et al., Water and Waste Engineering,* Vol. 2, *Water Purification and Wastewater Treatment and Disposal,* John Wiley & Sons, N.Y., 1968.
4. P.H.M. Guo *et al., J. Water Pollut. Control Fed., 53*:32, 1981.
5. R.H. Baker, Jr., *J. San. Engng. Div., Proc. Am. Soc. Civil Engrs., 88*:SA6, 75, 1962.

3.4-26   DESIGN PARAMETERS FOR ACTIVATED SLUDGE PROCESSES

Process modification	Sludge "Age," days	BOD removal, lb BOD$_5$/lb MLVSS[a] day	Volumetric loading, lb BOD$_5$/ 1,000 ft^3	Solids accumulation, lb/lb BOD removed	MLSS[b], mg/l	Detention time, hr	Volume ratio, return sludge/ influent	Sludge zone settling rate, ft/hr
Conventional	5–15	0.2–0.4(0.2–0.5)[2]	20–40	0.4[3]	1,500–3,000 (2,000)[3]	4–8(6)[3]	0.25–0.5	18[3]
Complete-mix	5–15	0.2–0.6	50–120		3,000–6,000	3–5	0.25–1.0	
Step-aeration	5–15	0.2–0.4	40–60		2,000–3,500	3–5	0.25–0.75	
Modified-aeration	0.2–0.5	1.5–5.0	75–150		200–500	1.5–3	0.05–0.15	
Contact-stabilization Contact unit	5–15	0.2–0.6(0.15–0.2)[2]	60–75	0.23[3]	(1,000–3,000) (3,000)[c,3]	(0.5–1.0) (0.25–0.5)[c,3]	0.25–1.0	12[3]
Solids stabilization unit					(4,000–10,000) (6,000)[d,3]	(3–6)(2–4)[d,3]		
Extended-aeration	20–30	0.05–0.15(0.01–0.07)[2]	10–25	0.14[3]	3,000–6,000 (4,000)[3]	18–36(24)[3]	0.75–1.50	6.5[3]
Kraus process	5–15	0.3–0.8	40–100		2,000–3,000	4–8	0.5–1.0	
High rate aeration	5–10	0.4–1.5(≥ 1)[2]	100–1,000		4,000–10,000 (450)[3]	0.5–2(3)[3]	1.0–5.0	
Pure oxygen systems	8–20	0.25–1.0	100–250	0.18[3]	6,000–8,000	1–3	0.25–0.5	10[3]
2 stage aeration		0.07–0.15[2]			{3,000 6,000}[3]	1.5[3] 7.5[3]		

[a] MLVSS – mixed liquor volatile suspended solids.
[b] MLSS – mixed liquor suspended solids.
[c] Contact unit.
[d] Solids stabilization unit.

REFERENCES

1. Metcalf and Eddy, Inc., *Wastewater Engineering*, McGraw-Hill, New York, 1973, p. 498.
2. B.L. Goodman, *Design Handbook of Wastewater Systems: Domestic, Industrial, Commercial*, Technomic, Westport, Conn., 1971, p. 31.
3. B.L. Goodman, *Manual for Activated Sludge Sewage Treatment*, Technomic, Westport, Conn., 1971, p. 12.

3.4-27 REPORTED REMOVAL OF PRIORITY POLLUTANTS IN FULL-SCALE ACTIVATED SLUDGE PLANTS[a]

Compound	Concentration influent (μg/l)[b]	Removal range (%)
Benzene	10—140	70—98
1,1,1-Trichloroethane	18—184	69—99
Chloroform	27—46	56—59
1,2-*trans*-Dichloroethylene	25	72
Ethylbenzene	10—44	77—98
Methylene chloride	36—282	36—74
Tetrachloroethylene	15—385	65—90
Toluene	15—73	70—99
Trichloroethylene	17—497	72—99
Phenol	18—20	94—99
Naphthalene	51	94
bis(2-Ethylhexyl)phthalate	10—217	31—64
Butyl benzyl phthalate	60	97
Di-*N*-butyl phthalate	10—18	67—70
Diethyl phthalate	11	82

[a] Constructed from Fate of Priority Pollutants in Publicly Owned Treatment Works, Interim Report, EPA-440-1-80-301, U.S. Environmental Protection Agency, Washington, D.C., 1980; and W. Eckenfelder, Jr., M.C. Goronszy, and T.P. Quirk, *CRC Crit. Rev. Environ. Control, 15*:111, 1985.

[b] Reported influent concentrations ≥10 μg/l.

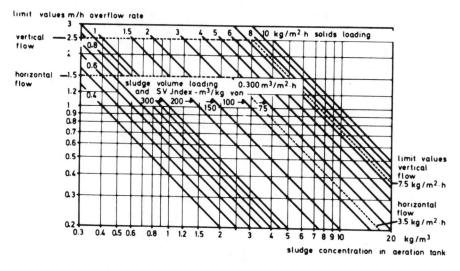

3.4-28   Design data for secondary sedimentation tank.

Source: K.H. Kalbskoph in *Water Quality Improvement by Physical and Chemical Processes,* E.F. Gloyna and W.W. Eckenfelder, Jr., Eds., University of Texas Press, Austin, 1970, p. 100. With permission.

3.4-29   DESIGN FACTORS FOR STABILIZATION BASINS

	Aerobic	Facultative	Anaerobic	Aerated
Depth, ft	0.6–1.0	2–5	8–10	6–15
Detention, days	2–6	7–30	30–50	2–10
BOD loading, lb/acre day	100–200	20–50	300–500	
BOD removal, %	80–95	75–85	50–70	55–90
Algae concentration mg/l	>100	10–50	nil	nil

REFERENCES

1. W.J. Oswald in *Advances in Biological Waste Treatment,* W.W. Eckenfelder and
   B.J. McCabe, Eds., Pergamon Press, New York, 1963, p. 357.
2. W.W. Eckenfelder and D.J. O'Connor, *Biological Waste Treatment,* Pergamon
   Press, New York, 1961.

3.4.3 DISINFECTION

3.4-30   REMOVAL OR DESTRUCTION OF BACTERIA BY DIFFERENT TREATMENT PROCESSES

Process	Removal (%)		
	Ref. 1	Ref. 2	Ref. 3
Coarse screens	0—5		
Fine screens	10—20	10—20	
Grit chambers	10—25		
Plain sedimentation	25—75	25—75	
Chemical precipitation	40—80	40—80	
Trickling filters	90—95	90—95	
Activated sludge	90—98	90—98	
Chlorination of treated sewage	98—99	98—99[a]	
Intermittent sand filtration		95—98	
Land treatment			94—99.7

[a] Chlorination of raw sewage or settled sewage 90—95%.

REFERENCES

1. Metcalf and Eddy, Inc., *Wastewater Engineering: Treatment Disposal, Reuse*, 2nd ed., revised by G. Tchobanoglous, McGraw-Hill, N.Y., 1979.
2. K. Imhoff and G.M. Fair, *Sewage Treatment*, John Wiley & Sons, N.Y., 1940.
3. C.E. Keefer, *Sewage Treatment Works*, McGraw-Hill, N.Y., 1940.

3.4-31   CHLORINE REQUIRED FOR DISINFECTION

Sewage or effluent	Dosage, mg/l			Dosing capacity to produce 2 mg/l residual[5]
	Enslow and Grune[1,2]	Rice[3]	Metcalf & Eddy[4]	
Raw sewage	6–12	20–25		
Septic	12–25	–	6–25	25
Settled sewage	5–10	15–20		
Septic	12–40		5–20	20
Chemical precipitation	3–10[6]	20	2–6	–
Trickling filter, standard, normal	3–5	12–15		
Trickling filter, standard, poor	5–10		3–15	15
High-rate trickling filter		15–20		
Activated sludge, normal	2–4	5–15		
Activated sludge, poor	5[6]–8		2–8	8
Sand filter, normal	1–3	5–10	–	
Sand filter, poor	3–5			6
Odor control, etc	–	10–20	–	–

REFERENCES

1. L.H. Enslow in *Modern Sewage Disposal*, L.H. Pearse, Ed., Federation of Sewage Works Associations, N.Y., 1938, p. 98.
2. W.N. Grune, *Water Sewage Works, 102*:350, 1955.
3. L.G. Rice, *Water Sewage Works, 97*:436, 1950.
4. Metcalf & Eddy, Inc., *Wastewater Engineering*, McGraw-Hill, New York, 1972.
5. *Standards for Sewage Works*, Upper Mississippi River Board of Public Health Engineers and Great Lakes Board of Public Health Engineers, 1954.
6. *Manual of Practice No. 11*, Water Pollution Control Federation, 1968.

3.4-32   CHLORINATION APPLICATIONS IN WASTEWATER COLLECTION, TREATMENT, AND DISPOSAL

Application	Dosage range, mg/l				Remarks
	Ref. 1	Ref. 2	Ref. 3	Ref. 4	
Collection					
Slime growth control	1—10	—	—	—	Control of fungi and slime-producing bacteria
Corrosion control (H$_2$S)	2—9[a]	2—9	0.6—27.6	—	Control brought about by destruction of H$_2$S in sewers
Odor control	2—9[a]	2—9	1.6—27.6	10—20	Especially in pump stations and long flat sewers
Treatment					
Grease removal	2—10	2—10	1—10	1.5	Added before preaeration
BOD reduction	0.5—2[b]	0.5—2[b]	0.5—10.9[b]		Oxidation of organic substances
Filter ponding control	1—10	1—10	0.5—5	—	Residual at filter nozzles
Filter fly control	0.1—0.5	0.1—0.5	3—6	3	Residual at filter nozzles, used during fly season
Sludge bulking control	1—10	1—10	0.7—7		Temporary control measure
Digester supernatant oxidation	20—140	20—140			
Digester and Imhoff tank foaming control	2—15	2—15			
Nitrate reduction					Conversion of nitrate to ammonia, depends on amount of nitrate present
Disposal					
Bacterial reduction	2—20				Plant overflows, storm water
Disinfection	—	1.5—6.25	—	1—40	Depends on nature of wastewater and on degree of treatment

[a] Per mg/l of H$_2$S.
[b] Per mg/l of BOD$_5$ destroyed.

REFERENCES

1. Metcalf and Eddy, Inc., *Wastewater Engineering,* McGraw-Hill, N.Y., 1972.
2. Metcalf and Eddy, Inc., 2nd ed., *Wastewater Engineering: Treatment, Disposal, Reuse,* revised by G. Tchobanoglous, McGraw-Hill, N.Y., 1979.
3. C.E. Keefer, *Sewage Treatment Works,* McGraw-Hill, N.Y., 1940.
4. H.E. Babbitt and E.R. Baumann, *Sewerage and Sewage Treatment,* 8th ed., John Wiley & Sons, N.Y., 1958.

3.4.4 ADVANCED OR TERTIARY TREATMENT PROCESSES

3.4-33 ALTERNATIVE PROCESSES OF WASTEWATER TREATMENT AND SOLIDS HANDLING

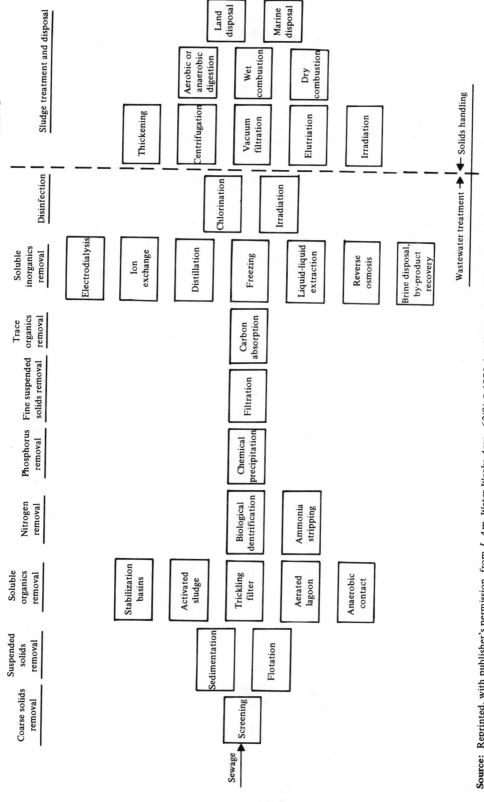

Source: Reprinted, with publisher's permission, from *J. Am. Water Works Assn., 62(9)* © 1970, by the American Water Works Association, Inc.

3.4-34 TYPICAL RESULTS — ADVANCED WASTEWATER TREATMENT PROCESSES[a,1]

Process	BOD$_5$ (mg/l)	COD (mg/l)	SS (mg/l)	PO$_4$ (mg/l)	NH$_3$ (mg/l)	TDS (mg/l)
Preliminary[b]						
Coagulation and sedimentation	67—83	62—83	87—96	67—83		
Plus mixed media filtration	77—90	69—90	98—99	83—96		
Plus activated carbon adsorption	97—98	91—95	98—99	83—96		
Plus ammonia stripping	97—98	91—95	98—99	83—96	47—95	
Primary						
Coagulation and sedimentation	67—83	62—83	89—96	67—83		
Plus mixed media filtration	77—90	69—90	98—99	83—96		
Plus activated carbon adsorption	97—98	91—95	98—99	83—96		
Plus ammonia stripping	97—98	91—95	98—99	83—96	47—95	
High-rate trickling filter						
Plus mixed media filtration	93—99	88—93	91—96			
Coagulation and sedimentation	95—97	89—93	95—98	75—92		
Plus mixed media filtration	96—98	90—94	99.6—100	92—99		
Plus activated carbon adsorption	99—99.7	95—98	99.6—100	92—99		
Plus ammonia stripping	99—99.7	95—98	99.6—100	92—99	47—95	
Conventional activated sludge				10—30[3]	30—50[3]	
Plus mixed media filtration	98—99	90—94	95—99			
Coagulation and sedimentation	98—99	90—94	96—99	75—92		
Plus mixed media filtration	99—99.7	91—95	99.6—100	92—99		
Plus activated carbon adsorption	99.7—100	97—99	99.6—100	92—99		
Plus ammonia stripping	99.7—100	97—99	99.6—100	92—99	47—95	
Algae harvesting	50—75[4]	40—60[4]		~50[4]	50—90[3,4]	
Ammonia stripping					80—93[3]	
					85—98[4]	
Anaerobic denitrification					65—95[3]	
Bacterial assimilation	75—95[4]	60—80[4]	80—95[4]	10—20[4]	30—40[4]	
Carbon adsorption	70—90[4]	60—75[4]	80—90[4]			
Chemical precipitation	70—90[4]	75—95[4]	60—80[4]	88—95[3]	5—15[4]	~20[4]
				88—95[4]		
Plus activated sludge	90—95[4]	85—90[4]	80—95[4]	30—40[4]	30—40[4]	~10[4]
Plus filtration				95—98[3]		
Electrochemical treatment	50—60[4]	40—50[4]	80—90[4]	80—95[3]	80—85[4]	
				80—85[4]		
Electrodialysis				30—50[4]	30—50[3,4]	~40[4]
Filtration						
Multimedia	50—70[4]	40—60[4]	80—90[4]			
Diatomite bed			95—99[4]			
Microstrainer	40—70[4]	30—60[4]	50—80[4]			95—99[4]
Flotation			60—80[4]			
Foam fractionation	~70[4]	60—70[4]	75—90[4]			
Freezing	95—99[4]	90—99[4]	95—98[4]			95—99[4]
Gas phase separation					50—70[4]	
Ion exchange	40—60[4]	30—50[4]		86—98[3]	80—92[3]	
				85—98[4]	80—90[4]	
Land application	90—95[4]	80—90[4]	95—98[4]	60—90[3]	60—80[4]	
Modified activated sludge					60—80[3]	
Oxidation (chlorine)	80—90[4]	65—70[4]			50—80[4]	
Reverse osmosis	95—99[4]	90—95[4]	95—98[4]	95—99[4]	65—95[3]	95—99[4]
					95—99[4]	
Sorption	~50[4]	~40[4]		90—98[3]		~10[4]
				~99[4]		
Distillation	98—99[4]	95—98[4]	~99[4]	~99[4]	90—98[4]	95—99[4]

[a] Based on raw wastewater values of 300 mg/l BOD$_5$;[1] 480 mg/l COD;[1] 230 mg/l SS;[1] 12 mg/l PO$_4$;[1] and 19 mg/l NH$_3$-N.[2]

[b] Preliminary treatment — grit removal, screen chamber, Parshall flume, overflow.

3.4-34   TYPICAL RESULTS — ADVANCED WASTEWATER TREATMENT PROCESSES[a,1]
(continued)

REFERENCES

1. R.L. Culp *et al., Handbook of Advanced Wastewater Treatment,* 2nd ed., Van Nostrand Reinhold, 1978.
2. C.E. Keefer, *Sewage Treatment Works,* McGraw-Hill, New York, 1940.
3. R. Eliassen and G. Tchobanoglous, *Environ. Sci. Technol.,* 3:539, 1969.
4. Metcalf and Eddy, Inc. *Wastewater Engineering,* McGraw-Hill, New York, 1972.

3.4-35 COMPARISON OF NUTRIENT REMOVAL PROCESSES

Process	Class	Removal efficiency, %	Estimated cost, $/mg[c]	Wastes to be disposed of	Remarks
Nitrogen					
Ammonia stripping	Chemical	80–98	9–25	—	Efficiency based on ammonia nitrogen only
Anaerobic denitrification	Biological	60–95	25–30	None	
Algae harvesting	Biological	50–90	20–35	Liquid and sludge	Large land area needed
Breakpoint chlorination	Chemical	90–98	—	None	Ammonia nitrogen only cost based on dosage
Conventional biological treatment	Biological	30–50 nitrogen / 10–30 phosphorous	30–100	Sludge	
Nitrogen and Phosphorous					
Ion exchange	Chemical	70–95	170–300	Liquid	Ammonia selective resins
Ion exchange	Chemical	80–92 nitrogen / 86–98 phosphorous	170–300	Liquid	Efficiency and cost depend on degree of pretreatment
Electrochemical treatment	Chemical	80–85	4–8[a]	Liquid and sludge[b]	
Electrodialysis	Chemical	30–50	100–250	Liquid	Cost based on 1–10 mil gal/day capacity, 1,000 ppm solids
Reverse osmosis	Physical	65–95	250–400	Liquid	
Distillation	Physical	90–98	400–1,000	Liquid	
and application	Physical	60–90 phosphorous[b]	75–150	None	Large land area needed
Phosphorous					
Modified activated sludge	Biological	60–80	30–100	Sludge	
Chemical precipitation	Chemical	88–95	10–70	Sludge	
Chemical precipitation with filtration	Chemical	95–98	70–90	Liquid and sludge	
Sorption	Chemical	90–98	40–70	Liquid and solids	Cost based on water treatment costs

[a] Power cost only; installation costs unavailable.
[b] Nitrogen efficiency depends on form of nitrogen.
[c] Based on Engineering News Record Construction Cost Index (ENRCC) of 1500.

Source: Reprinted with permission from R. Eliassen and G. Tchobanoglous, *Environ. Sci. Technol.*. *3*(6):539, 1969. Copyright 1969, American Chemical Society.

3.4-36   COMPILATION OF THE VARIOUS LIME PRECIPITATION SCHEMES FOR
THE REMOVAL OF PHOSPHORUS

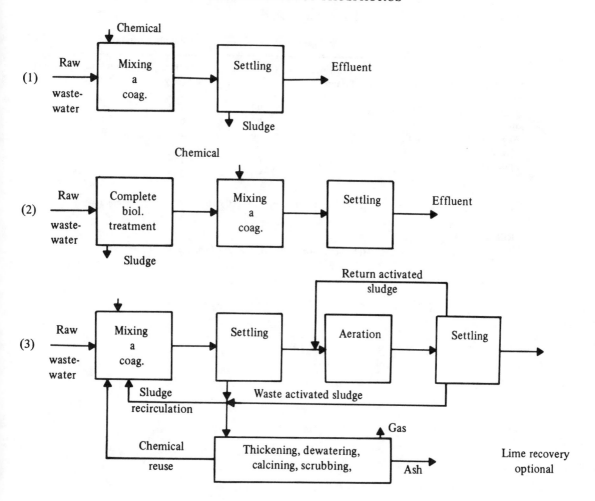

Source: J.B. Nesbitt, *J. Water Pollut. Control Fed., 41*(5):707, 1969. With permission.

3.4-37   NUTRIENT LOADING AND RETENTION IN OXIDATION PONDS

Located in Wisconsin

Site	Date	Inorganic nitrogen		Soluble phosphorus	
		Loading, lb/acre yr	Retention, %	Loading, lb/acre yr	Retention, %
Junction City	1957	2,427	97	402	94
Junction City	1959	3,760	85	3,680	80
New Auburn	April	4,000	6	767	0
New Auburn	August	4,600	98	1,350	58
Spooner	December	3,614	65	1,168	0
Spooner	August	3,430	93	1,680	66

Source: K.M. Mackenthun, *J. Water Pollut. Control Fed., 34*(10):1077, 1962. With permission.

REFERENCE

1. K.M. Mackenthun and C.D. McNabb, *J. Water Pollut. Control Fed., 33*(12):1234, 1961.

3.4-38 NITROGEN REMOVAL PROCESSES

Process	Principle	Drawbacks
Air stripping	NH_3° [solution] $\xrightarrow{air}$ $NH_3^\circ \uparrow$ [gas]	Excessive air requirements, precipitation of $CaCO_3$ by absorption of atmospheric CO_2
Ammonium – ion exchange	$NH_4^+ \rightarrow$ clinoptilolite $\cdot NH_4$ + regeneration with lime-NaCl solution	Requires air stripping to remove NH_3° from regenerant
Nitrate – ion exchange	$NO_3^- \rightarrow$ resin $\cdot NO_3$	Requires highly selective anion exchange resin
Reverse osmosis	Selective rejection of NH_4^+ and NO_3^- by a semipermeable membrane, pressure driving force	Membrane life, organic fouling
Electrodialysis	Selective transport of NH_4^+ and NO_3^- by a membrane, electrical potential driving force	Membrane life, organic fouling
Distillation	Evaporation of water and recondensation, separating out nonvolatile salts	High energy requirements; poor removal of NH_3°
Break-point chlorination	$2\ NH_3 + 3\ Cl_2 \rightarrow N_2^\circ \uparrow + 6\ HCl$	Excessive chemical costs; chloramine by-products
Chemical denitrification	$NO_3^- + Fe^{++} \xrightarrow{Cu} N_2^\circ \uparrow + Fe^{+++}$	By-product iron sludge
Biological denitrification	$NO_3^- \xrightarrow[\text{organisms}]{\text{organic C}} N_2^\circ \uparrow + CO_2$	Anaerobic process, requires external organic carbon source, e.g., methanol

Source: R.B. Dean *et al.,* Nitrogen removal from wastewaters, Adv. Waste Treatment and Water Reuse Symp., Session 2, Dallas, 1971.

3.4-39   GRANULAR ACTIVATED CARBON USED IN WASTEWATER TREATMENT

	Ref. 1	Ref. 2—8
Total surface area $(m^2/mg)^a$	950—1500	460—1600
Bulk density (lb/ft^3)	26	18—36
Particle density	1.3—1.4	1.4—1.5
Effective size (mm)	0.8—0.9	0.55—0.90
Uniformity coefficient	≤1.9	≤1.7—≤1.9
Mean particle diameter (mm)	1.5—1.7	1.07—1.35
Iodine number	>900	570—1530
Ash (%)	<8	5.4
Moisture (%)	<2	

[a]   E. Brunauer and Teller (BET) method, *J. Am. Chem. Soc., 60*:309, 1938.

REFERENCES

1. R.L. Culp and G.L. Culp, *Advanced Wastewater Treatment,* Litton Educational Publications, New York, 1971.
2. R.S. Joyce and V.A. Sukenik, Feasibility of Granular Activated-Carbon Adsorption for Wastewater Renovation, 2, PHS Publ. No. 999-WP-28, Public Health Service, U.S. Department of Health, Education, and Welfare, Cincinnati, Oct. 1965.
3. D.F. Bishop *et al., J. Water Pollut. Control Fed., 39*:188, 1967.
4. M.E. Flentje and D.G. Hager, *J. Am. Water Works Assn., 56*:191, 1964.
5. D.G. Hager and M.E. Flentje, *J. Am. Water Works Assn., 57*:1440, 1965.
6. M.A. Rollor *et al., J. Environ. Eng. Div., Proc. Am. Soc. Civil Eng., 108*:1361, 1982.
7. J. Fettig and H. Sontheimer, *J. Environ. Eng. Div., Proc. Am. Soc. Civil Eng., 113*:764, 1987.
8. R.M. Clark *et al., J. Environ. Eng. Div., Proc. Am. Soc. Civil Eng., 112*:744, 1986.

3.4-40 CHARACTERISTICS OF ACTIVATED CARBON SPECIFIED FOR WASTEWATER
TREATMENT

				Sorption data			
Identification	Source	Activating agent	Surface area, m^2/g	Methylene blue[h]	ABS number[i]	Phenol number	Eosin B[j], mg/l
Nuchar C1000®[a]	Pulp by-product	Oxidizing gases	1,000	1.5	32	18	0.11
Aqua Nuchar A®[a]	Pulp by-product	Oxidizing gases	600	140	105	14	2.9
Darco KB®[b]	Wood	Phosphoric acid	1,600	4	50	–	0.87
Hydrodarco® B[b]	Lignite	Steam	600–650	178	102	26	0.26
RB®[c]	Bituminous coal	925°C	1,250–1,400	0.3	28	32	0.16
Type C[c] pulverized	Bituminous	925°C	1,000	10	38	22	0.34
Norit® C[d]	Pine wood	Steam	600	260	260	15	20
Norit® EX[d]	Pine wood	Steam-acid	1,200	0.8	38	–	0.05
XH[e]	Petroleum	Steam	1,300	1.2	38	–	0.92
182[f]	Petroleum	Steam	1,180	75	51	–	1.3
E-65-2[f]	Coal	Steam	475	180	92	38	0.10
Ultra-fine grade[g] 441-2-PB	Subbituminous coal	Steam	950	230	112	20	2.7

[a] West Virginia Pulp and Paper.
[b] Atlas Chemical Industries.
[c] Pittsburgh Activated Carbon Co. (Calgon Co., subsidiary of Merck Co.)
[d] L. A. Solomon.
[e] Barneby Cheney.
[f] Union Carbide.
[g] FMC.
[h] Reported as the residual concentration of dye after 340 mg/l solution had been treated with 18 liters carbon for 1 hr.
[i] mg carbon necessary to reduce ABS from 5.0–0.5 mg in 1 liter.
[j] Residuals starting with 45 mg/l included as example of a negatively charged dye.

Source: D.F. Bishop *et al., J. Water Pollut. Control Fed.,* *39*(2):188, 1967. With permission.

3.4-41 PROPERTIES OF CARBON AT
VARIOUS POINTS IN REGENERATION
PROCESS

Property	Virgin carbon	Spent carbon	Carbon after one re-generation with steam
Apparent density	0.481	0.519–0.563	0.473–0.497
Iodine number	935	680–851	900–969
Ash, %	5.02	4.43–5.68	4.69–4.88

Source: A.F. Slechta *et al., J. Water Pollut. Control Fed.,*
39(5):791, 1967. With permission.

3.4-42   PLANKTON REMOVAL DURING TERTIARY TREATMENT
OF OXIDATION POND EFFLUENT

Bihourly composite May 9, 1968

Organism	Oxidation pond effluent, organisms/ml	Sand filter effluent		Microstrainer effluent	
		Organisms/ml	Removal, %	Organisms/ml	Removal, %
Ciliates	1,152	0	100	729	37
Green algae					
Euglena	1,845	1,363	26	1,737	6
Scenedesmus	2,925	1,080	73	1,656	43
Actinastrum	1,143	805	30	1,188	0
Diatoms					
Pennales	7,776	2,943	62	7,884	0

Source: B.T. Lynam *et al., J. Water Pollut. Control Fed., 41*(2)(Part 1):247, 1969. With permission.

REFERENCE

1. N. Youngsteadt, Metropolitan Sanitary District Greater Chicago Res. Rept., July 1968.

3.4.5   SLUGE HANDLING AND DISPOSAL

3.4-43   COMMON SLUDGE TREATMENT PROCESSES AND METHODS OF FINAL DISPOSAL

A. Concentration
    1. Clarifier thickening
    2. Separate concentration
        a. Gravity thickening
        b. Flotation
B. Digestion
    1. Aerobic
    2. Anaerobic
C. Dewatering
    1. Drying beds
    2. Lagoons
    3. Vacuum filtration
    4. Centrifugation
D. Heat drying and combustion
    1. Heat drying
    2. Incineration
        a. Multiple hearth
        b. Fluid solids
    3. Wet oxidation
E. Final sludge disposal
    1. Land fill
    2. Soil conditioning
    3. Discharge to sea

Source: P.L. McCarty, *J. Water Pollut. Control Fed.,* *38*(4):495, 1966. With permission.

3.4-44   REPORTED SLUDGE HEAT VALUES

Description	Combustibles, %	Ash, %	Sludge heat value, Btu/lb	Ref.
Grease and scum	88.5	11.5	16,750	2
Raw sewage solids	74.0	26.0	10,285	2
Fine screenings	86.4	13.6	8,990	2
Waste sulfite liquor solids	—	—	7,900	1
Primary sewage sludge	—	—	7,820	1
Activated sewage sludge	—	—	6,540	1
Semichemical pulp solids	—	—	5,812	1
Digested primary sludge	59.6	40.4	5,290	2
Grit	33.2	69.8	4,000	2

Source: D.L. Ford, in *Water Quality Improvement by Physical and Chemical Processes,* Vol. III, E.F. Gloyna and W.W. Eckenfelder, Jr., Eds., University of Texas Press, Austin, 1970, p. 347. With permission.

REFERENCES
1. R.A. Burd, *A Study of Sludge Handling and Disposal,* U.S. Department of the Interior, Federal Water Pollution Control Administration, 1968.
2. F. Sebastian and P. Cardinal, *Chem. Eng., 75:*(October 14, 1968) p. 112.

3.4-45   CHEMICAL COMPOSITION OF VARIOUS SEWAGE SLUDGES

Location	Type of sludge	Total nitrogen, %	Carbon, %	Carbon/ nitrogen ratio	Phos- phoric oxide, %	Acid soluble potassium, %	Ash, %
Washington, D.C.	Influent solids	2.39	43.69	18.3	1.09	–	37.59
	Digested sludge	2.06	28.59	13.9	1.44	0.14	52.83
Baltimore, Md.	Influent solids	2.23	47.09	21.1	1.29	–	24.16
	Activated sludge	2.36	30.37	12.4	11.01	–	29.70
	Heat dried digested sludge	3.05	36.53	12.0	2.97	–	39.73
Jasper, Ind.	Influent solids	2.90	42.31	14.6	1.62	–	32.29
	Activated sludge	3.51	23.01	6.6	2.81	–	52.43
	Digested sludge	5.89	22.95	3.9	3.49	–	36.96
Richmond, Ind.	Influent solids	3.80	28.21	7.4	5.19	–	40.94
	Activated sludge	3.02	44.04	14.6	3.64	–	31.37
	Digested sludge	2.24	26.36	11.8	4.34	–	50.09
Chicago, Ill.[a]	Raw sludge	2.70	46.62	17.3	2.71	–	28.24
	Activated sludge	4.98	28.62	5.7	5.58	–	34.82
	Heat dried sludge	5.56	29.41	5.3	6.56	–	37.42
Milwaukee, Wisc.[a]	Activated sludge	5.96	–	–	3.96	0.41	27.70
Rochester, N.Y.	Digested sludge	2.54	–	–	1.16	0.24	42.80
Des Moines, Iowa	Digested primary and activated sludge	1.81	–	–	3.31	0.40	61.40

[a] Undigested.

Source: R.A. Burd, *A Study of Sludge Handling and Disposal,* U.S. Department of the Interior, Federal Water Pollution Control Administration, 1968.

3.4-46 INFORMATION ON VARIOUS METALS IN SEWAGE SLUDGES AND MANURES

Element	Sweden (1968—1971)[a]			Michigan (1973)[b]			England, Wales (1972)[c]			U.S./Canada (1955—1973)[d] (range)	Swedish manure (mean)[e]	Canadia manure (mean)[f]
	Range	Mean	Median	Range	Mean	Median	Range	Mean	Median			
B	—	—	—	—	—	—	15—1,000	70	50	7—680	—	20.2
Cd	2.3—171	12.7	6.7	2—1,100	74	12	<60—1,500	<200	—	1—810	0.32	—
Co	2.0—113	15.2	10.8	—	—	—	2—260	24	12	—	2.4	1.0
Cr	20—40,615	872	86	22—30,000	2,031	380	40—8,800	980	250	Tr—19,600	5.2	—
Cu	52—3,300	791	560	84—10,400	1,024	700	200—8,000	970	800	100—11,700	41	16
Hg	<0.1—55	6.0	5.0	0.1—56	5.5	3.0	—	—	—	—	0.09	—
Mn	73—3,861	517	386	—	—	—	150—2,500	500	400	60—1,470	268	201
Ni	16—2,120	121	51	12—2,800	371	52	20—5,300	510	80	11—3,500	7.8	—
Pb	52—2,914	281	180	80—26,000	1,380	480	120—3,000	820	700	15—1,900	6.6	—
V	—	—	—	—	—	—	20—400	75	60	—	—	—
Zn	705—14,700	2,055	1,567	72—16,400	3,315	2,200	700—49,000	4,100	3,000	373—28,400	215	96
Ag	—	—	—	—	—	—	5—150	32	20	—	—	—
As	—	—	—	1.6—18	7.8	—	—	—	—	—	—	—
Ba	—	—	—	—	—	—	150—4,000	1,700	1,500	—	—	—
Mo	—	—	—	—	—	—	2—30	7	5	Tr—1,000	—	2.1
Sn	—	—	—	—	—	—	40—700	160	120	—	—	—
Be							1—30	5	3			
Bi							12—100	34	25			
Fe							6,000—62,000	24,000	21,000			
Ga							1—20	8	8			
La							30—150	72	60			
Li							10—150	45	40			
Sr							80—2,000	340	300			
Ti							<1,000—4,500	2,000	200			
Y							15—1,000	42	40			
Zr							30—3,000	310	150			

Note: Results are expressed in mg/kg DS.

a 93 treatment plants.[1]
b 57 sources.[2]
c 42 sources.[3]
d 15 treatment plants.[4]
e Reference 5.
f Reference 6.

3.4-46   INFORMATION ON VARIOUS METALS IN SEWAGE SLUDGES AND MANURES (continued)

Source: P.J. Matthews, *CRC Crit. Rev. Environ. Control, 14*:199, 1984.

REFERENCES

1. B. Berggren and S. Oden, Analysresultat Rorande Fundmetaller och Klorerade Kolvåten. I. Rötslam Fran Svenska Reningsverk 1968/71, Institutionen fur Markvetenskap Lant-brukshögskolan, 750 07, Uppsala, 1972.
2. P.A. Blakeslee, Monitoring Considerations for Municipal Wastewater Effluent and Sludge Application to the Land, Universities Workshop, U.S. Environmental Protection Agency, and U.S. Department of Agriculture, Champaign, Ill., July 1973.
3. M.L. Berrow and J. Webber, *J. Sci. Food Agric., 93*:93, 1972.
4. A.L. Page, Fate and Effects of Trace Elements in Sewage Sludge When Applied to Agricultural Lands, A Literature Review Study, Program Element No. B2043, prepared for Environmental Protection Agency, distributed by U.S. National Technical Information Service, Springfield, Va., January 1974.
5. A. Anderson, *Swed. J. Agric. Res., 7*:1, 1977.
6. A.J. Atkinson, G.R. Giles, and J.G. Desjardins, *Can. J. Agric. Sci., 34*:76, 1954.

3.4-47   VITAMIN CONTENT OF ACTIVATED SLUDGE

Vitamin	Contents in mg/100 g	
	Centrifuged	Dried
Thiamine	2.05	0.63
Riboflavin	2.16	1.3
Pyridoxine	trace	0.7
Nicotinic acid	13.6	10.0
Pantothenic acid	4.6	4.4
Biotin	158[a]	83.4[a]
Folic acid	182	140
B_{12}	159	100

[a] In micrograms/100 g.

Source: V. Kocher and V. Corti, *Schweiz. Z. Hydrol.,* *14*:333, 1952. With permission.

3.4-48   LIPIDS IN DRIED SEWAGE SLUDGE

Solvent	Domestic sewage, Madison, Wisc.		Packing house		Milorganite, dried activated sludge fertilizer, %	Activated sludge Milwaukee, Wisc. %
	Sample 1, %	Sample 2, %	Sample 1, %	Sample 2, %		
Petroleum ether	24.7	26.0	18.4	27.8	5.6	6.5
Ethyl ether	23.6	–	21.3	–	–	–
Isopropyl ether	31.1	–	30.3	–	–	–
Chloroform	42.5	40.5	38.8	46.0	9.6	10.3

Source: O.J. Knechtges, W.H. Peterson, and F.M. Strong, *Sewage Works J., 6*(6):1082, 1934. With permission.

3.4-49   ORGANIC FRACTIONS FOR VARIOUS SLUDGES

Percent of Dry Weight

Constituents	Plain settled	Digested	Activated	Raw activated	15%[a] oxidized	80%[a] oxidized
Organic matter	60–80	45–60	62–75	67.0	64.0	23.4
Total ash	20–40	40–55	25–38	33.0	36.0	–
Pentosans	1.00	1.6	2.1	7.2	2.4	0.0
Grease and fat[b]	7–35	3–17	5–12	21.0	19.3	1.8
Hemicellulose	3.20	1.6	–	–	–	–
Cellulose	3.80	0.6	7.0	–	–	–
Lignin	5.80	8.4	–	–	–	–
Protein	22–28	16–21	32–41	12.4	8.4	0.5

[a] Solids fraction only.
[b] Ether extract.

Source: G. Teletzke, Wet air oxidation of sewage sludges, *Process Biochemistry, 1,* 1966. With permission.

3.4-50 THERMAL DEATH POINTS OF ORGANISMS IN HEAT DRIED SLUDGE

Organisms	Conditions	Ref.
Coliform	60°C for 15–20 min	1
Salmonella group	60°C for 15–20 min	1
Shigella group	60°C for 15–20 min	1
Cysts *Endamoeba histolytica*	52°C for 1 min	2
Ova of *Ascaris*	56°C for 1 min	3
Larvae *Taenia saginata*	56° for 5 min	4
Enteric viruses	62°C for 30 min	5
Enteric viruses	71°C for 15 sec	5

REFERENCES

1. G.S. Wilson and C.B.E. Miles, *Principles of Bacteriology & Immunology,* 4th ed., Vol. 1, Williams and Wilkins, Baltimore, 1955.
2. M.F. Jones and W.I. Newton, *Am. J. Trop. Med., 30*:53, 1950.
3. L.O. Nolf, *Am. J. Hyg., 16*:288, 1932.
4. R.W. Allen, *J. Parasitology, 33*:331, 1947.
5. N.A. Clarke and S. L. Chang, *J. Am. Water Works Assn., 51*(10):1299, 1959.

3.4-51 COLIFORM CONTENT OF HEAT DRIED SEWAGE SLUDGE

Samples	Houston	Galveston	Chicago	Baltimore
Number	136	30	12	11
Number of analyses	174	43	18	18
Number positive (showing confirmations)	0	15	2	4
Number of analyses positive	0	18	2	8
Coliforms/g in positive samples, MPN				
Maximum observed	–	17	5	9
Minimum observed	–	2	5	2
Average	<2	7	5	4
Median of all analyses	<2	<2	<2	<2

Source: C.H. Connell *et al., J. Water Pollut. Control Fed., 35*(10):1262, 1963. With permission.

3.4-52   SPECIFIC GRAVITY OF RAW SLUDGE

Source	Ref. 1	Ref. 2[a]	Ref. 3
From sanitary sewerage system	1.02—1.03	1.02—1.03	
From combined sewerage system	1.05—1.07	1.05—1.07	
Septic tank, scum		0.97	
Septic tank sludge		1.02—1.03	
Plain sedimentation			1.02—1.05[b]
Undigested		1.02	
Digested, separate tank		1.04	
Digested, vacuum filtered		1.00	
Chemical precipitation		1.03	
Secondary tanks			1.025
Trickling filters		1.016—1.025	
Roughing filters			1.020
Activated sludge			
Wet sludge		1.005	1.005
Dewatered on vacuum filters		0.95	
Heat dried		1.25	
Extended aeration			1.015
Aerated lagoons, waste sludge			1.010
Filtration			1.005
Algae removal			1.005
Suspended growth denitrification			1.005

[a] Based on sewage flow of 100 gal (378 l) per capita-day and 300 ppm or 0.25 lb (114 g) per capita-day of suspended solids in sewage.

[b] Higher value results from addition of phosphorus removal chemicals.

REFERENCES

1. Metcalf and Eddy, Inc., *Wastewater Engineering*, McGraw-Hill, N.Y., 1972.
2. C.E. Keefer, *Sewage Treatment Works*, McGraw-Hill, N.Y., 1940.
3. Metcalf and Eddy, Inc., *Wastewater Engineering: Treatment, Disposal, Reuse*, 2nd ed., revised by G. Tchobanoglous, McGraw-Hill, N.Y., 1979.

3.4-53   TYPICAL CHEMICAL COMPOSITION
OF RAW AND DIGESTED SLUDGE

Percent dry basis

Item	Raw primary sludge	Digested sludge	Filter cake[a] Raw	Filter cake[a] Digested
Total dry solids	2.0–7.0	6.0–12.0		
Volatile solids	60–80	30–60	55–75	40–60
Grease and fats (ether soluble)	6.0–30.0	5.0–20.0	5–30	2–15
Protein	20–30	15–20	20–25	14–30
Nitrogen	1.5–4.0	1.6–6.0		
Phosphorus	0.8–2.8	1.5–4.0	1.4	0.5–3.5
Potash	0–1.0	0.0–3.0		
Cellulose	8.0–15.0	8.0–15.0	8–10	8–12
Iron (not as sulfide)	2.0–4.0	3.0–8.0		
Silica	15.0–20.0	10.0–20.0		
pH	5.0–8.0	6.5–7.5		
Alkalinity, mg/l	500–1,500	2,500–3,500		
Organic acids, mg/l	200–2,000	100–600		
Thermal content, Btu/lb	6,800–10,000	2,700–6,800		

[a] The median average annual rate of dry sludge produced in 7 cities from 1949–1953 was 32 lb/cap.

REFERENCES

1. Metcalf and Eddy, Inc., *Wastewater Engineering,* McGraw-Hill, New York, 1972.
2. J.M. Morgan and J.F. Thomson, *Water Sewage Works,* June 1, 1955, p. R293.
3. *Manual of Practice No. 2,* Water Pollution Control Federation, 1946.

3.4-54 DESIGN CRITERIA FOR AEROBIC DIGESTERS

Parameter	Values	
	Ref. 1	Ref. 2
Hydraulic detention time, days at 20°C[a]		
Activated sludge only	12—16	10—15
Activated sludge from plant operated without primary settling	16—18	12—18
Primary plus activated or trickling filter sludge	18—22	15—20
Solids loading, lb volatile solids per ft^3 per day (kg volatile solids per m^3 per day)	0.1—0.20 (1.6—3.2)	0.1—0.3 (1.6—4.8)
Oxygen requirements, lb/lb destroyed (kg/kg)		
Cell tissue[b]	≅2	≅2.3
BOD$_5$ in primary sludge	1.7—1.9	1.6—1.9
Energy requirements for mixing		
Mechanical aerators, hp/1,000 ft^3 (kw/10^{3}m^3)	0.5—1.0 (13—26)	0.76—1.52 (20—40)
Air mixing, scfm/1,000 ft^3 (m^3/10^{3}m^{3}min)	20—30 (20—30)	20—40 (20—40)
Dissolved oxygen level in liquid, mg/l	1—2	1—2

[a] Detention times should be increased for temperatures below 20°C. If sludge cannot be withdrawn during certain periods (e.g., weekends, rainy weather), additional storage capacity should be provided.
[b] Ammonia produced during carbonaceous oxidation oxidized to nitrate.

REFERENCES

1. Metcalf and Eddy, Inc., *Wastewater Engineering*, McGraw-Hill, N.Y., 1972.
2. Metcalf and Eddy, Inc., *Wastewater Engineering: Treatment, Disposal, Reuse*, 2nd ed., revised by G. Tchobanoglous, McGraw-Hill, N.Y., 1979.

3.4–55 TIME REQUIRED FOR 90% DIGESTION OF SEWAGE SLUDGE AT DIFFERENT TEMPERATURES

	Mesophilic digestion						Thermophilic digestion			
Temperature (°F)	50	60	70	80	90	100	110	120	130	140
Temperature (°C)	10	16	21	27	32	38	43	49	54	60
Digestion period (days)	75	55	41	28	24	24	26	16	16	18

Adapted from: K. Imhoff and G.M. Fair, *Sewage Treatment*, John Wiley & Sons, N.Y., 1940.

3.4-56 DIGESTION TANK CAPACITY

Type of plant	Tank space, ft^3/cap	
	Heated	Unheated
Imhoff tanks	–	3–4
Primary	2–3	4–6
Primary and low rate trickling filter	3–4	6–8
Primary and high rate trickling filter	4–5	8–10
Activated sludge	4–6	8–12

Source: *Manual of Practice No. 8*, Water Pollution Control Federation, 1959. With permission.

3.4-57   DIGESTION GAS PRODUCTION

Average ft^3/cap day

Imhoff tanks	0.55
Plain sedimentation	0.76
Trickling filters	0.92
Activated sludge	0.91

Source: *Manual for Sewage Plant Operators,* Texas Water and Sewage Works Assn., Austin, 1955. With permission.

3.4-58   ANALYSES OF SEWAGE GAS

Plant	H_2	CO_2	$SO_2 + SO_3$	O_2	N_2	CH_4	C_2H_6	H_2S
San. Dist., Chicago[1][a]		23.4	8.7	0.27	1.43	66.20	0.00	0.00
San. Dist., Chicago[1][a]		24.2	7.8	0.21	1.28	66.51	0.00	0.00
San. Dist., Chicago[1][b]		20.8	14.0	0.08	0.47	65.45	0.00	0.00
Aurora San. Dist.	2.6	31.8		0.4	0.9	64.2		0.01
Springfield, Ill.[2][c]	1.4	32.6		0.0	1.7	64.3		–
Newark, N.J.[2]	3.1	20.1	2.6[d]	0.4	0.7	71.8		
Aurora, Ill.[2][c]	3.5	32.3		0.6	2.3	61.2		
Salinas, Cal.[2]		29.3		2.5[e]	4.4	63.8		
Halle, Germany[2][c]		24.6	0.08–0.6[d]	0.6	1.6	72.9		
Decatur, Ill.[3][f]		16.8		0.1	14.1	69.0		
Calcumet, Chicago[3][f]		14.7		0.5	8.2	76.0		
Stuttgart, Germany[3][f]	4.7	14.0			4.7	75.5		
Milwaukee, Wisc.[3]		30.0		0.2	1.7	67.5		
Birmingham, U.K.[3]		18.1		0.4	3.2	77.0		
Baltimore, Md.[3]		26.5		0.2	2.8	70.5		

[a] Normal operation.
[b] Experimental high loading.
[c] Average of numerous tests.
[d] H_2S.
[e] Air.
[f] Imhoff.

REFERENCES
1. B. Lynam *et al., J. Water Pollut. Control Fed., 39*(4):518, 1967.
2. E.W. Steel, *Water Supply & Sewerage.* 4th ed., McGraw-Hill Book Co., Inc., New York, p. 587.
3. H.E. Babbitt and E.R. Baumann, *Sewage and Sewage Treatment,* 8th ed., John Wiley & Sons, New York, 1958.

3.4-59   SPECTROPHOTOMETRIC ANALYSIS[a] OF
PRIMARY AND DIGESTED SLUDGE BEFORE
AND DURING FAILURE

Element	Concentration in primary sludge slurry, mg/l		Concentration in digested sludge, mg/l	
	Before failure	During faliure	Before failure	During failure
Silver	0.0019	0.017	1.89	1.58
Aluminum	1.93-19.3	1.7-17.0	189-1,890	158-1,580
Boron	None	0.008	None	None
Barium	0.29	0.08	18.9	11.0
Calcium	19.3-193	17.0-170	189-1,890	158-1,580
Chromium	0.772	0.80	151.2	143
Copper	0.135	0.068	37.8	22.8
Iron	1.73	3.12	189-1,890	158-1,580
Magnesium	0.965	0.68	56.7	79.0
Manganese	0.097	0.068	9.45	6.3
Sodium	1.93-19.3	10.2	56.7	95
Nickel	0.019	0.034	5.67	1.58
Lead	None	0.051	9.45	15.8
Silicon	1.93-19.3	1.7-17.0	189-1,890	158-1,580
Tin	None	0.017	1.32	None
Strontium	0.039	0.034	5.67	1.58
Titanium	1.158	1.118	189	158
Vanadium	0.0019	None	0.38	None
Potassium	0.772	0.800	75.6	95

[a] Scanning spectrophotometry results are semiquantitative.

Source: H.R. Zablatzky *et al., J. Water Pollut. Control Fed., 40*(4):583, 1968.
With permission.

3.4-60   SUMMARY OF DIGESTER SUPERNATANT ANALYSIS DATA

Analysis	80% confidence	Median	Mean
Bicarbonate alkalinity, mg/l as $CaCO_3$	850–2,950	1,580	1,681
5 day BOD, mg/l	500–4,260	910	1,401
Total phosphate, mg/l as PO_4	46–380	132	211
mg/l as P	15–124	43	69
Ammonia nitrogen, mg/l	100–710	405	413
pH	6.6–8.0	7.3	7.3

Source: J.H. Masselli *et al.*, New Engl. Water Pollut. Control Comm., Boston, 1965.

3.4-61   COEFFICIENTS OF THERMAL CONDUCTIVITY FOR SEWAGE AND SLUDGE

Material	K
Water	4.0
Sewage	4.0–4.1
Raw sludge	4.5
Partly digested sludge	5.0
Digested sludge	5.1
Ice	13–17

Note: K in $Btu/hr/ft^2/°F/in.$

Source: H.A. Thomas, Jr., *Sewage Ind. Wastes,* 23(1):21, 1951. With permission.

3.4-62   HEAT TRANSFER COEFFICIENTS IN SEWAGE TREATMENT PLANT APPLICATIONS

Application	C $Btu/hr/ft^2/°F$
Hot water to outside fluid	
Water, sewage, thin supernatant	60–80
Thin sludge	30
Thick sludge	8–15
Vessel to ambient air	
Concrete tank in dry soil, top covered and flush with surface	0.10
Concrete tank in dry soil, top covered with ¼ of structure above ground	0.20
Concrete tank in wet soil, top covered with ¼ of structure above ground	0.40–0.75
Concrete tank in wet soil, top open with ½ of structure above ground[a]	1–2.5
Housed trickling filters	0.5–2.0
Open trickling filters[a]	2.0–6.0

[a] Heat loss variable, depending on wind, humidity, and rate of flow.

Source: H.A. Thomas, Jr., *Sewage Ind. Wastes,* 23(1):21, 1951. With permission.

3.4-63   GRAVITY THICKENING OF SLUDGES

Description	Loading, lb/ft² day	Influent suspended solids, %	Thickened suspended solids, %	Ref.
Domestic				
Trickling filter	–	–	8.0	3
Activated sludge	–	–	6.0	3
Activated sludge	–	0.8	3.5	4
Primary sludge	–	1.8	9.0	4
	24.2	0.4	4.5	4
	31.2	0.6	4.6	4
	24.9	0.4	4.9	4
	17.2	0.3	4.1	4
	13.2	0.2	3.8	4
	16.3	0.2	4.2	4
Primary and secondary	3.1	0.2	4.5	4
	7.8	0.6	6.3	4
	17.9	1.2	8.1	4
	37.5	0.7	4.0	4
	28.5	0.5	4.5	4
Trickling filter	8–10	–	7–9	3
Trickling filter and activated sludge	10–12	–	7–9	3
Pulp and paper				
53% primary 47% activated	25.0	2.0	4.0	5
67% primary 33% activated	25.0	2.0	6.0	5
100% activated	25.0	2.0	9.0	5
Brewery, domestic	–	2.0	8.0	1
Carbon processing	18–76	0.2–0.5	5.0	2
Lime softening	–	4.5	18.7	4

Source: D.L. Ford in *Water Quality Improvement by Physical and Chemical Processes*, E.F. Gloyna and W.W. Eckenfelder, Jr., Eds., University of Texas Press, Austin 1970, p. 197. With permission.

REFERENCES

1. D.L. Ford and W.W. Eckenfelder, unpublished data.
2. W.W. Eckenfelder and J.F. Malina, unpublished data.
3. H. Ludwig, unpublished data.
4. H. Edde and W.W. Eckenfelder, *J. Water Pollut. Control Fed.*, *40*(8):1, 1968.
5. W.W. Eckenfelder, unpublished report, 1965.

3.4-64 COEFFICIENTS FOR GRAVITY THICKENING OF SLUDGES

Sludge type	Average n
Lime-softening sludge	1.80
Pulp and paper sludge primary, secondary, and fly ash	0.70
Primary and trickling filter	0.80
Contact stabilization sludge	0.75
Primary and activated sludge	0.25
Ferric hydroxide	0.53

Note: A relationship has been suggested[1] for the design of a gravity thickener:

$$\frac{Cu}{Co} - 1 = \frac{B}{(ML)^n}$$

where

Cu = solids concentration of underflow in percent solids,

Co = solids concentration of the influent in percent solids,

ML = mass loading in lb solids/ft^2 day,

B, n = coefficients dependent on sludge characteristics.

Source: J. Barnard and W.W. Eckenfelder, Jr., in *Water Quality Improvement by Physical and Chemical Processes*, E.F. Gloyna and W.W. Eckenfelder, Jr., Eds., University of Texas Press, Austin, 1970, p. 414. With permission.

REFERENCE

1. H. Edde and W.W. Eckenfelder, *J. Water Pollut. Control Fed.*, *40*(8):1, 1968. With permission.

3.4-65 SLUDGE SOLIDS PRODUCED BY FLOTATION THICKENER

	Solids, %
Combined primary and activated sludge	6.1
Combined primary and activated sludge	7.4
Activated sludge only	4.9
Activated sludge only[a]	3.7
Primary + activated sludge 1:1	
Cannery waste in season	5.3
Cannery waste out of season	7.1

[a] Same as above, but higher volatile content.

Source: J. Barnard and W.W. Eckenfelder, Jr. in *Water Quality Improvement by Physical and Chemical Processes*, E.F. Gloyna and W.W. Eckenfelder, Jr., Eds., University of Texas Press, Austin, 1970, p. 411. With permission.

3.4-66   SLUDGE FILTRATION DATA

Type of sludge	Capacity, lb/hr ft²	Moisture in cake
Primary	3– 8	60–68
Chemically precipitated	3– 8	66–73
Digested	4–10	55–75
Digested, elutriated	5–10[a]	60–70
Activated	2/3– 2	80–84

Note: Mixtures give values equal to the weighted mean value of the individual component parts.

[a] Plant scale operating data.

Source: *Manual for Sewage Plant Operators,* Texas Water and Sewage Works Assn., Austin, Texas, 1955. With permission.

3.4–67   SLUDGE DRYING REQUIREMENTS FOR DEWATERING DIGESTED SLUDGE[a]

Type of sludge	Bed load, kg dry solids per m² per yr		Area, m²/1000 persons		
	Ref. 1	Ref. 2	Ref. 1	Ref. 2	Ref. 3
Primary	120	120—200	90	90—140	90
Trickling filter, standard rate[b]	100	100—160	140	110—160	120
Trickling filter, high rate[b]	—	—	—	—	140
Activated sludge	85	60—100	190	160—275	160
Chemical precipitation[b]	110	100—160	140	185—230	190
Intermittent sand filters	—	—	—	—	90

[a] Covered bed areas vary from 70 to 75% of those for open-bed areas.
[b] Includes primary sludge.

REFERENCES

1. E.W. Steel, *Water Supply and Sewage,* 4th ed., McGraw-Hill, N.Y., 1960.
2. Metcalf and Eddy, Inc., *Wastewater Engineering: Treatment, Disposal, Reuse,* 2nd ed., revised by G. Tchobanoglous, McGraw-Hill, N.Y., 1979.
3. H.E. Babbitt and E.R. Baumann, *Sewerage and Sewage Treatment,* 8th ed., John Wiley & Sons, N.Y., 1958.

3.4–68   RESULTS OF VACUUM FILTRATION OF SLUDGE[a]

Constituent	Raw		Digested	
	Ref. 1	Ref. 2, 3	Ref. 1	Ref. 2, 3
Volatile matter	63—72	55—75	43—70	40—60
Ash	28—37	25—45	30—57	40—60
Insoluble ash	—	15—30	—	30—45
Grease and fats	—	15—30	12.2	2—15
Protein	—	20—25	—	14—30
Ammonia nitrate	5.2—5.8	1—3	1.6—2.3	1.3—1.6
Phosphorus (P_2O_5)	2.4—3.6	1.4	1.0—2.5	0.5—3.5
Cellulose, etc.	—	8—10	—	8—12

[a]   Filter cake, dry basis (%).

REFERENCES

1. S.I. Zack, *Sewage Ind. Wastes,* 22:975, 1950
2. H.E. Babbitt and E.R. Baumann, *Sewerage and Sewage Treatment,* 8th ed., John Wiley & Sons, N.Y., 1958.
3. J.M. Morgan and J.F. Thomson, *Water Sewage Works,* June 1, 1955, p. R-293.

3.4-69   VACUUM FILTRATION UNIT PROCESS PERFORMANCE

Description	Influent suspended solids, %	Filter aid		Cake solids, %	Yield, lb/ft² hr	Ref.
		FeCl₃ (%)	CaO(%)			
Pulp and paper	6.0	–	–	20.0	10.7	1
	5.0	–	–	17.8	16.0	1
Carbon processing	2.8	–	–	13.8–17.3	1.3–4.3	2
	1.4	–	–	18.7	0.9	2
	4.9	–	–	17.6	3.5	2
	1.6	–	–	10–19.2	0.8–1.8	2
Domestic:						
Raw primary	–	2.7	8.2	18.2	6.3	3
	–	2.0	12.0	19.7	8.0	3
	–	3.5	8.7	15.3	6.3	3
	0.2–1.2	–	–	28–37	6–20	3
Activated	–	1.7	10.3	25.0	3.7	3
Trickling filter	–	1.8	4.7	33.1	10.4	3
Trickling filter	–	3.5	8.7	20.2	9.9	3
Trickling filter	–	2.2	6.7	25.2	8.8	3
Digested primary	0.2–1.5	–	–	26–34	4–15	4
Digested primary and activated sludge	0.5–2.0	–	–	24–32	4–8	4
	–	2.2	7.7	23.9	5.1	5
	–	11.6	0.0	23.3	1.4	5
	–	4.4	16.4	24.8	2.8	5
	–	3.2	10.3	23.7	5.5	5

Source: D.L. Ford in *Water Quality Improvement by Physical and Chemical Processes,* E.F. Gloyna and W.W. Eckenfelder, Jr., Eds., University of Texas Press, Austin, 1970, p. 354. With permission.

REFERENCES

1. W.W. Eckenfelder, unpublished data.
2. W.W. Eckenfelder and J.F. Malina, unpublished data.
3. E. Trubnick, *Water Sewage Works, 106:*10, 1959.
4. R.A. Burd, *A Study of Sludge Handling and Disposal,* U.S. Department of the Interior, Federal Water Pollution Control Administration, 1968.
5. G. Crawford and N. McDonald, *J. Water Pollut. Control Fed., 38*(2):271, 1966.

3.4-70   COEFFICIENT OF COMPRESSIBILITY
s OF VARIOUS SLUDGES

Type of sludge	Range in value of *s*
Digested	0.70–0.86
Activated	0.60–0.79
Raw	0.87[a]
Humus	0.80[b]

[a] One sample.
[b] Two samples.

Source: P. Crackley *et al., Sewage Ind. Wastes, 28*(8):963; 1956. With permission.

3.4-71   CENTRIFUGE PERFORMANCE ON SEWAGE SLUDGE

Bowl speed, rpm	Pool level	Feed rate, gal/min	Solids Feed, mg/l	Cake, %	Overflow, mg/l	Removal,[a] %
5,650	Deep	3	1,076	12.1	249	77
5,650	Deep	3	1,076	12.5	240	78
5,650	Deep	4	1,076	11.1	253	77
4,850[b]	Deep	2	1,252	11.5	265	79
4,850	Deep	3	1,153	10.4	255	78
4,850	Deep	4	1,153	9.7	294	75
4,850	Medium	1½	1,780	18.2	459	48

[a]   Percent recovery: $\dfrac{C\,(F - OF)}{F\,(C - OF)}$ x 100. C is the cake, F, the feed, and OF, the overflow.

[b]   Z-1 unit @ 4,850 rpm develops 3,000 G's; Z-3 unit @ 3,500 rpm develops 2,870 G's. Scale-up factor of the Z-1 to Z-3 is 5. (Results at 5,650 rpm cannot be duplicated with the Z-3 unit).

Source: C.G. Golueke and W.J. Oswald, *J. Water Pollut. Control Fed., 37*(4):471, 1965. With permission.

3.4-72   TEMPERATURES REQUIRED FOR DESTRUCTION OF ODORS FROM VOLATILE FATTY ACIDS

Acid	Temperature, °F
Acetic	1,370
Propionic	1,370
Butyric	1,425
Caproic	1,425

Source: C.N. Sawyer *et al., J. Water Pollut. Control Fed., 32*(12):1274, 1960. With permission.

3.4-73   BASIC CLASSES OF ORGANIC MATTER
IN RAW AND OXIDIZED SEWAGE SLUDGE

Item	Raw, g/l	Value for sludge oxidized at given temperature, g/l				
		150°C	175°C	200°C	225°C	250°C
COD	57.7	57.9	51.9	34.6	16.5	12.9
Protein[a]	6.2	6.1	3.9	2.5	1.2	Trace
Lipids[b]	10.5	10.7	8.9	3.6	0.6	0.5
Total sugars	0.0	0.6	0.9	0.0	0.0	0.0
Starch	3.6	3.0	1.2	0.6	0.1	0.0
Crude fiber	13.5	13.5	8.9	5.8	0.9	0.0
Volatile acid as acetic	1.3	3.0	4.9	6.5	8.1	7.7
Sum of basic classes	35.1	36.9	28.7	19.0	10.9	8.2
Observed volatile matter	35.5	42.0	31.4	19.0	5.1	3.1

Note: Oxidation by wet-air process.

[a]   Protein taken as nonammoniacal nitrogen x 6.25.
[b]   Lipids – petroleum ether extract of $MgSO_4 \cdot H_2O$ dried sample.

Source: G.H. Teletzke *et al., J. Water Pollut. Control Fed., 39*(6):994, 1967. With permission.

3.4.6 LAND APPLICATION OF EFFLUENT AND SLUDGE

3.4-74 TYPICAL WASTEWATER LAND APPLICATION RATES

Application method	Type of wastewater	Units	Application rate		Remarks
			Range	Avg	
Spreading basin	Clear water of good quality[a]	ft/day	1.6–10	4.2	Coarse soil
			1.7–3.6	2.2	Medium soil
			<1	0.16	Fine soil
	Primary effluent		0.05–0.50	0.15	Continuous inundation
	Treated effluent		0.2–6.0	1.5	Intermittent operation
Spray irrigation	Cannery wastes	gal/ft² day	0.2–2.0[b]	0.5	Applied at 0.2–0.6 in./hr
	Treated effluent		0.09–1.0[b]	0.2	
Leaching fields	Septic tank effluent	gal/ft² day	0.2–1.5	0.5	Continuous inundation
				1.5	Intermittent operation

[a] California data.
[b] Rate will vary with type of soil and type and degree of ground cover.

Source: © 1969 by the Dun-Donnelley Corp. Reprinted by special permission from the February 1969 issue of *Water and Wastes Engineering.*

3.4-75 COMPARISON OF DESIGN FEATURES FOR LAND TREATMENT PROCESSES

	Principal processes			Other processes	
Feature	Slow rate	Rapid infiltration	Overland flow	Wetlands	Subsurface
Application techniques	Sprinkler or surface[a]	Usually surface[a]	Sprinkler or surface	Sprinkler or surface	Subsurface piping
Annual application rate (m)	0.6—60	60—170	3—20	1—30	2—27
Field area required (ha)[b]	23—230	0.8—23	6—45	4—110	5—57
Typical weekly application rate (cm)	1—10	10—300	6—15[c] 15—40[d]	2—64	5—50
Minimum preapplication treatment provided in U.S.	Primary sedimentation[e]	Primary sedimentation	Screening and grit removal	Primary sedimentation	Primary sedimentation
Disposition of applied wastewater	Evapotranspiration and percolation	Mainly percolation	Surface runoff and evapotranspiration with some percolation	Evapotranspiration, percolation, and runoff	Percolation with some evapotranspiration
Need for vegetation	Required	Optional	Required	Required	Optional

[a] Includes ridge-and-furrow and border strip.
[b] Field area in hectares, not including buffer area, roads, or ditches for 3.8×10^3 m^3 day^{-1} flow.
[c] Range for application of screened wastewater.
[d] Range for application of lagoon and secondary effluent.
[e] Depends on the use of the effluent and the type of crop.

Source: S.R. Hutchins, M.B. Tomson, P.B. Bedient, and C.H. Ward, *CRC Crit. Rev. Environ. Control, 15*:355, 1985.

REFERENCE

1. H.J. Ongerth, D.P. Spath, J. Crook, and A.E. Greenberg, *J. Am. Water Works Assn., 65*:495, 1973.

3.4-76 FERTILIZING INGREDIENTS IN SLUDGE, VARIOUS MANURES, AND ORGANIC NITROGENOUS MATERIALS

Material	Fertilizing ingredients, %				
	Nitrogen N	Phosphoric acid	Potash	Organic matter	Moisture
Digested settled sludge	0.8–3.5	0.7–4.0[a]	0.26[e]	–	6.4,[e] 70[f]
Digested settled sludge with trickling filter sludge	1.0–4.5	tr.–4.0[a]	0.8–1.6[a]	30–60[b,c]	
Digested activated sludge	2.0–4.8	1.0–3.6[a]	0.13[e]	–	
Heat dried activated sludge	4.0–7.0	1.7–2.5	0.13[f]	–	10
Commercial pulverized					
Sheep manure	1.2–2.5	1.0–2.0	2.0–4.0	48	64
Cattle manure	1.6–2.1	1.0	1.0–2.2	66	79
Poultry manure	1.9–4.0	2.5–3.7	0.8–1.3	64	68
Animal tankage	5–10	7–16	–	–	
Blood	9–13	0.5–14	–	–	
Fish scrap	6.5–10	5–10	–	–	
Cottonseed meal	5–8	2–3	1–2	–	
Castor pomace	5–6	2	1	–	
Vacuum filtered sludge[d]	3.2	2.26	0.28	–	

[a] Dry basis.
[b] Estimated from other tables.
[c] Volatile matter.
[d] Ash, 51.3% and volatile, 48.7% on dry basis. Grease, 12.2%.
[e] Dried.
[f] From drying beds.

REFERENCES

1. R.E. Leaver, *Sewage Ind. Wastes,* 28(3):323, 1956.
2. H.E. Babbitt and E.R. Baumann, *Sewerage and Sewage Treatment,* John Wiley & Sons, New York, 1958.
3. *Manual of Practice No. 2,* Fed. Sewage Works Assn., Champaign, Ill., 1946.
4. D.R. Bloodgood, *Water Sewage Works,* January 1955, p. 44.

3.5   RECYCLING AND REUSE

3.5.1   *INDUSTRIAL*

3.5-1   WATER REUSE IN THE U.S.

Type	1971 volume		No. of plants
	m^3	10^9 gal	
Irrigation and agriculture	3×10^8	77	338
Industrial	2×10^8	54	14
Recreational	11×10^6	3	5
Nonpotable domestic	$<4 \times 10^6$	<1	1
Groundwater augmentation	2.6×10^7	0.07[a]	9

[a] 1975.

Source: F.M. Middleton, News of Environmental Research in Cincinnati, U.S. Environmental Protection Agency, p. 2, July 1977.

3.5-2   RECYCLING OF INDUSTRIAL WATER

Source	Recycle ratios[a]
Auto	2.62
Beet sugar	1.48
Coal preparation	14.91
Corn and wheat milling	1.22
Distilling	1.51
Food processing	1.19
Machinery	5.5
Meat	4.03
Petroleum	7.62
Pulp and paper	3.02
Soaps and detergents	3.08
Steel	1.60
Tanning – leather	1.04
Textiles	1.3
Natural gas transmission	2.32

[a] Recycle ratios
$$= \frac{\text{freshwater intake required without recycle}}{\text{actual freshwater intake using recycle}}$$
or:
= average number of times a gallon of water was recycled.

Source: *Water in Industry*, National Association of Manufacturers and the Chamber of Commerce of the United States, January 1965.

3.5-3 WATER REUSE IN MANUFACTURING

Selected Regions, 1964

Industry description	Total U.S.	Chesa-peake Bay Region	Ohio River Region	Eastern Great Lakes Region	Southeast Region	Arkansas White Red Region	Western Gulf Region	California Region
All manufacturers	2.2	2.1	1.6	1.6	2.7	8.6	2.9	3.7
Paper and allied products	2.9	6.0	3.3	1.9	2.8	5.8	–	4.5
Chemicals and allied products	2.0	1.3	1.7	1.3	2.2	11.7	1.9	3.6
Petroleum and coal products	4.4	–	7.4	3.3	12.6	19.6	6.5	4.1
Primary metals	1.5	1.3	1.3	1.4	2.8	–	2.2	11.1

Note: Recirculation rate = gross water used/water intake.

Source: Reprinted, with publisher's permission, from *J. Am. Water Works Assn.*, 60(10):1129, © 1968, by the American Water Works Association, Inc.

3.5.2   AGRICULTURAL

3.5-4   METHODS OF MANURE DISPOSAL

I. Crop utilization – sufficient land available.
  A. Storage up to 6 months in colder climates.
    1. Open pond.[1]
      a. Mostly liquid.
    2. Underground tank, mixed.[1]
      a. For liquid or manure or both.
    3. Manure pile.
      a. For manure with bedding. Few problems.
  B. Compost.
    1. Feasible for very large operators with proximity to urban society.
  C. Final disposal.
    1. Fresh manure spread on land.
      a. Highest fertilizing value.
    2. Spread on land after storage.
      a. Good fertilizing value.
    3. Run-off.
      a. Snow melt and storm water; wash surface or ground.
    4. Evaporate and infiltrate.
      a. From storage area.
II. New ways (modern factories) – possible extensive treatment or destruction.
  A. Preparation.
    1. Lagoons.*
      a. Anaerobic, either indoor or outdoor.
        1. Odor problems when stirred or cleaned.
      b. Anaerobic-aerobic.
        1. Odor problems. Dilution water required.
      c. Anaerobic-aerated.
        1. Odor problems. Less dilution water required than for (b).
    2. Oxidation ditch.*
      a. Dilution water required.

    3. Septic tank.
      a. Needs frequent cleaning.
    4. Imhoff tank.
      a. Needs frequent cleaning.
    5. Dry.
      a. Natural.
        1. Needs proper climate conditions.
      b. Artificial.
        1. Expensive, final disposal not solved.
    6. Settling and trickling filters.**
    7. Activated sludge.**
    8. Compost.**
    9. Package plants.**
    10. Digestion.**
      a. Anaerobic.
      b. Aerobic.
  B. Final Disposal.
    1. Incineration.
    2. To surface watercourse (for liquid only).
      a. Not acceptable.
    3. Evaporation and seepage into ground (for liquid only).
      a. Satisfactory, if proper soil conditions are available.
    4. Tile beds (for liquid only).
      a. Satisfactory if the soil is permeable.
      b. Need large area.
    5. Town sewers.
      a. Has worked satisfactorily.
      b. Some modifications necessary at sewage treatment plant.
    6. Pelletizing and selling (solids, only).
      a. Cannot compete with commercial fertilizers.

*Sensitive to temperature, antibiotics. Final disposal not solved.
**Expensive. Very sensitive biological processes. Affected by antibiotics, loads, temperature, nutrients. Final disposal not solved.

Source: R. Laak, *Water Sewage Works, 117*(4):134, 1970. Courtesy of Scranton Gillette Communications, Inc.

REFERENCE
1. Liquid manure, Ontario Dept. of Agric. and Food, Ontario, Canada, December, 1967.

3.5-5   POSSIBLE SCHEMES FOR ANIMAL WASTE TREATMENT

Operation	Purpose	Remarks
Solids disposal or treatment		
Anaerobic digestion	Reduction of solids	Must be followed by satisfactory ultimate disposal
Aerobic digestion	Reduction of solids	Must be followed by satisfactory ultimate disposal
Incineration	Solids disposal	Operated to produce only inorganic residual (15–20%) of original mass
Primary sedimentation	Removal of settleable solids	Cattle waste must be diluted at least ten-fold for good sedimentation
Biological conversion Anaerobic	Conversion of non-settleable solids to biological solids CO_2, H_2O, CH_4, NH_3 and other organics	Operation must usually be followed by additional biological processes and sometimes degasification steps
Biological conversion Aerobic	Conversion of non-settleable solids to biological solids CO_2, NH_3, NO_2 and NO_3	Operation followed by sedimentation facilities for removal of biological solids
Lime addition	PO_4 removal and increased BOD removal	
In line anaerobic	Denitrification	Followed by sedimentation

Source: R.W. Okey *et al.*, "Relative Economics of Animal Waste Disposal by Selected Wet and Dry Techniques," *Animal Waste Management*, Cornell University Conf. Agric. Waste Manage., January 13–15, 1969, p. 370. With permission.

3.5-6 POSSIBLE TREATMENT UNITS FOR LIVESTOCK WASTES

	Dairy cattle, 100 head	Beef cattle		Hogs, 1,000 head	Chickens	
		1,000 head	10,000 head		10^4 birds	10^5 birds
Anaerobic lagoon						
Loading – 200 lb VS[a]/1000 ft³ day						
Volume, ft³	4,000	40,000	400,000	3,600	2,400	24,000
Surface area, acres	0.009	0.09	0.9	0.008	0.005	0.055
Detention time, days	30	–	–	5	18	18
Loading – 50 lb VS[a]/1,000 ft³ day						
Volume, ft³	16,000	160,000	1,600,000	14,400	9,600	96,000
Surface area, acres	0.037	0.37	0.37	0.033	0.022	0.22
Detention time, days	120	–	–	22	72	72
Combined anaerobic-aerobic system						
Anaerobic unit – 50 lb VS[a]/1000 ft³ day						
Volume, ft³	16,000	160,000	1,600,000	14,400	9,600	96,000
Surface area, acres	0.037	0.37	3.7	0.033	0.022	0.22
Oxidation pond						
Surface area, acres	1.6	16.5	165	5	2.5	25
Oxidation ditch						
Surface area, acres	0.014	0.14	1.4	0.021	0.029	0.29
Land disposal						
Thin spreading						
Area, acres	0.46	4.6	46	0.46	0.23	2.3
Sub-soil injection						
100 ton wet solids/acre						
Area, acres	0.035	0.3	3	0.02	0.013	0.13
10 ton wet solids/acre						
Area, acres	0.35	3	30	0.2	0.13	1.3
Spraying, .05 in./acre						
Area, acres	7.3	–	–	3.7	7.3	73
High rate anaerobic digestion, 20 lb VS[a]/ft³ day						
Volume, ft³	4,000	40,000	400,000	3,600	2,400	2,400
Annual cost per year per animal	$1,820	$18,200	$182,000	$1,650	$1,100	$11,000
Wet oxidation						
High pressure						
Construction cost/unit	$25,000	$250,000	$1,500,000	$22,500	$15,000	$150,000
Total cost/day	$5	$50	$500	$4.5	$3	$30
Low pressure						
Total cost/day	$1.50	$15	$150	$1.35	$0.90	$9
Incineration and composting[b]						
Construction costs	$1,500	$15,000	$150,000	$1,350	$900	$9,000
Total costs/day	$2.50	$25	$250	$2.25	$1.50	$15
Cost per animal capacity/yr	$9.20	$9.20	$9.20	$0.83	$0.55	$0.55

[a] VS – volatile solids.
[b] The same figures apply for both incineration and composting.

Source: *Pollution Implications of Animal Wastes — A Forward Oriented Review,* U.S. Department of the Interior, 1968, p. 127.

3.5-7   PERFORMANCE OF ANAEROBIC LAGOONS

Influent quality, mg/l BOD	Effluent quality, mg/l BOD	BOD reduction, %	Ref.
530	160	70	1
360	85	77	2
520	100	81	3
1,100	160	85	4
15,000	1,500	90	5
1,380	130	90	6

Source: *Pollution Implications of Animal Wastes – A Forward Oriented Review,* U.S. Department of the Interior, 1968, p. 86.

REFERENCES
1. C.D. Parker, H.L. Jones, and W.S. Taylor, "Purification of Sewage in Lagoons," *Sew. Ind. Wastes, 22:*760, 1950.
2. C.D. Parker, H.L. Jones, and N.C. Greene, "Performance of Large Sewage Lagoons at Melbourne, Australia," *Sew. Ind. Wastes, 31:*133, 1959.
3. C.D. Parker, "Sewage Treatment by Lagoons," *J. Sanit. Eng. Div.,* Am. Soc. Civ. Eng., *84,* 1958.
4. F.W. Sollo, "Pond Treatment of Meat Packing Plant Wastes," *Proc. 15th Purdue Ind. Wastes Conf.,* 1961, p. 386.
5. J.E. Etzel, "Industry's Idea Clinic," *J. Water Pollut. Control Fed., 36:*943, 1964.
6. A.J. Steffen and M. Bedker, "Operation of Full-Scale Anaerobic Contact Treatment Plant for Meat Packing Waste," *Proc. 16th Purdue Ind. Wastes Conf.,* 1962, p. 423.

3.6 BIOASSAY AND TOXICITY

3.6.1 AQUATIC INSECTS

3.6-1 METAL TOXICITY—BIOASSAY DATA[a]

Metal	Insect	96 hr TL_m, mg/l	50% survival days	50% survival mg/l	Physical and chemical data of the 16 mg/l test water after two weeks C°	DO mg/l	pH	Alkalinity, mg/l	Acidity, mg/l	Hardness, mg/l	N, mg/l	Metal, mg/l	Mg metal absorbed by insects
Cu^{++} from $CuSO_4 \cdot 5H_2O$	Acroneuria	8.3			18.5	8.0	6.8	54	20	40	4.9	12.3	0.06
	Ephemerella	0.32 (48 hr)	14	32.0	18.5	8.0	6.9	42	6	40	1.5	10.4	0.14
	Hydropsyche		14	32.0	18.5	8.0	6.8	40	4	46	0.9	14.5	0.22
Zn^{++} from $ZnSO_4 \cdot 7H_2O$	Acroneuria		10	16.0	18.5	8.0	7.6	46	12	50	6.5	5.5	0.01
	Ephemerella		11	32.0	18.5	8.0	7.6	30	8	54	1.7	7.9	0.12
	Hydropsyche		14	32.0	18.5	8.0	7.6	30	8	52	1.1	4.7	0.18
Cd^{++} from $CdSO_4 \cdot 8H_2O$	Acroneuria	2.0			18.5	8.0	7.3	60	12	52	7.2	14.1	0.42
	Ephemerella		10	32.0	18.5	8.0	7.0	56	8	54	1.6	15.2	0.05
	Hydropsyche		>14	64.0	18.5	8.0	7.0	54	6	56	1.5	14.0	0.13
Pb^{++} from $PbSO_4$	Acroneuria		7	16.0	18.5	8.0	7.3	60	12	54	7.6	2.2	0.05
	Ephemerella		7	32.0	18.5	8.0	7.0	56	6	52	3.1	3.0	0.12
	Hydropsyche		9	16.0	18.5	8.0	7.1	42	6	54	3.5	3.2	0.08
Fe^{++} from $FeSO_4$	Acroneuria	0.32			18.5	8.0	7.7	72	14	48	14.1	1.2	0.02
	Ephemerella				18.5	8.0	8.2	54	6	48	0.95	1.4	0.01
	Hydropsyche		7	16.0	18.5	8.0	8.1	46	6	50	1.28	1.6	0.01
Ni^{++} from $NiSO_4 \cdot 6H_2O$	Acroneuria	33.5			18.5	8.0	7.0	54	10	40	6.0	6.4	0.04
	Ephemerella	4.0	>14	64.0	18.5	8.0	7.0	46	6	42	1.5	7.2	0.03
	Hydropsyche				18.5	8.0	7.0	42	6	48	1.1	13.0	0.02
Co^{++} from $CoSO_4 \cdot 7H_2O$	Acroneuria	16.0	8	32.0	18.5	8.0	7.2	66	12	50	12.3	9.5	0.01
	Ephemerella				18.5	8.0	6.9	46	6	50	1.2	10.6	0.03
	Hydropsyche				18.5	8.0	7.0	46	6	46	1.3	14.7	0.01
Cr^{+++} from $CrCl_3 \cdot 6H_2O$	Acroneuria	2.0	7	32.0	18.5	8.0	6.8	50	8	50	9.2	4.7	0.03
	Ephemerella	64.0			18.5	8.0	6.0	46	14	50	4.7	7.7	0.01
	Hydropsyche		7	32.0	18.5	8.0	6.4	46	14	42	2.2	5.2	0.01
Hg^{++} from $HgCl_2$	Acroneuria	2.0			18.5	8.0	7.8	52	8	46	6.5	—	—
	Ephemerella	2.0			18.5	8.0	7.6	46	6	42	1.6	—	—
	Hydropsyche	2.0			18.5	8.0	7.6	46	6	42	1.4	—	—

a DO concentration of 8.0 mg/l maintained.

Source: S.L. Warnick and H.L. **Bell**, *J. Water Pollut. Control Fed.*, *41*(2):280, 1969. With permission.

3.6.2   TOXICITY

3.6-2   CHEMICAL TOXICITY TO FISH AND FISH FOOD

Substance	Concentration		Test organism	Effect	Ref.
	ppm	As			
Chromic acid	0.3	Cr	*Daphnia magna*[a]	Toxic	1
Hydrochloric acid	62	HCl	*Daphnia magna*	Toxic	1
Nitric acid	107	HNO_3	*Daphnia magna*	Toxic	1
Sulfuric acid	88	H_2SO_4	*Daphnia magna*	Toxic	1
Strong acids	to pH 5.0	–	Fish	Toxic	2
Cadmium chloride	0.01	Cd	Goldfish	Kills in 8–18 hr	3
Cadmium sulfate	513	Cd	Minnows	Kills in 3 hr	4
Copper sulfate	0.8	Cu	Goldfish	Kills in 24–96 hr	5
Copper sulfate	0.04	Cu	*Daphnia magna*	Toxic	1
Sodium chromate	0.1	Cr	*Daphnia magna*	Toxic	6
Potassium dichromate	36	Cr	Goldfish	No effect in 108 hr	5
Potassium dichromate	180	Cr	Goldfish	Kills in 3 days	5
Chromate, ion	20	Cr	Trout and minnows	Kills in 8 days	7
Chromate, ion	50	Cr	Sunfish, bluegills	Not toxic in month	8
Chromate, ion	0.01	Cr	Microflora	Toxic	9
Ferric chloride	34	Fe	Goldfish	Kills in 1–1.5 hr	5
Ferrous sulfate	37	Fe	Goldfish	No effect in 100 hr	5
Ferrous sulfate	368	Fe	Goldfish	Kills in 2–10 hr	5
Nickel chloride	4.5	Ni	Goldfish	Kills in 200 hr	5
Lead nitrate	63	Pb	Goldfish	Kills in 80 hr	5
Stanuous chloride	.626	Sn	Goldfish	Kills in 4–5 hr	5
Zinc sulfate	25	Zn	Trout	Kills in 133 min	10
Zinc, ion	0.3	Zn	Fish	Kills some fresh water fishes	11
Sodium cyanide	0.3	CN	Minnows, catfish, carp	No effect in 24 hr	12
Potassium cyanide	0.04–0.12	CN	Goldfish	Kills in 3–4 days	5
Cyanogen chloride	0.08	CNCl	Fish	Critical	13
Potassium ferrocyanide	948	CN	Minnows, goldfish	Not lethal	14
Potassium ferricyanide	848	CN	Minnows, goldfish	Not lethal	14
Ammonia	2.5	NH_3	Goldfish	Kills in 1–4 days	5
Ammonia	2–7	NH_3	Fish	Lethal	2
Hydrogen sulfide	10	H_2S	Goldfish	Kills in 96 hr	5
Sulfide, ion	3	S	Trout	Kills in 5 min	15
Sulfide, ion	0.5–1.0	S	Fish	Critical	2
Potassium cyanate	264	KCNO	Trout fingerlings, adult minnows	No effect in 24 hr	12
Sodium hydroxide	156	NaOH	*Daphnia magna*	Toxic	6
Trisodium phosphate	52	Na_3PO_4	*Daphnia magna*	Toxic	6
Chlorine	0.05–1.0	Cl	Fish	Critical	2

[a] *Daphnia magna* is a representative fish food organism commonly found in streams.

3.6-2   CHEMICAL TOXICITY TO FISH AND FISH FOOD (continued)

REFERENCES

1. R.H. Beaton and C.C. Furnas, *Ind. Eng. Chem., 33*:1500, 1941.
2. B.F. Dodge and D.C. Reams, American Electroplaters' Society Research Report, Serial No. 14, American Electroplaters' Society, Jenkintown, Pa., 1949, p. 29.
3. D.L. Belding, *Trans. Am. Fish. Soc., 57*:100, 1927.
4. M.K. Bezzubets and V.N. Vozhdayeva, *J. Gen. Chem. Ind. (USSR), 18*:17, 1941; *Chem. Zentr.,* 1942, II, 2833; *C.A., 38*:2777, 1944.
5. G.E. Barnes, *Water Sewage Works, 94*:8, 1947; *Proc. 3rd Ind. Waste Conf.,* Purdue University, Lafayette, Ind., 1947, p. 179.
6. D.E. Bloodgood and F.J. Losson, Jr., *Proc. 3rd Ind. Waste Conf.,* Purdue University, Lafayette, Ind., 1947, p. 196.
7. D.E. Bloodgood and A. Strickland, *Water Sewage Works, 97*:28, 1950.
8. B.F. Dodge and D.C. Reams, American Electroplaters' Society Research Report, Serial No. 14, American Electroplaters' Society, Jenkintown, Pa., 1949, p. 17.
9. B.F. Dodge and D.C. Reams, American Electroplaters' Society Research Report, Serial No. 14, American Electroplaters' Society, Jenkintown, Pa., 1949.
10. K.E. Carpenter, *Brit. J. Exp. Biol., 4*:378, 1927.
11. G.H. Clevenger and H. Morgan, *Mining Sci. Press, 113*:413, 1916; *C.A., 10*:2861, 1916.
12. F.L. Coventry, *et al., Ecology, 16:*60, 1935.
13. Doudoroff and M. Katz, *Sewage Ind. Wastes, 22:* 1432, 1950.
14. J. Denis, British Patent 250,824 (1924). *C.A., 21*:1321, 1927. C.R. Hoover, Biennial Report State Water Comm. Connecticut, *5*:45, 1934. M. Strell, *Gesundh.-Ing., 62*:546, 1939. *C.A., 34*:560, 1940. G.E. Barnes and M.M. Braidech, *Eng. News Record, 129*:496, 1942.
15. J.G. Dobson, *Sewage Works J., 19*:1007, 1947.

3.6–3   TOXIC EFFECTS OF CYANIDE

Subject	Toxic conc (mg/l)	Remarks
Acute toxicity	0.05[2]	
Chronic toxicity	0.1[2]	
Brown trout	0.0057—0.0112[4]	HCN based on spawning data
Sunfish	0.18[1]	Concentration killing 50% of fish in 24 hr
Daphnia	1.8[1]	50% of organisms immobilized in 48 hr
BOD	0.04[1]	95% recovery
	0.4[1]	60% recovery
Sludge digestion	25[1]	No adverse effect in 24 days; 3 ppm initial retarding effect for 6 days
	50[1]	10% reduction in gas production
Effluent discharge limit	0.025[3]	Function of type of cyanide discharged

REFERENCES

1. W.W. Eckenfelder, *Industrial Water Pollution Control,* McGraw-Hill, N.Y., 1966.
2. L.A. Klapow and R.H. Lewis, *J. Water Pollut. Control Fed., 51:*2054, 1979.
3. D.T. Lordi *et al., J. Water Pollut. Control Fed., 52:*597, 1980
4. W.M. Koenst *et al., Env. Sci. Tech., 11:*883, 1977

3.7   NUTRIENTS, LAKES, EUTROPHICATION

Data in this section are concerned with monitoring aspects of algae growth in lakes and the loading of nutrients.

3.7.1   INTRODUCTION

3.7-1   METHODS OF QUANTITATIVE ASSESSMENT OF EUTROPHICATION

Hypolimnetic oxygen
    Dissolved oxygen concentration
    Rate of consumption
Biological productivity
    Standing crop
    Volume of algae
    Transparency
    Chlorophyll in epilimnion
    Oxygen production
    Carbon dioxide utilization
Nutrient levels
    Nitrogen
    Phosphorous
    Nitrogen-phosphorous ratios

Source: C.N. Sawyer, *J. Water Pollut. Control Fed.,* *38*(5):737, 1966. With permission.

3.7-2   INDICATORS OF TROPHIC STATUS

Physical	Chemical	Biological
Transparency (Secchi disc)	Sediment type	Algal bloom frequency
Morphology	Oxygen supersaturation in epilimnion	Algal species variety
Mean depth, etc	Hypolimnetic oxygen deficit	Characteristic algal groups
	Conductivity	Littoral vegetation
	Dissolved solids	Fish species and biomass
	Nutrient concentrations at spring maximum	Characteristic zooplankton
	Chlorophyll level	Bottom fauna
		Primary production

Source: P.L. Brezonik and H. D. Putnam, Eutrophication: Small Florida lakes as models to study the process, *Proc. 17th Southern Water Resources Pollut. Control Conf.,* University of North Carolina, Chapel Hill, 1968, p. 315. With permission.

3.7-3   PLANKTON OF OLIGOTROPHIC
AND EUTROPHIC LAKES

Parameter	Oligotrophic	Eutrophic
Quantity	Poor	Rich
Variety	Many species	Few species
Distribution	To great depths	Trophogenic layer
Diurnal migration	Extensive	Limited
Water–blooms	Very rare	Frequent
Characteristic algal groups or genera	Chlorophyceae	Cyanophyceae
	Desmids	*Anabaena*
	Staurastrum	*Aphanizomenon*
	Diatomaceae	*Microcystis*
	Tabellaria	Diatomaceae
	Cyclotella	*Melosira*
	Chrysophyceae	*Fragilaria*
	Dinobryon	*Stephanodiscus*
		Asterionella

Source: C.N. Sawyer, *J. Water Pollut. Control Fed., 38*(5):737, 1966. With permission.

3.7-4   RELATIVE COMPARISON OF THE BIOLOGICAL
COMMUNITIES IN OLIGOTROPHIC
AND EUTROPHIC LAKES

Parameter	Oligotrophic lakes	Eutrophic lakes
Algal bloom	Rare	Frequent
Algal species variety	Many	Variable to few
Characteristic algal group	–	Blue-green
		Anabaena
		Aphanizomenon
		Microcystis
		Oscillatoria rubescens
Growth of aquatic plants in littoral regions	Sparse	Abundant
Characteristic zooplankton	*Bosmina obtusirostris*	*B. longirostris*
	B. coregoni	*D. cucullata*
	Diaptomus gracilus	
Bottom fauna	Tanytarsus	Chironomids
Fish	Deep-dwelling, cold-water fish as trout, salmon, cisco	Surface-dwelling, warm water fish as pike, perch, bass

Source: E.G. Fruh *et al., J. Water Pollut. Control Assn., 38*(8):1237, 1966. With permission.

3.7-5   CLASSIFICATION OF THE GREAT LAKES

Lake	Morphometry mean depth, m	Transparency avg Secchi disc depth, m	Total dissolved solids, mg/l	Specific conductance, μmho at 18°C	Dissolved oxygen
Oligotrophic	>20	High	<100	<200	High, all depths all year
Superior	148.4	10	60	78.7	Saturated, all depths
Huron	59.4	9.5	110	168.3	Saturated, all depths
Michigan	84.1	6	150	225.8	Near saturation, all depths
Eutrophic	<20	Low	>100	>200	Depletion in hypolimnion <70% saturation
Ontario	86.3	5.5	185	272.3	50 to 60% saturation in deep water in winter
Erie (average)	17.7	4.5	180	241.8	–
Erie (central basin)	18.5	5.0	–	–	<10% saturation hypolimnion
Erie (eastern basin)	24.4	5.7	–	–	40 to 50% saturation hypolimnion

Source: A.M. Beeton, *Limnol. Oceanogr., 10*:240, 1965. With permission.

3.7.2   *NUTRIENT LOADS*

3.7-6   APPROXIMATE QUANTITIES OF ALGAL GROWTH MATERIALS

From Sources Either In or Entering U.S. Surface Waters

Source	Organic matter	Nitrogen	Phosphorus
	Quantities available, mil lb/yr		
Natural			
Air	–	15–300[a]	–
Rainfall (direct into surface water)	4,200–7,400 (COD)	30–590	2–17
Aquatic plants	N.A.	0–1,070	0–107
Waterfowl, fish, bottom fauna, and the like	N.A.	N.A.	N.A.
Muds under water bodies	N.A.	N.A.	N.A.
Runoff from			
Forest land (including commercial forests)	N.A.	990–2,250	243–587
Other land	N.A.	N.A.	N.A.
Man generated sewage			
Domestic			
Human and food wastes			
Washing wastes	5,200 (COD)[b]	1,330[b]	137–166[b]
			250–280[b]
Industrial			
Food processing wastes	N.A.	N.A.	N.A.
Other wastes	N.A.	N.A.	N.A.
Runoff from			
Urban land	5,500 (COD)	200	19
Cultivated land			
Fertilized	17,900 (COD)[c]	2,040[c]	110–380
Unfertilized	N.A.	N.A.	N.A.
Land on which animals are kept	N.A.	420	170

N.A. – Not available.
[a] Approximately half the quantity entering from rainwater.
[b] Data based on the assumption of 70% removal of COD and 30% removal of nutrients in treatment plant.
[c] Estimated.

Source: Reprinted with permission from F.A. Ferguson, *Environ. Sci. Technol.,* 2(3):188, 1968. Copyright 1968, American Chemical Society.

3.7-7   AMOUNT OF NUTRIENTS DEPOSITED BY PRECIPITATION (kg ha^{-1} year^{-1})[a]

State	Site	NH$_4^+$N	NO$_3^+$N	Total N[b]	Period	PO$_4^{+3}$P	Period	SO$_4^{-2}$S	Cl$^-$	Period
Arizona	Tombstone		3.5	3.5	1975/78	5.8	1978	4.1	1.7	1975/78
Florida	Gainesville									
	(I)	1.2	2.3	3.5	1976/77	0.41[c]	1976/77	8.0	11.8	1976/77
	(II)	1.5	2.8	4.3		1.02[c]		9.0	22.7	
Iowa	Ames	6.0	6.0	12.0	1971/73	1.41	1980	16.8	(2.2)	1971/73
		(7.4)	(3.0)	(10.4)				(7.1)		
	Atlantic	(5.7)	(5.0)	(10.7)	(1980)	0.87		(6.0)	(4.8)	(1980)
	Boone	6.0	7.2	13.2		0.48		15.6	(1.1)	
		(7.1)	(4.0)	(11.1)				(6.1)		
	Charles City	7.2	6.0	13.2				13.2		
	Creston	6.7	6.0	12.7				16.7		
	Eldora	5.0	4.8	9.8				17.3		
	Fairfax	(7.6)	(5.5)	(12.1)		0.60		(17.1)	(2.3)	
	Guthrie Center	7.1	7.2	14.3				16.1		
	Sigourney	(3.6)	(8.9)	(12.5)				(16.1)	(12.6)	
	Storm Lake	(8.6)	(5.5)	(14.1)		1.37		(5.7)	(3.4)	
	Tripoli	(8.8)	(5.4)	(14.2)		1.28		(15.3)	(5.1)	
Michigan	Antrim								9.0	4/59-4/61
	Clinton								10.3	
	Houghton Lake	2.1	3.2	5.3		0.31[c]	1973/74			
	Ingham							14.0		
	Kalamazoo							13.5		
	Ottawa National Forest					0.46[c]		8.1		
	Pellston	3.1	4.0	7.1		0.25[c]		6.0		
	Saginaw					0.21[c]		27.7[d]		
Minnesota	Cloquet Forest Center					0.22[c]	1975	2.1		1975
	Lamberton			13.6	1975/78					
	Mercell Experimental Station					0.41	1971/73	5.6		1978/79
	Morris			5.8						
	Staples			6.3						
	Waseca			11.1						
Nebraska	Alliance								5.7	1959/60
	Auburn	4.9	3.6	8.5	1970/72	0.60	1970/72		13.1	
	Clay Center	4.9	3.6	8.5	1970/72	0.60	1970/72			
	Concord	13.7	5.8	19.5		1.20				
	Hasley							7.3		1959/60
	Lincoln							9.2		1959/60
	Mead	8.0	5.2	13.2	1970/72	0.40	1970/72			
	North Platte	4.4	3.3	7.7	1970/72	0.10	1970/72			
	Pierce							4.6		1959/60
	Red Cloud							6.1		1959/60
	Scotts Bluff	3.9	1.8	5.7	1970/72	0.40	1970/72			
	Valentine							4.6		1959/60
New York	Aurora	3.3	4.3	7.6	1970/71	0.60	1972	16.4		1970/71
	Geneva	3.5	4.8	8.3		0.05		11.5		
	Ithaca	3.2	4.2	7.4		0.05		13.0		
Wisconsin	Drummond					0.16[c]				

[a] Compiled from M.A. Tabatabai, *CRC Crit. Rev. Environ. Control, 15*:65, 1985.
[b] Total inorganic nitrogen.
[c] Total phosphorus.
[d] Saginaw Bay.

3.7-8   ESTIMATE OF NATIONWIDE NUTRIENT CONTRIBUTIONS FROM VARIOUS SOURCES

	Nitrogen		Phosphorus	
Source	1,000,000 lb/yr	Usual concentration in discharge, mg/l	1,000,000 lb/yr	Usual concentration in discharge, mg/l
Domestic waste	1,100–1,600	18–20	200–500	3.5–9
Industrial waste	>1,000	0–10,000	b	b
Rural runoff				
Agricultural land	1,500–15,000	1–70	120–1,200	0.05–1.1
Nonagricultural land	400–1,900	0.1–0.5	150–750	0.04–0.2
Farm animal waste	>1,000	b	b	b
Urban runoff	110–1,100	1–10	11–170	0.1–1.5
Rainfall[a]	30–590	0.1–2.0	3–9	0.01–0.03

[a] Considers rainfall contributed directly to water surface.
[b] Insufficient data available to make estimate.

Source: Reprinted, with publisher's permission, from *J. Am. Water Works Assn., 59*(3) © 1967, by the American Water Works Association, Inc.

3.7-9   NUTRIENTS IN AGRICULTURAL RUNOFF

Land area or type of runoff	Nitrogen Type	Nitrogen lb/mi² yr	Phosphorus Type	Phosphorus lb/mi² yr	Ref.
Woodlands		19		1.9	1
Crop and pasture		38		26	1
Rural	Total	830–1,900	Total	125–180	2
Agricultural				250	3
Agricultural	Nitrite and nitrate	1,260	Phosphate	320	4
Agricultural	Kjeldahl	195			4
Agricultural	Total	3,620	Soluble	250	5
Agricultural			Total	340	5
Agricultural		4,500		258	6
Manured		1,900		640	1

Source: R.C. Loehr and S.A. Hart, *Crit. Rev. Environ. Control,* *1*(1):69, 1970.

REFERENCES

1. J.W. Biggar and R.B. Corey, *Agricultural Drainage and Eutrophication,* paper presented at the Int. Symp. Eutrophication, Madison, Wisconsin, June 1967.
2. R.O. Sylvester, *Trans. Sem. Algae Metropolitan Wastes,* U.S. Dept. of Health, Education and Welfare, Cincinnati, Ohio, 1961.
3. *Lake Erie Report: A Plan for Water Pollution Control,* Federal Water Pollution Control Administration, Washington, D.C., 1968.
4. N.A. Jaworski, O. Villa, and L.J. Hetling, *Nutrients in the Potomac River Basin,* Tech. Rep. #9, Chesapeake Tech. Support Lab., Federal Water Pollution Control Administration, Annapolis, Md., 1969.
5. S.A. Witzel, *et al.,* The Effect of Farm Wastes on the Pollution of Natural Waters, Paper 69-428, presented at the Annual Meeting, Am. Soc. Agric. Eng., Lafayette, Ind., 1969.
6. C.N. Sawyer, *J. New Engl. Water Works Assn.,* *61*:109, 1947.

3.7-10   PHOSPHORUS ENTERING SURFACE WATERS WITH RUNOFF FROM CULTIVATED LAND[a]

	Size of site, acres	Phosphorus entering surface water, lb/acre yr Range	Phosphorus entering surface water, lb/acre yr Average
Kaskaskia River, Ill.	3,320,000	0.02–0.76	0.35
Madison, Wisc.	140,000	N.A.[b]	0.40
San Joaquin Valley, Cal.	358	0.20–0.90	0.55
Yakima Valley, Wash.	265,000	0.9–8.9	1.33
Coshocton, Ohio	1.45	N.A.	1.2

[a] Includes runoff through drain tile under the crop land.
[b] NA – not available.

Source: Reprinted with permission from F.A. Ferguson, *Environ. Sci. Technol.,* *2*(3):188, 1968. Copyright 1968, American Chemical Society.

3.7-11   REPRESENTATIVE CONCENTRATION RANGE OF PHOSPHORUS IN WATER

	Total phosphorus, mg/l as P	Present as soluble orthophosphate, %	Ref.
Domestic wastewater	5–20	15–35	3, 4
Effluents from secondary treatment plants	3–10	50–90	5
Agricultural drainage	0.05–1.0	15–50	3
Lakes, unpolluted	0.01–0.04	10–30	1
Lakes, eutrophic	0.03–1.5	5–20	5
U.S. rivers	0.01–1.0		5
Oceans mean value	0.07		5
Rainwater	0.004–0.03		
Forested streams	0.02–0.10		2

REFERENCES

1. G.E. Hutchison, *A Treatise on Limnology,* Vol. 5, John Wiley & Sons, New York, 1957.
2. R.O. Sylvester, *Nutrient Control Content of Drainage Water from Forested, Urban, and Agricultural Areas,* U.S. Public Health Service, 1961, p.80.
3. Task Group Report, *Sources of Nitrogen and Phosphorus in Water Supplies, J. Am. Water Works Assn., 59*:344, 1967.
4. E.F. Gloyna and W.W. Eckenfelder, *Water Quality Improvement by Physical and Chemical Processes,* University of Texas Press, Austin, 1970.
5. C.M. Weiss, *J. Am. Water Works Assn., 61*(8):387, 1969.

3.7-12   NUTRIENT CONTENT OF DRAINAGE WATER FROM FORESTED, URBAN, AND AGRICULTURAL AREAS

	Mean nutrient concentrations, ppb			
	Total phosphorus	Soluble phosphorus	Nitrates	Total Kjeldahl nitrogen
Urban street drainage[a]	208	76	527	2,010
Streams from forested areas[b]	69	7	130	74
Subsurface irrigation drains[c]	216	184	2,690	172
Surface irrigation drains[c]	215	162	1,250	205

[a] From major highways, arterial and residential streets from thirty min. to several hr after a rainstorm had commenced.

[b] From three streams containing large reservoirs, roads, and some logging, but no human habitation.

[c] From the Yakima River Basin irrigation return flow drains.

Source: Algae and metropolitan wastes, Transaction of seminar on algae and metropolitan wastes, held at Cincinnati, Ohio, April 27–29, 1960, under sponsorship of Division of Water Supply and Pollution Control and R.A. Taft Sanitary Engineering Center, Cincinnati, 1961.

3.7-13   NITROGEN BALANCE FOR HARVESTED CROP AREA

U.S., 1930

Items	Nitrogen, lb/acre yr
Additions	
Rain and irrigation	4.7
Seeds	1
Fertilizers	1.7
Manures	5.2
Symbiotic nitrogen fixation	9.2
Nonsymbiotic nitrogen fixation	6
Total additions	27.8
Losses	
Harvested crops	25.1
Erosion	24.2
Leaching	23
Total losses	72.3
Net annual loss	44.5

Source: F.E. Allison in *Advances in Agronomy,* A.G. Norman, Ed., Academic Press, New York, 1955, p. 213. With permission.

Section 4
Solid Wastes

4.1 SOURCES

4.1.1 DOMESTIC

4.1-1 URBAN SOLID WASTE CONSTITUENTS

Source	Waste	Composition	Means of treatment or disposal
Households, restaurants, institutions, stores, markets	Garbage	Wastes from preparation, cooking, and serving of food; market wastes from handling, storage, and sale of food	Grinding, incineration, landfill, composting, hog feeding
	Rubbish	Paper, cartons, boxes, barrels, wood, excelsior, tree branches, yard trimmings, wood furniture, bedding, dunnage, metals, tin cans, metal furniture, dirt, glass, crockery, minerals	Salvage, incineration, landfill, composting, dumping
	Ashes	Residue from fires	Landfill, dumping
Streets, sidewalks, alleys, vacant lots	Street refuse	Sweepings, dirt, leaves, catch basin dirt, contents of litter receptacles, bird excreta	Incineration, landfill, dumping
	Dead animals	Cats, dogs, horses, cows, marine animals	Incineration, rendering, explosive destruction
	Abandoned vehicles	Unwanted cars and trucks left on public property	Salvage, dumping
Factories, power plants	Industrial wastes	Food processing wastes, boiler house cinders, lumber scraps, metal scraps, shavings	Incineration, landfill, salvage
Urban renewal, expressways	Demolition wastes	Lumber, pipes, brick masonry, asphaltic material and other construction materials from razed buildings and structures; bat guano, pigeon excreta	Incineration, landfill, dumping, salvage
New construction, remodeling	Construction wastes	Scrap lumber, pipe, concrete, other construction materials	Incineration, landfill, dumping, salvage
Households, hotels, hospitals, institutions, stores, industry	Special wastes	Hazardous solids and liquids, explosives, pathologic wastes, radioactive wastes	Incineration, landfill, burial, salvage
Sewage treatment plants, lagoons, septic tanks	Sewage treatment residue	Solids from coarse screening and grit chambers, sludge	Incineration, landfill, composting, fertilizing

Source: *Municipal Refuse Disposal,* Committee on Refuse Disposal, Am. Publ. Works Assn., Public Administration Service, Chicago, 1961. With permission.

4.1-2 COMMERCIAL TREATMENT OF INDUSTRIAL AND MUNICIPAL SOLID WASTES

Solid waste	Waste-producing industry	Waste-processing industry	Waste-separation process	Recovered useful products
Garbage and refuse	Residential, commercial, and industrial assemblages	Municipalities	Hand sorting, magnetic separation	Nonferrous scrap, ferrous scrap, glass, plastics, rubber, rags, paper
Waste wood	Pulp	Paper	Gravity separation by heavy media	Wood, bark
Bark	Pulp	Fertilizer	Screening	Sized bark
Bagasse	Sugar	Paper	Screening	Bagasse fiber, pith
Wastepaper	Residential, commercial, and industrial assemblages	Paper	Hand sorting, gravity separation by cyclones, mechanical sorting, screening, magnetic separation, flotation	Paper fiber
Grinding wastes	Tool		Gravity separation by heavy liquids, flotation	Diamonds
Fly ash	Electric power	Electric power	Magnetic separation, air classification, screening	Ferropozzolan, purified pozzolan
Slag	Steel	Steel	Magnetic separation, screening	Sized slag
Foundry wastes	Foundry	Foundry	Magnetic separation, screening, gravity separation by shaking tables, air classification	Molding sand, metals, alloys
Nonferrous metal scrap	Atomic power, electric power, automobile	Atomic power, electric power, automobile	Magnetic separation, screening	Alloys, metals
Nonferrous metal residues	Smelting	Smelting	Screening, gravity separation by jigs and shaking tables, air classification, heavy-media separation	Metals

Source: *Solid Waste Processing*, U.S. Department of Health, Education, and Welfare, Public Health Service, 1969.

4.1.2   *INDUSTRIAL AND COMMERCIAL*

4.1-3   WASTE PRODUCED BY PROCESSING MINERALS AND FUELS IN 1965

Thousands of Short Tons

Industry	Mine waste (other than surface mine overburden)	Mill tailings	Washing plant rejects	Slag	Processing plant wastes	Total
Alumina	b				5,350[a]	5,350
Anthracite			2,000			2,000
Bituminous coal	12,800		86,800			99,600
Copper	286,600	170,500		5,200		462,300
Iron and steel	117,599	100,589		14,689	1,000	233,877
Lead and zinc	2,500	17,811	970			21,281
Phosphate rock	72		54,823	4,030	9,383	68,308
Other	c	c	c	c	c	229,284[d]
Total	419,571	288,900	144,593	23,919	40,233	1,122,000

[a] Of this total 1.7 million tons are discharged directly into the Mississippi River annually by two alumina production plants in Louisiana. Therefore, annual accumulation is 3.65 million tons.
[b] Total not available, but quantities are negligible and are included in washing plant wastes.
[c] Not available.
[d] Wastes of remaining mineral mining and processing industries—20 per cent of total wastes generated.

Source: *Wealth out of Waste*, U.S. Bureau of Mines, 1968.

4.1-4   HOW PACKAGING RELATES TO ASPECTS OF SOLID WASTE PROBLEMS

Aspect	Basic problem	Packaging contribution	Possible solution
Waste collection	Collection is labor-intensive, thus costly	Proportion of packaging materials in municipal waste growing	Automation of collection On-site volume reduction and disposal
Waste processing	Disposal technology is relatively backward Insufficient support of research and development Land available for waste disposition is dwindling	Packaging materials are usually nondegradable by natural processes	Retooling of financial support for waste processing Development of new disposal technology Modify packaging materials to make them more degradable
Aesthetic blight from littering	Public carelessness and or indifference	Packaging is major component of litter	Intensive anti-litter publicity Rigorous anti-litter law enforcement Economic incentives for returning containers
Soil and groundwater pollution from decomposition of organics	Inadequate waste processing Poor selection of disposal sites	Packaging plays indirect role; some materials may contain organic residues	Relocation of adequate sites Replacement of dumps by incinerators
Air pollution from waste combustion	Existence of burning dumps Poorly operated or designed incinerators	Role of packaging same as for solid waste in general	Elimination of burning dumps and inadequate incinerators Research and development on improved combustion equipment High quality pollution abatement equipment for incinerators
Loss of potentially valuable raw materials	Solid waste has low value because of contamination by intermixing Low cost of virgin material reduces demand for secondary materials	Many high value raw materials combined in single package Exploitation of concept of throwaway containers	New technology for low cost automated separation of heterogeneous solid waste Incentives for wider industrial use of secondary materials Modifications in package design which would utilize more homogeneous materials

Source: Reprinted with permission from A. Darnay, *Environ. Sci. Technol., 3*(4):331, 1969. Copyright 1969, American Chemical Society.

4.1-5   POLYETHYLENE FILM CONSUMED IN PACKAGING, 1961–66

Millions of Pounds

End use	1961	1962	1963[a]	1964	1965	1966
Food packaging						
Candy	10	11	13	15	16	20
Bread, cake	30	55	57	60	80	85
Crackers, biscuits	1	1	3	5	5	8
Meats, poultry	5	7	10	12	16	20
Fresh produce	95	100	112	125	145	160
Snacks	1	1	1	2	2	2
Noodles, macaroni	5	5	5	6	6	7
Cereals	2	2	3	3	3	4
Dried vegetables	5	5	6	8	8	10
Frozen foods	5	7	10	12	14	19
Dairy products	1	1	5	7	8	12
Other foods	10	15	10	5	7	13
Frozen food bags, household wrap	10					
Total food uses	180	210	235	260	310	360
Nonfood packaging						
Shipping bags, liners	40	45	58	70	80	95
Rack and counter				35	40	50
Textiles	35	35	40	45	60	90
Paper	15	15	17	20	25	30
Laundry, dry cleaning	35	35	42	50	60	75
Miscellaneous	35	40	48	20	40	30
Total nonfood uses	160	170	205	240	305	370
Total packaging uses	340	380	440	500	615	730

[a] Estimated by Midwest Research Institute.

Source: *Modern Packaging Encyclopedia*, William C. Simms, ed., McGraw-Hill, Inc., Vol. 40, No. 13A, September 1967; Vol. 39, No. 4A, December 1965; Vol. 38, No. 3A, November 1964.

4.1-6   CONSUMPTION OF FILM IN PACKAGING

1966 and 1976

Type of film	Quantity, millions of pounds		10-year rate of change, per cent
	Actual, 1966	Forecast, 1976	
Cellophane	395	320	–2.0
Polyethylene film	730	2,030	10.7
Other plastic film	192	560	11.3

Source: Midwest Research Institute, Kansas City, Mo. With permission.

4.1.3 AGRICULTURAL

4.1-7 AGRICULTURAL WASTES

	Source Units[a]	Multiplier	ton/year
Apricots	310 acres	1.5420 ton/acre yr[b]	478
Cherries	53 acres	1.4265 ton/acre yr[b]	76
Vegetables, berries, and seed crops	3,614 acres	3.0 ton/acre yr[c]	10,842
Greenhouse and nursery	125 acres	25.0 ton/acre yr[d]	3,125
Feedlot cattle manure	840 head	13.21 ton/head yr[e]	11,096
Total agricultural waste			25,617

[a] Basic crop acreage and livestock inventory are from Alameda County Agriculture Commission Report (1967). Data were allocated to subcounty areas with help of Alameda County Agriculture Extension Office.

[b] Apricot and cherry orchard waste multipliers determined by estimate of trees per acre from University of California Agricultural Extension Service and per tree waste multipliers from FMC Corporation Santa Clara Study.

[c] State of California Department of Public Health, *Status of Solid Waste Management in California*, 1968.

[d] Professor Harry C. Kohn, Department of Environmental Horticulture, Davis Campus of the University of California. (Verbal Communication.)

[e] The manure multiplier for cattle is from Taiganides, E. Paul, Table 4, *Agricultural Solid Wastes*. Three of the cattle manure multipliers in his table averaged to 72.4 lb/day.

REFERENCE

1. C.G. Golueke and P.H. McGauhey, *Comprehensive Studies of Solid Wastes Management*, Sanitary Engineering Research Laboratory, University of California, June 1970, p. 50.

4.1-8 PRODUCTION OF WASTES BY LIVESTOCK

U.S., 1965

Livestock	Population, millions	Annual production of solid wastes, million tons	Annual production of liquid wastes, million tons
Cattle	107	1004.0	390.0
Horses[a]	3	17.5	4.4
Hogs	53	57.3	33.9
Sheep	26	11.8	7.1
Chickens	375	27.4	–
Turkeys	104	19.0	–
Ducks	11	1.6	–
Totals		1138.6	435.4

[a] Horses and mules on farms as work stock.

Source: *Wastes in Relation to Agriculture and Forestry,* U.S. Department of Agriculture, Government Printing Office, March 1968.

4.1-9 POPULATION EQUIVALENTS[a] OF ANIMAL WASTES

BOD_5 [b] Basis

Animal	Ref.
Chickens (4–5 lb)	
11.8	1
11.3	2
12	3
11	4
10–20	5
Swine (100 lb)	
0.6	1
0.33	6
Cattle (1,000 lb)	
Dairy 0.13	7
Beef 0.17	7

[a] Equivalent animals per capita, i.e., 11.8 chickens contribute the BOD equivalent to one person per day.

[b] BOD (Biological Oxygen Demand) is a measure of the water pollution potential of an organic waste. It corresponds to the amount of oxygen required by the bacteria which consume the organic waste.

Source: *Pollution Implications of Animal Wastes—A Forward Oriented Review,* U.S. Department of the Interior, Federal Water Pollution Control Administration, July 1968, p. 44.

REFERENCES
1. E.P. Taiganides and T.E. Hazen, "Properties of Farm Animal Excreta," *Trans. Am. Soc. Agric. Eng., 9,* 1959.
2. J.N. Dornbush and J.R. Anderson, "Lagooning of Livestock in South Dakota," *Proc. 19th Ind. Waste Conf.,* Purdue Univ., 1964, p. 317.
3. S.A. Hart, "Fowl Fecal Facts," *Nat. Symp. Poultry Ind. Wastes,* May 1963.
4. "Population Equivalents of Chickens," *Public Works,* July 1966, p. 156.
5. E.A. Cassell and A. Anthonisen, "Studies on Chicken Manure Disposal: Part I," New York State Department of Health, 1966.
6. E.P. Taiganides *et al.,* "Properties and Pumping Characteristics of Hog Wastes," *Trans. Am. Soc. Agric. Eng., 7:*123, 1964.
7. S.A. Witzel *et al.,* "Physical, Chemical and Bacteriological Properties of Farm Wastes (Bovine Animals)," *Proc. Natl. Symp. Anim. Wastes Manage., Am. Soc. Agric. Eng.,* October 1963.

4.1.4   RECREATIONAL

4.1-10   WASTE GENERATION RATES FOR RECREATION SITES

Recreation site	Average rate of waste generation, 90% confidence interval
Campgrounds (lb/camper day)	1.26 ± 0.08
Campgrounds (lb/visitor day)	0.92 ± 0.06
Family picnic area (lb/picnicker)	0.93 ± 0.16
Group picnic area (lb/picnicker)	1.16 ± 0.26
Organization camps (lb/occupant day)	1.81 ± 0.39
Job Corps Civilian Conservation Corps Camps	
Kitchen waste (lb/corpsman day)	2.44 ± 0.63
Administrative and dormitory waste	
(lb/corpsman day)	0.70 ± 0.66
Resort areas	
Rented cabins (with kitchens)	
(lb/occupant day)	1.46 ± 0.31
Lodge rooms (without kitchens)	
(lb/occupant day)	0.59 ± 0.64
Restaurants (lb/meal served)	0.71 ± 0.40
Overnight lodges in winter sports areas	
(wastes from all facilities) (lb/visitor day)	1.87 ± 0.26
Day lodge in winter sports areas	
(lb/visitor day)	2.92 ± 0.61
Recreation residences (lb/occupant day)	2.13 ± 0.54
Observation sites (lb/incoming axle)	0.05 ± 0.03
Visitor centers (lb/visitor)	0.02 ± 0.008
Swimming beaches (lb/swimmer)	0.04 ± 0.01
Concession stands (lb/patron)	0.14 (1 site)
Administrative residences (lb/occupant day)	1.37 ± 0.35

Source: *Solid Waste Management in Recreational Forest Areas,* Environmental Protection Agency, 1971, p. 2.

4.2 COMPOSITION

4.2.1 DOMESTIC

4.2-1 COMPOSITION OF REFUSE (%)

Component	Estimated annual average[1]	Municipal, 1969[2]	Avg Municipal, 1963[3],a	Municipal, 1965[4],a	Municipal, 1971[5-12]	Municipal, 1966[13]	Residential, 1966–1969[14]	Municipal, 1965[15]	Example of refuse composition, 1967[16]
Paper products	60	35	42	54	55	9.40		10.33	
Newspaper									14
Cardboard					12				7
Corrugated boxes						23.38		23.92	
Magazine paper						6.80		7.48	
Brown paper					11	5.57		6.12	
Mail						2.75		3.03	
Food cartons						2.06		1.27	
Tissue paper						1.98		2.18	
Wax cartons						0.76		0.84	
Plastic coated paper						0.76		0.84	
Miscellaneous					32				25
Subtotal	60	35	42	54	55	53.46	43.8	56.01	46
Food waste	8.5	8.0	10	15	14				12
Vegetable						2.29		2.52	
Meat			2			2.29		2.52	
Fried fats						2.29		2.52	
Citrus rinds						1.53		1.68	
Subtotal	8.5	8.0	12	15	14	8.40	18.2	9.24	12
Plastics	3.5		0.7	1	1	0.76		3.50	4.0
Rubber	0.4		0.6	1		0.38		0.42	4.0
Leather	0.4		0.5	1		0.38		0.42	4.0
Subtotal	4.3		1.8		1	1.52	3.0	4.34	4.0
Dirt and vacuum cleaner catch	0.5		1.0			2.29		2.52	
Paints, oils, and removers	0.3		0.8			0.76		0.84	
Wood		1.5	2.4	2	4	2.29	2.5	2.52	7
Glass and ceramics	8.0	6.0	6.0	2.5	9	7.73	9.0	8.50	10
Subtotal	8.8	7.5	10.2	4.5	13	13.07	11.5	14.38	17

4.2-1   COMPOSITION OF REFUSE (%) (continued)

Component	Estimated annual average[1]	Municipal, 1969[2]	Avg Municipal, 1963[3],[3a]	Municipal, 1965[4],[3a]	Municipal, 1971[5-12]	Municipal, 1966[13]	Residential, 1966–1969[14]	Municipal, 1965[15]	Example of refuse composition, 1967[16]
Metals	8.0	8.0	8.0	7.0	9	6.85		7.53	8
Ferrous					7.5				
Nonferrous					1.5				
Subtotal	8.0	8.0	8.0	7.0	9	6.85	9.1	7.53	8
Plants and grass	6.5				5				10
Evergreens			7			1.53		1.68	
Flowers and lawn grass		5.0	5	10		3.06		3.36	
Ripe tree leaves		5.0	12	10		2.29		2.52	
Subtotal	6.5				5	6.88	7.9	7.56	10
Rags	0.5	0.5		2.0		0.76		0.84	
Textiles		2.0		1.0					
Synthetics	0.5	2.5	0.6	3.0					3
Subtotal			0.6	5.5		0.76	2.7	0.84	3
Ashes, stone, and dust			10.0						
Street sweepings			3.0						
Furniture and boxes	1.5								
Organics	1.0								
Construction	2.5	6.0		5.5					
Subtotal		6.0	13.0	5.5			3.7		
Chemical moisture		28				9.05			
House and garden chemicals	0.4								
Miscellaneous	0.5		0.6					0.10	
Subtotal	0.9	28	0.6			9.05		0.10	
TOTAL	100	100	100.2	100	100	99.99	99.9	100	100

4.2-1   COMPOSITION OF REFUSE (%) (continued)

REFERENCES

1. *Rail Transport of Solid Wastes—A Feasibility Study,* Interim Report: Phase One. U.S. Department of Health, Education, and Welfare, Public Health Service, 1969, p. 42.
2. F.H. Miller, *Conversion of Organic Solid Wastes into Yeast*, U.S. Department of Health, Education, and Welfare, Public Health Service, 1969, p. 61.
3. E. Kaiser, in "Characteristics of Municipal Refuse" by J.M. Bell, Proc. Natl. Conf. Solid Wastes Res., Am. Publ. Works Assn. Res. Found., December 1963, p. 37.
4. "Composition of Municipal Refuse and Properties of Typical Combined Refuse," California Waste Management Study, Department of Public Health, August 1965, p. II-27.
5. N.L. Drobny *et al., Recovery and Utilization of Municipal Solid Waste,* Environmental Protection Agency, Solid Waste Management Office, 1971.
6. C.A. Rogers, "Refuse Quantities and Characteristics," in Proc. Natl. Conf. Solid Waste Res., Am. Publ. Works Assn., 1963, p. 17.
7. M. Bell, "Characteristics of Municipal Refuse," Proc. Natl. Conf. Solid Waste Res., Am. Publ. Works Assn., 1963, p. 28.
8. E.R. Kaiser and C.D. Zeit, *The Composition and Analysis of Domestic Refuse at the Oceanside Plant,* Personal communication to N.L. Drobny, August 1968.
9. J.Houser, Fairfield Engineering Company, Personal communication to N.L. Drobny, August 1967.
10. J.E. Heer, Jr., University of Louisville, Kentucky, Personal communication to N.L. Drobny, September 5, 1967.
11. W. Galler, University of North Carolina, Personal communication to N.L. Drobny, July 1967.
12. Bureau of Solid Waste Management, Unpublished data.
13. E.R. Kaiser, *Chemical Analyses of Refuse Components,* 1966.
14. *Incinerator Guidelines,* U.S. Department of Health, Education, and Welfare, Public Health Service, 1969, p. 6.
15. E.R. Kaiser, *Chemical Analyses of Refuse Components,* Proc. Am. Soc. Mech. Eng., Nov. 7–11, 1965.
16. E.R. Kaiser, "Refuse Reduction Processes," in Surgeon General's Conf. Solid Waste Manage., U.S. Department of Health, Education, and Welfare, Public Health Service, 1967, p. 94.

4.2-2   REFUSE BREAKDOWN BY PHYSICAL COMPONENTS

Per Cent by Weight

	Europe[1]		U.S.A.[2]					
Component	Berlin, 1953–58	Six towns in England and Scotland, 1954–55	Chandler, Ariz., 1953		Philadelphia, 1957		Chicago, 1956–58	
Dust under 5/16 in.	40.84	37.22	Fines in a		Ashes, dirt	16.4	Ashes	18.7
Cinders under 3/4 in.	0.83	13.99	pulverized	11.0				
Cinders over 3/4 in.	1.00	12.01	mixture					
Putrescible matter	23.16	6.43	Garbage	21.8	Garbage	5.0	Garbage	4.8
			Grass, leaves	1.3			Grass	9.6
Paper, cardboard	12.97	12.02	Paper	42.7	Paper	54.4	Paper	56.5
Metal containers	4.22	3.83	Tin cans	8.7	Metal	8.4	Metal	7.9
Other metals	1.34	1.05	Other metals	1.1				
Textiles, rags	1.76	1.28	Rags	1.9	Rags	2.6	–	–
Bottles and jars	2.17	3.29	Glass	7.5	Glass	8.5	Glass	6.2
Broken glass	3.34	1.68						
Bones	0.84	0.33	–	–	–	–	–	–
Combustible debris	0.95	2.39	Wood	2.3	Wood	2.4	–	–
			Rubber	0.7	Rubber	0.9		
			Plastics	0.4	Plastics	0.2		
			Leather	0.3	Leather	0.6		
Noncombustible debris	6.58	4.48	Stone and ceramic	0.3	Ceramics	0.6	–	–

Source: C.A. Rogus, *Public Works, 93*(4):99, April 1962. With permission.

REFERENCES
1. "Dustless Refuse Collection" by R.F. Millard, Cleansing Superintendent, Cardiff City Council, England.
2. From Am. Publ. Works Assn. manuals "Refuse Collection Practice" and "Municipal Refuse Disposal."

4.2-3   CHARACTERISTICS OF BULKY WASTES[a]

Residential, per cent

Burnable	70
shreddable	70–75
Unburnable	30
shreddable	20

Commercial, per cent

Burnable	65
shreddable	95
Unburnable	35
shreddable	5

Density of various components, lb/yd^3

Household bulky	166.4
Tree cuttings	229.3
Wood	356.0
Construction wastes, burnable	430.8
Mixed construction waste	269.2
Rubbish and street dirt	423.5
Overall density	321.1

[a] Bulky wastes are those which cannot be collected in normal packer trucks; are not accepted at incinerators without preparation for burning because they are too large to fit into charge openings, or do not burn within the residence time afforded by grate travel, or burn too slowly, or burn with black smoke, or produce noxious gases; and cannot be compacted at landfills. Oversize items have a bulk in excess of 10 ft^3 or a single dimension exceeding 5 ft. Slow-burning timber usually is in excess of 12 in. in diameter. Large loads of grass clippings as well as large loads of rejected agricultural crops markedly lower furnace temperature and thus come under the category of problem wastes delivered in bulk. Demolition debris are considered bulky wastes.

REFERENCE

1. J.J. Baffa and N. Bartilucci, *Bulky and Demolition Wastes,* John J. Baffa Consulting Engineers, New York, 1969.

4.2-4 TYPICAL CHARACTERISTICS OF REFUSE

Weight
Loose combustible refuse, 200 lb/yd³
Compacted ashes, 1,200 lb/yd³
Uncompacted ordinary refuse, 200–300 lb/yd³
In collection, 400–500 lb/yd³
In fill, 700–1,000 lb/yd³

Btu value
9,000 to 10,000 Btu/lb dry combustibles
Ordinary refuse

Item	Weight, %
Combustible	45
Ash	25
Water	30
Btu/lb	4,500

Stability[a]

| | Lb/capita[b] | |
Component produced	Per year	Per calendar day
Garbage	200	0.5
Rubbish	1,300	3.6
Trash	400	1.1
Total	1,900	5.2

[a] The trend is from the former ratio of 50:50 mixed garbage and rubbish to a present one of 85–90% rubbish and 10–15% garbage.
[b] According to a field survey, production is from 700–1,500 lb/capita yr.

Source: "Refuse Quantities and Characteristics," in *Municipal Refuse Collection and Disposal,* Office for Local Government, New York State Executive Department, 1964, p. 6.

4.2-5   COMPOSITION AND ANALYSIS OF AVERAGE MUNICIPAL REFUSE

	Per cent of total refuse	Proximate analysis[a] as received basis, weight per cent				Ultimate analysis, dry basis, weight per cent								Btu per lb	
		Moisture	Volatile matter	Fixed carbon	Noncombustibles[b]	Carbon	Total hydrogen	Available hydrogen	Oxygen	Nitrogen	Sulfur	Noncombustibles[b]	Ratio C:(H)	Dry basis	Dry, ash-free basis
Rubbish, 64%															
Paper, mixed	42.0	10.24	75.94	8.44	5.38	43.41	5.82	(0.28)	44.32	0.25	0.20	6.00	155	7,572	8,055
Wood and bark	2.4	20.00	67.89	11.31	0.80	50.46	5.97	(0.672)	42.37	0.15	0.05	1.00	75	8,613	8,700
Grass	4.0	65.00	—	—	2.37	43.33	6.04	(0.83)	41.68	2.15	0.05	6.75	52	7,693	8,250
Brush	1.5	40.00	—	—	5.00	42.52	5.90	(0.75)	41.20	2.00	0.05	8.33	56.7	7,900	8,600
Greens	1.5	62.00	26.74	6.32	4.94	40.31	5.64	(0.77)	39.00	2.00	0.05	13.00	52.4	7,077	8,135
Leaves, ripe	5.0	50.00	—	—	4.10	40.50	5.95	(0.31)	45.10	0.20	0.05	8.20	131	7,069	7,700
Leather	0.3	10.00	68.46	12.44	9.10	60.00	8.00	(6.56)	11.50	10.00	0.40	10.10	9.1	8,850	9,850
Rubber	0.6	1.20	83.98	4.94	9.88	77.65	10.35	(10.35)	—	—	2.0	10.00	7.5	11,330	12,600
Plastics	0.7	2.00	—	—	10.00	60.00	7.20	(4.40)	22.60	—	—	10.20	13.6	14,368	16,000
Oils, paints	0.8	0.00	—	—	16.30	66.85	9.65	(9.00)	5.20	2.00	—	16.30	7.43	13,400	16,000
Linoleum	0.1	2.10	64.50	6.60	26.80	48.06	5.34	(3.00)	18.70	0.10	0.40	27.40	16	8,310	11,450
Rags	0.6	10.00	84.34	3.46	2.20	55.00	6.60	(2.70)	31.20	4.62	0.13	2.45	20.4	7,652	7,844
Sweepings, street	3.0	20.00	54.00	6.00	20.00	34.70	4.76	(0.36)	35.20	0.14	0.20	25.00	96	6,000	8,000
Dirt, household	1.0	3.20	20.54	6.26	70.00	20.62	2.57	(2.07)	4.00	0.50	0.01	72.30	10	3,790	13,650
Unclassified	0.5	4.00	—	—	60.00	16.60	2.45	(0.166)	18.35	0.05	0.05	62.50	100	3,000	8,000
Food wastes, 12%															
Garbage	10.0	72.00	20.26	3.26	4.48	44.99	6.43	(2.845)	28.76	3.30	0.52	16.00	15.8	8,484	10,100
Fats	2.0	0.00	—	—	0	76.70	12.10	(10.70)	11.20	0	0	0	7.2	16,700	16,700

4.2-5   COMPOSITION AND ANALYSIS OF AVERAGE MUNICIPAL REFUSE (continued)

Per cent of total refuse	Proximate analysis[a] as received basis, weight per cent				Ultimate analysis, dry basis, weight per cent								Btu per lb	
	Moisture	Volatile matter	Fixed carbon	Noncombustibles[b]	Carbon	Total hydrogen	Available hydrogen	Oxygen	Nitrogen	Sulfur	Noncombustibles[b]	Ratio C:(H)	Dry basis	Dry, ash-free basis
Noncombustibles, 24%														
Metallics 8.0	3.00	0.5	0.5	96.0	0.76	0.04	(0.02)	0.2	—	—	99.0	51	124	12,000
Glass and ceramics 6.0	2.00	0.4	0.4	97.2	0.56	0.03	(0.02)	0.11	—	—	99.3	34	65	8,000
Ashes 10.0	10.00	2.68	24.12	63.2	28.0	0.5	(0.40)	0.8	—	0.5	70.2	70	4,172	14,000
Composite refuse, as received 100.0	20.73	—	—	—	28.00	3.50	(0.71)	22.35	0.33	0.16	24.93	39.4	6,203	9,048

a Based on ASTM methods of analysis of coal and coke, as adapted for refuse.
b Noncombustibles — ash, metal, glass and ceramics.

Source: J.M. Bell, "Characteristics of Municipal Refuse," *Proc. Natl. Conf. Solid Waste Res.*. Am. Publ. Works Assn., Chicago, February 1964. With permission.

4.2-6   ORGANIC ANALYSIS OF COMPOSITE REFUSE

	Per cent
Moisture	20.7
Cellulose, sugar, starch	46.6
Lipids (fats, oils, waxes)	4.5
Protein, 6.25N	2.1
Other organic (plastics)	1.2
Ashes, metal, glass	24.9

Source: J.M. Bell, "Characteristics of Municipal Refuse," *Proc. Natl. Conf. Solid Waste Res.,* Am. Publ. Works Assn., Chicago, February 1964. With permission.

4.2-7   REFUSE ANALYSIS: INORGANIC CONSTITUENTS

Item	Glass and ceramics	Dirt	Metals	Wood products	Food wastes	Plastics
Sulfur oxides, SO_2, SO_3	x	x	x	x	x	
Silicon dioxide, SiO_2	x	x	x			
Magnesium oxide, MgO	x	x	x			
Chromium oxide, Cr_2O_3	x	x	x			
Iron oxide, Fe_2O_3	x	x	x			
Sodium oxide, Na_2O	x	x				
Calcium oxide, CaO	x	x				
Aluminum oxide, Al_2O_3	x	x				
Potassium oxide, K_2O	x	x				
Boron oxide, B_2O_3	x					
Lead oxide, PbO	x					
Tin oxides, SnO_2, SnO	x					
Titanium oxide, TiO_2	x					
Zirconium oxide, ZrO_2	x					
Beryllium oxide, BeO	x		x			
Nickel oxide, NiO			x			
Copper oxides, CuO, Cu_2O			x			
Manganese oxide, MnO			x			
Cadmium oxide, CdO			x			
Zinc oxide, ZnO			x			
Chlorides, Cl (acid and salts)						x
Fluorides, F (acid and salts)						x
Ammonia, NH_3				x	x	x
Nitrogen oxides, NO_x				x	x	x

Source: *Special Studies for Incinerators for the Government of the District of Columbia,* U.S. Department of Health, Education, and Welfare, Public Health Service, 1968.

4.2-8   REFUSE ANALYSIS: ORGANIC CONSTITUENTS

	Sources					
Item	Wood and wood products	Food wastes	Plants and grass	Plastics	Rubber	Pressurized cans
Carbohydrates	x	x	x			
Lipids (fats)	x	x	x			
Acrylonitrile-butadiene-styrene polymers				x		
Cellulose acetate				x		
Cellulose acetate butyrate				x		
Cellulose nitrate				x		
Melamine formaldehyde				x		
Polyethylene				x		
Polyvinyl dichloride				x		
Urea formaldehyde				x		
Urethane				x		
Polymethyl methacrylate				x		
Polypropylene				x		
Polystyrene				x		
Polyvinyl acetate				x		
Polyvinyl chloride				x		
Halogenated hydrocarbons				x	x	x
Polynuclear hydrocarbons				x	x	

Source: *Special Studies for Incinerators for the Government of the District of Columbia,* U.S. Department of Health, Education, and Welfare, Public Health Service, 1968, p. 5.

4.2-9   REFUSE ANALYSIS: PLASTICS

Type	Source
Acrylonitrile-butadiene-styrene	Shoe heels, appliances
Cellulose acetate	Pens, handles, frames, combs, toys
Cellulose acetate butyrate	Pens, handles, frames, combs
Cellulose nitrate	Pens, pencils
Melamine formaldehyde	Bottlecaps, buttons
Polyethylene	Film, flexible bottles, containers
Polyvinyl dichloride	Bottles, toys
Urea formaldehyde	Bottlecaps, buttons, dinnerware
Urethane	Coatings, laminates, adhesives
Polymethyl methacrylate	Buttons
Polypropylene	Fibers, packaging, films, appliances
Polystyrene	Combs, buttons, containers, toys, housewares
Polyvinyl acetate	Records
Polyvinyl chloride	Films, bottles, toys

GENERAL REFERENCES
1. Hunter-Wagner Engineering Division, "How We Get Our Plastics," *Plastic World Flow Chart of Major Plastic Materials,* Black, Swolls, and Bryson, Kansas City.
2. F.J. Spencer, "Progress in Polymers Today," *Hydrocarbon Processing, 45*(7), July 1966, p. 83.

4.2-10   REFUSE ANALYSIS: LEATHER

Composition	Per cent
Collagen (Amino acids)	95
Glycogen (Animal starch)	}
Glycerides (Natural fats—esters of glycerin)	5

GENERAL REFERENCES
1. A. Rogers, *Industrial Chemistry,* C.C. Furnas, ed., 2 volumes, D. Van Nostrand Company, Inc., 1942.
2. E.R. Riegel, *Industrial Chemistry,* J.A. Kent, ed., Reinhold Publishing Corporation, 1962, p. 963.

4.2-11   REFUSE ANALYSIS: RAGS

Cotton	Cellulose, sulfur
Nylon	Diamine, diacarboxylic acid, caprolactam
Silk	Complex polyamides
Orlon and acrilan	Acrylonitrile
Dynel	Copolymer of acrylonitrile and vinyl chloride
Dacron	Complex polyester
Rayon	Cellulose acetate, ethyl cellulose, viscose rayon
Wood	Protein–complex polyamide

GENERAL REFERENCES
1. A. Rogers, *Industrial Chemistry*, C.C. Furnas, ed., 2 volumes, D. Van Nostrand Company, Inc., 1942.
2. E.R. Riegel, *Industrial Chemistry*, J.A. Kent, ed., Reinhold Publishing Corporation, 1962, p. 963.

4.2-12   REFUSE ANALYSIS: PAINTS, OILS, REMOVERS

White lead	Methylene chloride
Titanium dioxide (TiO_2)	Maleic anhydride
Zinc oxide (ZnO)	Phthalic anhydride
Zinc sulfide	Polystyrene
Calcium chloride ($CaCl_2$) and other hygroscopic salts	Phenol
	Cresols
	Cresylic acid
Chromium (Cr)	Xylenols
Alkyd and phenol aldehydes	Acrylates
	Polyamides
Acetone	Urea
Methanol	Vinyl
Benzene	

GENERAL REFERENCES
1. A. Rogers, *Industrial Chemistry*, C.C. Furnas, ed., 2 volumes, D. Van Nostrand Company, Inc., 1942.
2. E.R. Riegel, *Industrial Chemistry*, J.A. Kent, ed., Reinhold Publishing Corporation, 1962, p. 963.

4.2-13   REFUSE ANALYSIS: RUBBER

Composition—natural	Per cent	Composition (types)—synthetics
Proteins—amino acids	2.0	Butadiene polymers
Fatty acids, esters	1.0	Chloroprene polymers
Quebrachital	1.0	Isobutene polymers
Inorganic salts	0.4	Organic polysulfides
Rubber hydrocarbons	35.0	Plasticized vinyl chloride
Water	60.0	Butadiene+styrene
Sulfur	0.6	Butadiene+acrylonitrile
		Ethylene dichloride
		Isobutylene+isoprene
		Isobutylene+butadiene

GENERAL REFERENCES

1. A. Rogers, *Industrial Chemistry*, C.C. Furnas, ed., 2 volumes, D. Van Nostrand Company, Inc., 1942.
2. E.R. Riegel, *Industrial Chemistry*, J.A. Kent, ed., Reinhold Publishing Corporation, 1962, p. 963.

4.2-14   REFUSE ANALYSIS: DIRT AND VACUUM CLEANER CATCH

Probable chemical constituents	Per cent
Oxygen (O_2)	47.0
Silicon (Si)	27.0
Aluminum (Al)	8.0
Iron (Fe)	5.0
Calcium (Ca)	2.8
Magnesium (Mg)	2.8
Potassium (K)	2.8
Sodium (Na)	2.8
80 miscellaneous elements	1.7

REFERENCE

1. *Special Studies for Incinerators for the Government of the District of Columbia*, U.S. Department of Health, Education, and Welfare, Public Health Service, 1968.

4.2-15   REFUSE ANALYSIS: GLASS AND CERAMICS

	Per cent
Composition of typical glass	
Silicon dioxide (SiO_2)	67.0–96
Boron oxide (B_2O_3)	1.0–16
Aluminum oxide (Al_2O_3)	1.0–4
Sodium oxide (Na_2O)	4.0–18
Calcium oxide (CaO)	0.3–13
Potassium oxide (K_2O)	0.1–12
Lead oxide (PbO)	15.0
Sulfur trioxide (SO_3)	0.4–0.7
Arsenic oxide (As_2O_3)	0.5–1

Composition of ceramics
Silicon dioxide (SiO_2)
Boron oxide (B_2O_3)
Aluminum oxide (Al_2O_3)
Sodium oxide (Na_2O)
Calcium oxide (CaO)
Potassium oxide (K_2O)
Titanium dioxide (TiO_2)
Chromium oxide (Cr_2O_3)
Beryllium oxide (BeO)
Zirconium oxide (ZrO_2)
Tin oxide (Sn_2O)
Magnesium oxide (MgO)
Fluorides

REFERENCE
1. Abstracted from A. Rogers, *Industrial Chemistry,* C.C. Furnas, ed., 2 volumes, D. Van Nostrand Company, Inc., 1942.

4.2-16   REFUSE ANALYSIS: METALS

Silicon (Si)	Cadmium (Cd)
Carbon (C)	Zinc (Zn)
Nickel (Ni)	Bismuth (Bi)
Chromium (Cr)	Sulfur (S)
Magnesium (Mg)	Tungsten (W)
Copper (Cu)	Mercury (Hg)
Aluminum (Al)	Arsenic (As)
Tin (Sn)	Vanadium (V)
Iron (Fe)	Antimony (Sb)
Manganese (Mn)	Phosphorus (P)
Molybdenum (Mo)	Beryllium (Be)

REFERENCE
1. Abstracted from A. Rogers, *Industrial Chemistry,* C.C. Furnas, ed., 2 volumes, D. Van Nostrand Company, Inc., 1942.

4.2-17   DENSITY OF SOLID PACKAGING MATERIALS

Material	Specific gravity	Density, lb/ft^3	ft^3/ton material
Aluminum	2.70	168	11.9
Steel	7.70	480	4.1
Glass	2.50	156	12.8
Paper	0.7–1.15	44–72	45.4–27.7
Cardboard	0.69	43	46.5
Wood	0.60	37	54.0
All plastics (average)	NA	71	28.1
Polyethylene	0.94	59	33.8
ABS	1.03	64	31.2
Acrylic	1.18	64	27.0
Polypropylene	0.90	56	35.7
Polystyrene	1.05	65	30.7
PVC	1.25	78	25.6
PVDC	1.65	103	19.4

NA – not available

Source: Compiled by Midwest Research Institute, Kansas City, Mo. With permission.

4.2-18   ANNUAL QUANTITIES OF RAW REFUSE PER CAPITA

Class	Weight, lb	% by weight	Uncompacted volume, yd^3	% by volume	Specific weight, lb/yd^3
Garbage	150	9.5	0.15	4.0	1,000
Rubbish and all combustibles (except garbage)	1,000	62.5	3.15	81.0	320
All noncombustibles	450	28.0	0.60	15.0	750
	1,600	100	3.90	100	410

Source: S. Wegman, "Planning a New Incinerator," Proc. 1964 Natl. Incin. Conf., Am. Soc. Mech. Eng., May 18–20, 1964. With permission.

4.2-19   AVERAGE SOLID WASTE COLLECTED

Pounds per Capita per Day

Solid wastes	Urban	Rural	National
Household	1.26	0.72	1.14
Commercial	0.46	0.11	0.38
Combined	2.63	2.60	2.63
Industrial	0.65	0.37	0.59
Demolition, construction	0.23	0.02	0.18
Street and alley	0.11	0.03	0.09
Miscellaneous	0.38	0.08	0.31
Totals	5.72	3.93	5.32

Source: *The National Solid Wastes Survey,* U.S. Department of Health, Education, and Welfare, Public Health Service, 1968, p. 13.

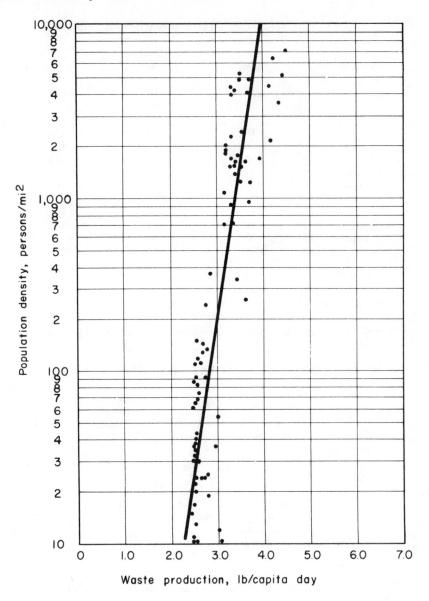

4.2-20   Relationship between per capita waste production and population density.[a,b]

[a] Data from 78 communities in 2 rural New York state counties.
[b] Correlation of data with the curve:

Variation from curve, %	Data within limits, %
±10	75
±15	95
±20	98

Source: G. Westerhoff, *Public Works, 101*(2):87, February 1970. With permission.

4.2.2   COMMERCIAL AND INDUSTRIAL

4.2-21   PHYSICAL CHARACTERISTICS OF REFUSE AT RECREATIONAL AREAS

Per Cent by Weight

	Rock Creek Park[a]	National Park Service,[b] range
Garbage	31	20–40
Paper	29	20–29
Glass	17	12–24
Metal	17	15–31
Plastic	1	–
Miscellaneous	5	–
Total	100	–

[a] Analysis of 857 pounds of refuse collected from picnic areas September 6, 1966, Rock Creek Park, Washington, D.C.
[b] Results of studies of samples from 6 camp ground and picnic areas in 1954.

Source: L. Weaver, *Public Works, 98* (4):128, April 1967. With permission.

4.2-22   COMPOSITION OF CAMPING AND RECREATION RESIDENCE SOLID WASTE COMPARED WITH URBAN RESIDENCE VALUES

Per Cent of Total

	Camping	Recreation residence	Urban residence
Food waste	24–50	24	15–30
Other combustibles	12–40	41	16–71
Noncombustibles	21–45	35	17–29

Source: T.J. Sorg and H.L. Hickman, Jr., *Sanitary Landfill Facts*, U.S. Government Printing Office, 1970.

4.2-23   COMPONENTS OF LITTER BY
PROPORTION OF TOTAL ITEMS LITTERED

Item	Per cent
Beverage containers	20.4
Other containers	4.4
Paper	59.5
Miscellaneous	15.7

Source: *A National Study of Roadside Litter.* Keep America Beautiful Inc., 1969. With permission.

4.2-24   CHARACTERISTICS
OF SURFACE REFUSE[a]

Average Value from Test Areas in Chicago

Area	Litter[b]	Dust and dirt[b]	BOD[c]	COD[c]
Industrial	11.15	7.76	2.88	17.1
Commercial	5.34	3.36	7.87	39.4
Multiple family	3.39	2.21	3.32	39.7
Single family	2.57	0.70	5.06	40.2

[a] Prepared by American Public Works Association.
[b] lb/day/100 ft of curb.
[c] mg/g.

Source: Reprinted with permission from *Environ. Sci. Technol., 3*(6):527, 1969. Copyright 1969, American Chemical Society.

4.2-25   REFUSE ANALYSIS: MISCELLANEOUS

Item	Composition
Developers	Nitrophenols, nitrobenzene, hydroquinone
Dyestuff	Carbazole, phenanthraquinone, anthraquinone, naphthalene sulfonic acids, salicylic acid, benzaldehyde, toluides, xylenes, halogenated benzene, aniline salts, dimethylaniline, aniline
Insecticides	Crude naphthalene, nitronapthalene, halogenated benzene
Preservatives	Anthranilic acid
Flavorings and perfumes	Benzoic acid, benzaldehyde

GENERAL REFERENCES

1.   A. Rogers, *Industrial Chemistry*, C.C. Furnas, ed., 2 volumes, D. Van Nostrand Company, Inc., 1942.
2.   E.R. Riegel, *Industrial Chemistry*, J.A. Kent, ed., Reinhold Publishing Corporation, 1962, p. 963.

4.2-26   PAPER WASTES

Type of paper	Constituents	Fillers, binders, and coatings	Ash,[a] %
Newspapers	Lignin, hemi-cellulose, pentosans	Rosin, alum, casein	3.5
Brown kraft paper	Lignin, hemi-cellulose, pentosans	Gum, starch, clay, rosin, alum, resin	6.5
Corrugated boxes	Lignin, pentosans, β and γ cellulose, hemi-cellulose	Clay, starch, glue, TiO_2	7.8
Books and magazines	β and γ cellulose, lignin	Clay, starch, rosin, casein, TiO_2, $CaCO_3$, satin white	28.0
Writing papers	β and γ cellulose, hemi-cellulose, lignin, pentosans	Rosin, clay, alum, starch, satin white, resin	
Glassine and grease papers	Hemi-cellulose, lignin, pentosans, β and γ cellulose	Glycerine, clay, starch, wax	6.0
Tissue papers	Lignin, pentosans	Starch	0.7
Paper food containers	Lignin, hemi-cellulose, β and γ cellulose, pentosans	Rosin, clay, starch, alum, wax	7.8
Paper boards	Lignin, pentosans, hemi-cellulose	Clay, rosin, wax, starch, resin	7.5

[a] Average values.

Source: *Rail Transport of Solid Wastes, A Feasibility Study,* U.S. Department of Health, Education, and Welfare, 1969, p.47.

REFERENCES

1.   Ralph W. Komler, *Varieties of Paper and Paperboard,* Waste Paper Utilization Council.
2.   James P. Casey, *Pulp & Paper,* Interscience Publishers, Inc.
3.   E. Sutermeister, *Chemistry of Pulp and Papermaking.* John Wiley and Sons, 1920.

4.2-27 CHEMICAL COMPOSITION OF FOOD WASTES

Food type	Organic					Inorganic		Example
	Proteins $(-R-CO-NH-R')_n$	Fat $\begin{matrix}H\\CO-COR\\HCO-COR'\\HCO-COR''\\H\end{matrix}$	Carbohydrates $CH_2OH-(HC-OH)_n-CHO$	Vitamins A,B,C	Water	Mineral matter, major constituents	Ash	
Dairy products[1]	0.5–3.3%	3.6–12.8%	0	Trace	15.0–88%	Calcium	0.72–0.76%	Milk, butter, cheese
Fish[2]	14.2–22.9%	0.3–12.8%	0 or Trace	Trace	64.6–84.2%	Calcium, phosphorus	1.0–1.7%	Cod, perch, trout, white fish
Fruit[3]	0.3–1.0%	0	3.5–14.9%	Trace	65.0–94.9%	Potassium, iron, phosphorus, copper, manganese	0.3–0.6%	Apples, grapes, oranges, peaches
Meat and meat products[4]	2.2–28.8%	1.2–92.8%	0–5.0%	Trace	40.0–75.0%	Sodium, chlorine	0.2–7.3%	Beef, pork, lamb
Vegetables[5]	1.0–6.7%	0.1–1.2%	2.9–27.9%	Trace	68.5–95.4%	Calcium, iron, phosphorus	0.58–1.53%	Beans, potato, carrot, spinach
Breakfast cereals[6]	6.3–15.5%	0.2–2.4%	71.0–83.0%	Trace	4.3–10.4%	Calcium, iron, phosphorus	0.4–4.2%	Corn flakes, puffed rice
Cereal grains[7]	9.9–12.4%	1.7–8.6%	62.3–71.9%	Trace	8.4–12.2%	Calcium, iron, phosphorus	1.7–2.4%	Wheat, corn, rye, barley

Source: *Rail Transport of Solid Wastes, A Feasibility Study.* U.S. Department of Health, Education, and Welfare, 1969, p.48.

REFERENCES
1. Gerrard, *Meat Technology*, 1951, p. 282.
2. Jacobs. *The Chemical Analysis of Food and Food Products*, 1951. p. 676.
3. Blanck, *Handbook of Food and Agriculture*, 1955. p. 561.
4. Gerrard, *Meat Technology*, 1951, p. 289.
5. Jacobs. *The Chemical Analysis of Food and Food Products*, 1951, p. 559.
6. Kent-Jones and Amos, *Modern Cereal Chemistry*, 1957, p. 388.
7. Jacobs. *The Chemical Analysis of Food and Food Products*, 1951, p. 498.

4.2-28 CHEMICAL PROPERTIES AND COMBUSTION DATA FOR PAPER, WOOD, AND GARBAGE

	Sulfite paper[a]		Average wood[1]		Douglas fir[2]		Garbage[b]	
Analysis, %								
Carbon	44.34		49.56		52.30		52.78	
Hydrogen	6.27		6.11		6.30		6.27	
Nitrogen			0.07		0.10			
Oxygen	48.39		43.83		40.50		39.95	
Ash	1.00		0.42		0.80		1.00	
Gross heating value (dry basis), Btu/lb	7,590		8,517		9,050		8,820	

Constituent (based on 1 lb)	Cubic feet	Pounds	Cubic feet	Pounds	Cubic feet	Pounds	Cubic feet	Pounds
Theoretical air	67.58	5.165	77.30	5.909	84.16	6.433	85.12	6.507
Theoretical air 40% saturated @ 60°F	68.05	5.188	77.84	5.935	84.75	6.461	85.72	6.536
Flue gas with theoretical air								
CO_2	13.993	1.625	15.641	1.816	16.51	1.917	16.668	1.935
N_2	53.401	3.947	61.104	4.517	66.53	4.918	67.234	4.976
H_2O formed	11.787	0.560	11.487	0.546	11.84	0.563	11.880	0.564
H_2O (air)	0.471	0.023	0.539	0.026	0.587	0.028	0.593	0.029
Total	79.652	6.155	88.771	6.905	95.467	7.426	96.375	7.495
Flue gas with % excess air as indicated								
0	79.65	6.16	88.77	6.91	95.47	7.43	96.38	7.50
50.0	113.44	8.74	127.42	9.86	137.55	10.64	139.24	10.77
100.0	147.23	11.32	166.07	12.81	179.63	13.86	182.00	14.04
150.0	181.26	13.91	204.99	15.79	222.01	17.09	224.86	17.21
200.0	215.28	16.51	243.91	18.75	264.38	20.12	267.72	20.58
300.0	283.33	21.70	321.75	24.68	349.13	26.58	353.44	27.12

[a] Sulfite paper constituents:

Cellulose	$C_6H_{10}O_5$	84%
Hemicellulose	$C_5H_{10}O_5$	8
Lignin	$C_6H_{10}O_5$	6
Resin	$C_6H_{10}O_5$	2
Ash	$C_{20}H_{30}O_2$	1

[b] Estimated on dry basis.

Source: *Interim Guide of Good Practice for Incineration at Federal Facilities,* U.S. Department of Health, Education, and Welfare, National Air Pollution Control Administration, 1969, p. 14.

REFERENCES
1. R.T. Kent, *Mechanical Engineer's Handbook,* 11th ed., John Wiley and Sons, 1936, p. 6.
2. R.T. Kent, *Mechanical Engineer's Handbook,* 12th ed., John Wiley and Sons, 1961, p. 2.

4.2-29   CHEMICAL COMPOSITION OF
PATHOLOGICAL WASTE AND COMBUSTION DATA

Ultimate analysis (whole dead animals)

Constituent	As charged, % by weight	Ash-free combustible, % by weight
Carbon	14.7	50.80
Hydrogen	2.7	9.35
Oxygen	11.5	39.85
Water	62.1	—
Nitrogen	Trace	—
Mineral (ash)	9	—

Dry combustible empirical formula — $C_5 H_{10} O_3$

Combustion data
(based on 1 lb of dry ash-free combustible)

Constituent	Quantity, lb	Volume, scf
Theoretical air	7.028	92.40
40% saturated at 60°F	7.059	93
Flue gas with theoretical air 40% saturated		
CO_2	1.858	16.06
N_2	5.402	73.24
H_2O formed	0.763	15.99
H_2O air	0.031	0.63
Products of combustion total	8.054	105.92
Gross heat of combustion		8,820 Btu/lb

Source: J. Danielson, *Air Pollution Engineering Manual,* U.S. Department of Health, Education, and Welfare, National Center for Air Pollution Control, 1967, p. 460.

4.2-30   FUEL PROPERTIES OF SOLID WASTE MATERIALS

Per Cent by Weight

Material	Moisture	Volatile combustible matter	Fixed carbon	Ash (fixed solids)	High heat value, Btu/lb
		As received basis			
Anthracite coal	4	7	78	11	12,800
Subbituminous coal	15	36	42	7	10,000
Lignite	35	29	27	9	6,700
Wood(air-dried)	17	69.5	13	0.5	5,500
Sewage sludge(fresh)	75	16	6	3	2,250
Sewage sludge(digested)	75	13	1	11	1,380
Chicken manure(fresh)	75	17	3	5	1,380
Chicken manure(air-dried)	20	54	7	19	4,400
		Moisture-free basis			
Anthracite coal	88	7	81	12	13,300
Subbituminous coal	92	42	50	8	11,800
Lignite	86	44	42	14	10,300
Wood(air-dried)	99	83	16	1	6,600
Sewage sludge(fresh)	76	65	11	24	8,980
Sewage sludge(digested)	56	52	4	44	5,510
Chicken manure(fresh)	76	67	9	24	5,500

Source: *Management of Farm Animal Wastes,* Am. Soc. Agric. Eng., Proc. Natl. Symp. Anim. Waste Manage., May 5–7, 1966. With permission.

4.2-31   ANALYSIS OF PACKAGING WASTE

Weight Per Cent

	Weight	H_2O[a]	C	H	O	N	S	Cl	Inert	Btu/lb
Paper and paperboard	49.1	8.0	42.3	5.6	40.5	0.14	0.14	–	3.32	7,551
Plastics	3.0	8.0	69.9	11.3	4.3	–	–	6.5	–	15,770
Wood	15.7	12.0	44.0	5.6	36.15	0.25	0.10	–	1.9	7,783
Textiles	0.3	10.0	41.5	5.8	37.7	2.0	0.2	–	2.8	7,232
Miscellaneous	3.2	10.0	45.0	6.8	34.0	0.14	0.08	–	3.98	8,728
Glass	16.0	5.0	0.49	0.63	0.32	0.03	0.00	–	93.53	80
Metal	12.7	10.0	4.1	0.56	3.85	0.05	0.01	–	81.43	668[b]
Combined analysis, weighted		8.47	31.94	4.37	27.43	0.13	0.09	0.20	27.37	5,797
Heat from partial oxidation of metal										103
Total										5,900

[a] Assumed moisture contents.
[b] Organics only.

Source: Proc. Natl. Conf. Packaging Wastes, September 22–24, 1969, Environmental Protection Agency, 1971, p. 184.

One thousand pounds of mixed packaging waste of the composite analysis given in 1.1–47 will yield approximately the following products of incineration:

Dry flue gas	9,075 lb
Water vapor	590
Fly ash, 4.5 C + 4.5 ash	9
Residue	293
Total	9,967 lb

The dry flue gas may be expected to consist of:

CO_2	1,072 lb
SO_2, SO_3	1
CO	10
Hydrocarbons, as CH_4	5
HCl	2
Organic acids, as CH_3CO_2H	1
Aldehydes, as HCHO	1
NO, NO_2	1
O_2	1,127
N_2	6,855
Total	9,075 lb

The average gas temperature leaving the secondary chamber would be 1680°F, assuming 10% heat loss through the setting.

The fly ash carried out of the secondary chamber with the gases is a mixture of carbon and ash, about 50% of each. The residue is largely glass and metal, with about 16.5 pounds of ash and 13.5 pounds of unburned carbon.

Source: Proc. Natl. Conf. Packaging Wastes, September 22–24, 1969, Environmental Protection Agency, 1971, p. 187.

4.2-32   DISTRIBUTION OF PACKAGING OUTPUT,
1958–63[a]

Per Cent

End use	Consumer packaging expenditures	Industrial/ commercial packaging expenditures
Corrugated board[b]	45.0	55.0
Fold/san boxes	73.7	26.3
Set-up boxes	68.1	31.9
Wrappers	75.7	24.3
Labels	93.3	6.7
Fiber cans	100.0	c
Metal cans	88.3	11.7
Metal collapsible tubes	89.3	10.7
Aerosol packages	97.1	2.9
Aluminum foil	84.1	15.9
Closures	91.8	4.3
Glass containers	95.7	8.2
Polyethylene	92.7	7.3
Plastic jars	88.1	11.9
Cellophane	95.0	5.0
All packaging	77.1	22.9

[a] Expressed as a per cent of a five-year average dollar value of packaging, 1958–63.
[b] Estimated by Midwest Research Institute.
[c] Minimal.

Source: U.S. Department of Commerce, Business and Defense Services Administration, *Containers and Packaging,* *20*(2):8, July 1967. Modified by Midwest Research Institute.

4.2-33   PRINCIPAL INDUSTRIAL WASTE COMPONENTS

Source	Waste	Characteristics	Composition	Means of treatment or disposal
Food and kindred product industries[2]	Fruit, vegetable and citrus[5]		Hull, rinds, cores, seeds, vines, leaves, tops, roots, trimmings, pulps, peelings, hydrochloric acid[16] (used in processing)	Screening, lagooning, soil absorption, spray irrigation reclamation[5,6]
Canning[2]	Cobs, shells, stalks, straws[2,6]	High in suspended solids (liquid waste), colloidal and dissolved organic matter[2,15]		
Vegetable oil refining			"Still pitch" – tarry residue, fatty acids, sodium hydroxide, trichlorethylene[6]	Reclamation[6]
Dairy[4,16]	Dilutions of whole milk, separated milk, buttermilk whey[16]	High in dissolved organic matter, mainly protein, fat, lactose[16]	N, CaO, K_2O, P_2O_5, Fe, Cl, SiO_2[23]	Aeration, trickling filter, activated sludge[16]
Slaughtering of animals, rendering of bones and fats, residues in condensates, grease, wash water[16,24]	Manure, paunch manure, blood, flesh, fat particles, hair, bones, oil, grease[24]		N, NH_3, NH_2, NO_3, NaCl[24]	Reclamation, screening, trickling filters,[16] chlorination[15]
Breweries and distilleries[5,16]	Spent grain, spent hops, yeast, alkalis, amyl alcohol, dissolved organic solids containing nitrogen and fermented starches[5]	High in dissolved organic solids, containing nitrogen and fermented starches or their products[16]	Amyl alcohol[10] (from processing)	Recovery, centrifugation and evaporation, trickling filtration, stock feeds,[16] fertilizer[5]
Pharmaceutical[16]	Microorganisms, organic chemicals[16]	High in suspended and dissolved organic matter, including vitamins[16]	Aniline, phenols[10]	Evaporation, incineration, stock feeds[16]
Textile mill products[16]	Textiles, i.e., cotton, wool, and silk[15]	Highly alkaline, colored, high BOD and temperature, high suspended solids[15,16]	H_2SO_4, NaOH, aniline, chlorine,[10] starch, malt, tin and iron salts, dyes, bleach, fibers, minerals[15]	Neutralization, precipitation, trickling filtration, aeration, recovery[6,16]

4.2-33   PRINCIPAL INDUSTRIAL WASTE COMPONENTS (continued)

Source	Waste	Characteristics	Composition	Means of treatment or disposal
Cooking of fibers, desizing of fabrics[16]		Same as textile mill products	For complete list of chemicals used in textile industry, see Reference 21	
	Rayon, other man-made materials, i.e., Acrilan, Dynel, Orlon, Nylon, etc[21]	Acidic, alkaline, inorganic[15]	Sulfides and polysulfides, colloidal sulfur, NaOH, H_2SO_4, $ZnSO_4$, HCl, $NaHSO_4$, H_2S, $CaSO_4$;[15] acrylonitrile, phenol, HNO_3;[10] ammonia, adiponitrile, hexamethylenediamine, sodium carbonate, alcohols, ketones[25]	Reclamation, neutralization, trickling filtration, lagooning[6,15]
Laundry[15,16]		High turbidity and alkalinity[15,16]	Spent soaps, synthetic detergents, bleaches, dirt, grease[5,15]	Screening, precipitation, flotation, adsorption[16]
Lumber and wood products (forest, mills, factories)[5,6,15,16]	Pulp and paper[5,16]	High or low pH; colored; high suspended, colloidal, and dissolved solids; inorganic fillers[16]	Sawmill usage (sawdust, shavings, wood chips), wood flour,[6] soda, sulfate, sulfite[15]	Reclamation, incineration, soil conditioning[6]
	Organic, inorganic, toxic, suspended and dissolved solids of lignin, resins, soda ash, fiber, adhesives, ink, fats, soaps, tallow[15]		Sodium lignate, sodium resinate, complex organosulfur compounds, some fiber in relatively dilute solutions,[18] sulfites,[9] mercaptans, sulfides, disulfides, sulfates, terpenes, carbohydrates, CaO, SO_2, N, PO_4[15]	Reclamation, settling, lagooning, biological treatment aeration[6,16]
Chemical plants (general)		Toxic[10,15]	Acrylonitrile, aniline,[a] amyl alcohol, carbon disulfide, carbon tetrachloride, chlorine, hydrogen cyanide, hydrochloric acid, phenol, sulfuric acid, toluene, xylene, dinitrobenzene, dimethyl sulfate, ethylene, chlorohydrin, ben-	Reclamation, lagooning and all other known methods of treatment[14]

4.2-33 PRINCIPAL INDUSTRIAL WASTE COMPONENTS (continued)

Source	Waste	Characteristics	Composition	Means of treatment or disposal
Chemical plants (*continued*)			zene, metallic compounds of lead, arsenic, mercury[18]	
	Fumes and/or dust[36]		Arsenic[36]	
	Particulate clouds and dusts[7]		Mn, V, Cd, Be, Fe, Zn, and their oxides[7]	
	Weed killer[32]		2-4-D[32]	Sewage[32]
	Cyanide waste[13]	Toxic to aquatic life[13]	Cyanides[13]	Ponding[13]
	Plastics, synthetic resins		Acrolein, acrylonitrile, formaldehyde, phenols, trichlorethylene[9]	Reclamation, incineration[6]
Aircraft manufacturing industry[34]	Cd and hexavalent[34]	Traces of metals[34]	Cd and hexavalent[34]	Leaching pits[34]
Waste treatment plants[29]	Well-digested sludge[29]	Blackish, amorphous, nonplastic material[29]	Mg, Ca, Zn, Cr, Sn, Mn, Fe, Cu, Pb[29]	Anaerobic decomposition of organic waste solids[29]
Petroleum industry[16]	Spent chemical[31]	Liquid wastes with oil, acid and alkaline solutions, inorganic salts, organic acids and phenols, etc[31]	Clays, H_2SO_4, H_3PO_4[31]	Streams[31]
Drilling		Oil, brine, chemicals[15]	Sodium, calcium, magnesium, chlorine, SO_4, bromine[15]	Separation, evaporation, lagooning[15]
Storage	Muds, salt, oils, natural gas[16]			Separation, evaporation, lagooning[15]
Distillation[16]	Acid sludges, miscellaneous oils[16]	Insoluble organic and inorganic salts, sulfur compounds, sulfonic and naphthenic acids, insoluble mercaptides, oil-water emulsions, soaps, waxy emulsions, metal oxides, phenolic compounds[16,20]	Na_2CO_3, $(NH_4)_2S$, Na_2S, sulfates, acid sulfates, H_2S, NaOH, NH_4OH, $Ca(OH)_2$, $(NH_4)_2SO_4$, NH_4Cl, phenols[16]	Settling, filtration, reclamation, evaporation[15]

4.2-33   PRINCIPAL INDUSTRIAL WASTE COMPONENTS (continued)

Source	Waste	Characteristics	Composition	Means of treatment or disposal
Treating[16]		See "Distilla-tion"[16,20]	See "Distilla-tion"; also lead, copper, calcium[16]	Reclamation, settling, filtration, evaporation, neutralization[15]
Recovery[16]		See "Distilla-tion"; also organic esters[16,20]	See "Distilla-tion"; also iron[16]	See "Treating"[15]
Leather and leather prod-ucts[15,16]	Tanneries[15]	Organic and inor-ganic, high BOD-lime sludge, hair, fleshing, tan liquor, bleach liquor, salt, blood, dirt, chrome[15]	Chromium, sulfu-ric acid, nitrogen, $CaCO_3$, P_2O_5, K_2O, Fe^{22}	Sedimentation, lagooning[15]
Energy pro-ducing in-dustry[3,8,11]	Fly ash[8]	Hollow spheres of fused or partially fused silicate glass or as small solid spheres of fused silicates, iron ox-ides or silica, un-burned carbon and mineral[8]	Silicates, iron oxide, silica[8]	Sold for use in concrete,[8] landfills, etc
Pulverized coal-fired plants; stoker-fired, cyclone-fired plants; wet-bottom pulverized coal-fired plants				
Electrical industry[27]	Ash[27]	Dust[27]	Silicates and aluminates of Fe, Cu, Mg with small percentages of Na, K[27]	
Metal finishing industry[5,12,17,19,33]	Pickling and washing liquors[5]	Toxic, waste waters[5]	Cu and Cu alloys[5]	Sewage[5]
	Acid wastes[17]	Harmful to aquatic life,[17] salts of metals[12,17,19,33]	Cu, Ni, Zn, Cr, $Fe^{10,12,17,19}$	Sewage[12,17,19,33]
Rubber and mis-cellaneous plastic products[16]	Rubber[16]	High BOD, odor, high suspended solids, variable pH, high chlorides[16]	Sulfuric acid, tri-chlorethylene, xy-lene, amyl alcohol, aniline, benzene, chromium formal-dehyde[10]	Aeration, chlo-rination, sulfona-tion, biological treatment[16]
Washing of latex; coagulated rubber; exuded impurities from crude rubber; rejects, cuttings, mold flashings, trims, excess ex-trusions[6,16]	Scraps from molding, extrusion, rejects, trimming, finishing[6]			

4.2-33   PRINCIPAL INDUSTRIAL WASTE COMPONENTS (continued)

Source	Waste	Characteristics	Composition	Means of treatment or disposal
Explosives,[16] washing TNT and guncotton for purification, washing and pickling of cartridges[16]		TNT, colored, acid, odorous, and contains organic acids and alcohol from powder and cotton, metals, acid, oils and soaps[16]	H_2SO_4, HNO_3, NO_2, SO_3, picric acid, TNT isomers, copper, zinc, nitrogen, toluene[15]	Dilution, neutralization, lagooning, flotation, precipitation, aeration, chlorination[16]
Phosphate and phosphorus[16]	Washing, screening, floating rock, condenser bleed-off[16]	Clays, slimes, tallows, low pH, high suspended solids[16]	Phosphorus, silica, fluoride	Settling, clarification (mechanical), lagooning[16]
Fertilizers			Nitrogen, phosphorus, potassium, sulfuric acid, traces of other chemicals	
Coke byproducts	Slag from ovens, ammonia still waste, spent acids and phenols	Suspended solids, volatile suspended solids, organic and NH_3-N, phenol, cyanide acids, alkalis[16]	Ammonia, benzene, H_2SO_4, phenol[16,20]	Discharged to sewers, dumped, incineration[20]
Industrial, not otherwise identified[1,21,26,30]	Inorganic industrial waste or stabilization[30]	Metals and compounds thereof[30]	Na, K, Ca, chlorides, sulfates, bicarbonates, nitrates, phosphates, fluorides, borates, chromates, etc[30]	
	Metallic fumes and dusts[26]		Pb, V, As, Be and compounds thereof[26]	
	Industrial wastes[1]	Mineral fines[1]	Chromates, heavy metals[1]	Underground aquifers[1]
	Laboratory wastes[35]		Metallic ions, phenolics, cyanides, oils, synthetic fibers, pharmaceuticals, rubber chemicals[35]	Landfill or dump[35]
	Industrial wastes	Toxic metals[10]	Pb, Be[10]	
Insecticides	Washing and purification of products[16]	High organic matter, toxic, acidic[16] Chlorinated hydrocarbons: toxaphene, benzene, hexachloride, DDT, aldrin, endrin, dieldrin, lindane,	Carbon, hydrogen, chlorine, carbon disulfide, carbon tetrachloride	See Chemical plants (general)

TABLE 4.2-33   PRINCIPAL INDUSTRIAL WASTE COMPONENTS (continued)

Source	Waste	Characteristics	Composition	Means of treatment or disposal
Insecticides (*continued*)		chlordane, methoxychlor, heptachlor[28]		
		Organic phosphorus compounds: parathion, Malathion, phosdrin, tetraethyl pyrophosphate[28]	Phosphorus, oxygen, carbon, hydrogen, carbon disulfide, carbon tetrachloride[28]	
		Other organic compounds[28]	Carbonates, dinitrophenols, organic sulfur compounds, organic mercurials, rotenone, pyrethrum, nicotine, strychnine[28]	
		Inorganic substances[28]	Copper sulfate, arsenate of lead, compounds of chlorine and fluorine, zinc phosphide, thallium sulfate, sodium fluoroacetate[28]	

[a] Most common and troublesome toxics.

Source: *Solid Waste/Disease Relationships,* U.S. Department of Health, Education, and Welfare, Public Health Service, 1967, p. 161.

REFERENCES

1. C.F. Gurnham, *Principles of Industrial Waste Treatment*, John Wiley & Sons, 1955.
2. *Fruit Processing Industry*, prepared through cooperation of Natl. Can. Assn. and Natl. Technical Task Comm. Ind. Wastes, U.S. Department of Health, Education, and Welfare, 1959.
3. *Commercial Laundering Industry*, Am. Assn. Text. Chem. and Color. in cooperation with the Am. Inst. Laund., U.S. Department of Health, Education, and Welfare, 1956.
4. *Milk Processing Industry*, Dairy Ind. Comm. in cooperation with the Natl. Task Comm. Ind. Wastes, U.S. Department of Health, Education, and Welfare.
5. G.M. Fair and J.C. Geyer, *Elements of Water Supply and Waste-Water Disposal*, John Wiley & Sons, 1965.
6. C.H. Lipsett, *Industrial Wastes and Salvage; Conservation and Utilization*, Atlas Publishing Company, 1951, 1963.
7. H.L. Green and W.R. Lane, *Particulate Clouds: Dusts, Smokes and Mists*, D. Van Nostrand Company, Inc.
8. M.J. Snyder, "Properties and Uses of Fly Ash," *Battelle Tech. Rev., 36*(2):168, February 1964.
9. L. Klein *et al., River Pollution. II. Causes and Effects*, Butterworths, 1962, p. 24.
10. J.B. Skinner, "Tabulating the Toxics," *Chem. Eng., 69*(12):183, June 11, 1962.
11. J.D. Watt and D.J. Thorne, "Composition and Pozzolanic Properties of Pulverised Fuel Ashes. I. Composition of Fly Ashes," *J. Appl. Chem.*, December 15, 1965, p. 585.
12. J.L. Kelch and A.K. Graham, "Electrometric System for Continuous Control of Reduction of Hexavalent Chromium in Plant Wastes," *Plating, 36*(1):1028, May 1949.
13. "Dilution," *Plating, 36*(1), May 1949.
14. A.C. Hyde, "Chemical Plant Waste Treatment by Ten Methods," *J. Water Pollut. Control Fed., 37*(11): 1486, November 1965.
15. E.B. Besselievre, *"Industrial Waste Treatment,"* McGraw-Hill Book Company, 1952.
16. N.L. Nemerow, *Theories and Practices of Industrial Waste Treatment*, Addison-Wesley Publishing Company, Inc., 1963.
17. "Metal Finishing Wastes," *Metal Progress, 83*(6):157, May 1963.
18. C.C. Porter and F.W. Bishop, "Treatment of Paper Mill Wastes in Biochemical Oxidation Ponds," *Ind. Eng. Chem., 42*(1):102, January 1950.
19. C.A. Walker and J.A. Tallmadge, "Metal Finishing Waste Reduction," *Chem. Eng. Prog., 55*(5):73, May 1959.
20. *Ohio River Pollution Control*, Part 2 Supplements, *An Industrial Waste Guide to the Canning Industry*, Report of the U.S. Public Health Service, 78th Congress, 1st Session, House Document No. 266, August 27, 1943.

TABLE 4.2-33   PRINCIPAL INDUSTRIAL WASTE COMPONENTS (continued)

21.   "The BOD of Textile Chemicals," *American Dyestuff Reporter*, Rhode Island Section Subcommittee on Stream Pollution, May 23, 1955, p. 355.
22.   H.B. Hommon, "Studies on the Treatment and Disposal of Industrial Wastes. 3. Purification of Tannery Wastes," Public Health Bulletin No. 100, 1919.
23.   H.B. Hommon, "Studies on the Treatment and Disposal of Industrial Wastes. 4. Purification of Creamery Wastes," Public Health Bulletin No. 109, 1921.
24.   D.D. Gold, "Summary of Treatment Methods for Slaughterhouse and Packing-house Wastes," *Eng. Exper. Stat. Bull. No. 17, Univ. of Tenn.*, 1953.
25.   M.E. Batz, "Deep Well Disposal of Nylon Waste Water," *Chem. Eng. Prog., 60*(10):85, October 1964.
26.   L.A. Chambers, "Where Does Air Pollution Come From?" *Proc. Natl. Conf. Air Pollut.*, November 18–20, 1958.
27.   A.A. Orning, "Industrial Sources of Air Pollution. 3. Electric Power Plants (Coal Fuel)," *Proc. Natl. Conf. Air Pollut.*, November 18–20, 1958.
28.   "Use of Pesticides," Wiesner Report, *Science, 2*:43, May 15, 1963.
29.   R.H.L. Howe, "Chemical Values of Digested Sludge and Activated Sludge for Chemical Wastes Degradation and Stabilization," *Water Sewage Works*, November 30, 1965, p. R-219.
30.   J.E. McKee, "The Impact of Industrial Wastes (on the Water Quality Equation)," Santa Ana River Basin Water Pollution Control Board, October 17, 1963.
31.   C. Phillips, Jr., "Treatment of Refinery Emulsions and Chemical Wastes," *Ind. Eng. Chem., 46*(2):300, 1954.
32.   "Ground Water Pollution in California Points to Industrial Waste Discharges," *Engineering News-Record, 137*:785, December 12, 1946.
33.   R. Lamb, "A Suggested Measure of Toxicity Due to Metals in Industrial Effluents, Sewage and River Water," *Int. J. Air Water Pollut., 8*(3-4):243, March-April 1964.
34.   "Underground Waste Disposal and Control," Task Group Report, *J. Am. Water Works Assoc.*, 1957, p. 1334.
35.   R.F. Weston, "Laboratory Waste Disposal," *Arch. Environ. Health, 10*:550, April 1965.
36.   A.N. Currie, "The Role of Arsenic in Carcinogenesis," *Br. Med. Bull., 4*:402, 1948.

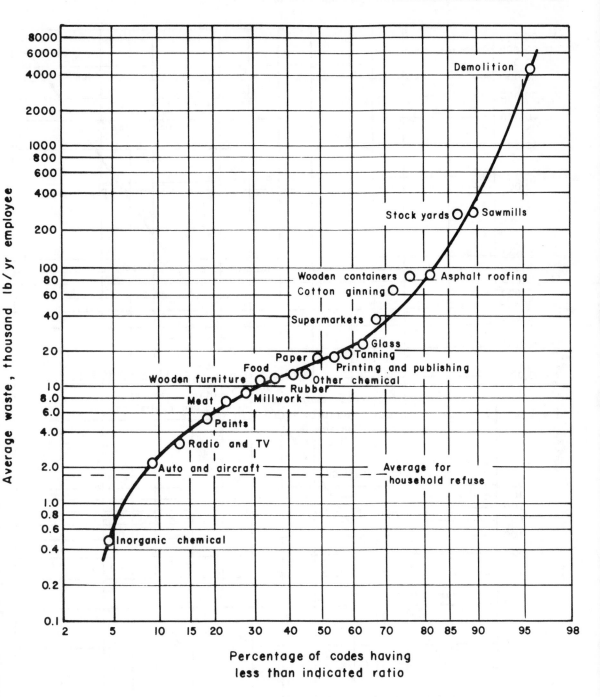

4.2-34   Distribution of waste for disposal: employee ratios in 21 SIC code groups.

Source: *Technical-Economic Study of Solid Waste Disposal Needs and Practices Industrial Inventory (Volume II),* U.S. Department of Health, Education, and Welfare, Public Health Service, 1969, p. 8.

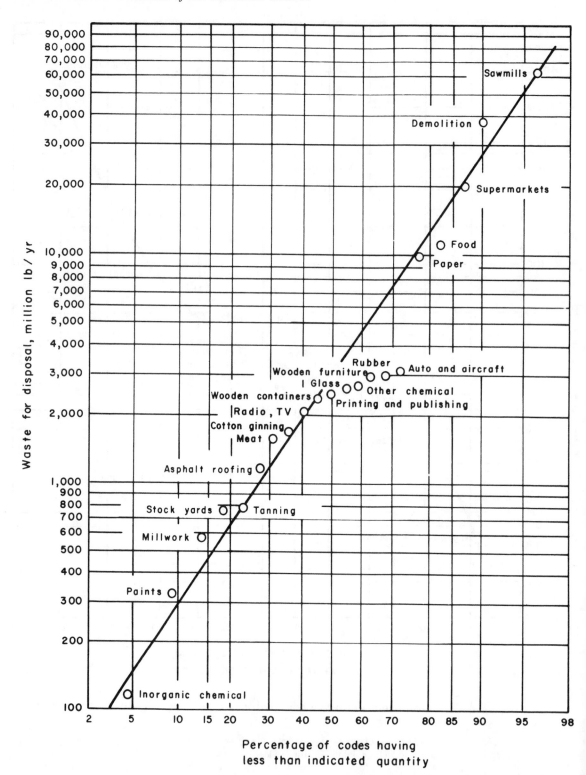

4.2-35   Distribution of waste for disposal among 21 SIC code groups.[a]

[a] Standard Industrial Classification.

Source: *Technical-Economic Study of Solid Waste Disposal Needs and Practices Industrial Inventory (Volume II),* U.S. Department of Health, Education, and Welfare, 1969, p. 7.

4.2.3   *AGRICULTURAL*

4.2-36   PER CAPITA ANIMAL CONTRIBUTION OF INDICATOR MICROORGANISMS

Animals	Average weight, feces/day, wet weight, g	Average indicator density/g of feces			Average contribution/ capita day		
		Mois-ture, %	Fecal coliform, million	Fecal strepto-cocci, million	Fecal coliform, million	Fecal strepto-cocci, million	Ratio
Man	150	77	13.0	3.0	2,000	450	4.4
Duck	336	61	33.0	54.0	11,000	18,000	0.6
Sheep	1,130	74	16.0	38.0	18,000	43,000	0.4
Chicken	182	72	1.3	3.4	240	620	0.4
Cow	23,600	83	0.23	1.3	5,400	31,000	0.2
Turkey	448	62	0.29	2.8	130	1,300	0.1
Pig	2,700	67	3.3	84.0	8,900	230,000	0.4

Sources: E.E. Geldreich, *Sanitary Significance of Fecal Coliforms in the Environment*, U.S. Department of the Interior, Federal Water Pollution Control Administration, 1966, and B.A. Kenner, H.F. Clark, and P.W. Kabler, "Fecal Streptococci: I. Cultivation and Enumeration of Streptococci in Surface Waters," *Appl. Microbiol., 9:*15, 1961.

4.2-37 ANIMAL WASTE DEFECATION

Per 1,000 Lb Liveweight

	Manure production					BOD production				Fertilizer nutrients			
		TS		VS									
lb/d	% wb	lb/d	% db	lb/d	lb/d	lb BOD[f]/lb TS	lb BOD/lb VS	BOD/COD[g] %	N, % db	P_2O_5, % db	K, % db	Ref.
Dairy cattle												
72	12.5	9.0	80[a]	7.2[a]	1.84	—	—	—	—	—	—	1
—	—	10.4	80.3	—	—	—	0.183	18.3	3.7	1.1	3.0	3
—	—	—	80[a]	—	—	0.102	0.129	8.2[a]	—	—	—	5
—	—	—	71.5[a]	—	—	0.278	0.388[a]	12.7[a]	2.8	1.04	0.34	6
105	9.0	9.4	—	—	—	—	—	—	—	—	—	b
—	—	—	—	—	—	—	0.232	—	5.5[a]	—	—	8
—	—	6.8	85[a]	—	1.32	—	—	22.8[a]	—	—	—	10
Average 88	—	9.0[c]	80.0[c]	5.7	—	—	0.233[c]	16.0[c]	4.0[c]	1.1[c]	1.7[c]	
Beef cattle												
—	—	—	73.2[a]	—	—	0.195[a]	0.267[a]	13.3	12.5[a]	1.52	0.44	6
—	—	3.6	86.5[a]	—	1.02	—	—	31.3[a]	7.2[a]	—	—	10
—	—	—	—	—	—	—	0.236	—	—	—	—	8
Average —	—	—	80.0[c]	—	—	—	0.252[c]	—	9.8[c]			
Poultry, hens												
64[a]	27.2	17.4[a]	70.3	12.2[a]	—	—	0.338	29.8[a]	23.3[a]	—	—	2
—	—	16.5[a]	77.5	12.8[a]	—	—	0.288	26.0	5.4	4.6	2.1	3
54	24.1	18.4	73.8	13.6	—	—	0.381	—	6.9	—	—	c
Average 59[c]	—	17.4[c]	74.0[c]	—	—	—	0.338[c]	28.0[c]	11.5[c]			

52	10.5	5.5	81.3	4.5	3.1	0.57	38.3	0.696	3.35	–	–	d
–	–	–	78.5	6.3	–	–	26.7[a]	0.320	4.0	3.1	1.4	3
49[a]	–	–	–	–	–	–	36.2[a]	–	–	–	–	4
–	–	–	–	–	–	0.262	19.3	0.302	–	–	–	5
–	15.4	–	85.0	–	–	0.450[a]	30.8[a]	0.382[a]	5.9[a]	–	–	7
50	17.0	8.5	83.0	7.0	–	–	45.0	0.540	7.0	–	–	8
–	–	–	–	–	–	–	–	0.270	–	–	–	e
–	–	–	80.3	4.5	–	–	41.2[a]	–	–	1.9	1.4	9
Average 50	–	–	82.0	5.9	–	–	33.0	0.363	5.6	2.5	1.4	
Sheep												
–	–	–	85.0[c]	–	–	0.074	6.2	0.087	–	–	–	5
37	–	8.4	–	–	–	–	–	0.116	–	–	–	b
–	–	–	79.0	–	–	0.104	–	0.101	–	–	–	e
Average 37[c]	–	8.4	82.0	–	–	–	–	–	–	–	–	

Note: lb/d = pounds per day; wb = wet basis; db = dry basis; TS = Total solids; VS = Volatile solids.

a Indicates value was calculated on the basis of data cited in the reference.
b W.B. Roller. Personal communication. Ohio Agric. Res. Center, 1968.
c E.P. Taiganides. Personal communication. Agric. Eng. Dept., Ohio State University, 1963.
d J.C. Converse. Personal communication. Agric. Eng. Dept., University of Illinois, 1970.
e E.P. Taiganides. Personal communication. Agric. Eng. Dept., Ohio State University, 1967.
f BOD (Biological Oxygen Demand) is a measure of the water pollution potential of an organic waste. It corresponds to the amount of oxygen required by the bacteria which consume the organic waste.
g COD (Chemical Oxygen Demand) is a measure of the water pollution potential of an organic waste. It corresponds to the amount of oxygen required for complete oxidation of the organic matter.

4.2-37   ANIMAL WASTE DEFECATION (continued)

REFERENCES

1. A.C. Dale and D.L. Day, "Some Aerobic Decomposition Properties of Dairy Cattle Manure," *Trans. Am. Soc. Agric. Eng., 10:*546, 1967.
2. J.R. Dornbush and J.R. Anderson, "Lagooning of Livestock Wastes in South Dakota," in *Proc. 1964 Ind. Waste Conf.,* Purdue Univ., 1965, p. 317.
3. S.A. Hart and M.E. Turner, "Lagoons for Livestock Manure," *J. Water Pollut. Control Fed., 37:*1578, 1965.
4. R.L. Irgens and D.L. Day, "Laboratory Studies of Aerobic Stabilization of Swine Waste," *J. Agric. Eng. Res., 11:*1, 1966.
5. E.A. Jeffrey, W.C. Blackman, and R.L. Ricketts, "Aerobic and Anaerobic Digestion Characteristics of Livestock Wastes," *Eng. Ser. Bull.,* 57, University of Missouri, 1964.
6. D.D. Jones, J.C. Converse, and D.L. Day, "Aerobic Digestion of Cattle Wastes," *Trans. Am. Soc. Agric. Eng., 11:*757, 1968.
7. C.K. Spillman, "Characteristics and Anaerobic Digestion of Swine Waste," M.S. thesis, University of Illinois, 1960.
8. E.P. Taiganides *et al.,* "Properties and Pumping Characteristics of Hog Wastes," *Trans. Am. Soc. Agric. Eng., 7:*123, 1964.
9. T.L. Willrich, "Primary Treatment of Swine Wastes by Lagooning," in Management of Farm Animal Wastes, *Am. Soc. Agric. Eng.,* 1966, p.70.
10. S.A. Witzel *et al.,* "Physical, Chemical and Bacteriological Properties of Bovine Animals," in *Management of Farm Animal Wastes, Am. Soc. Agric. Eng.,* 1966, p. 10.

4.2-38   CHARACTERISTICS OF ANIMAL MANURES

		lb/ton manure								
Animal	Moisture, %	N	P	K	S	Ca	Fe	Mg	Volatile solids	Fat
Dairy cattle	79	11.2	2.0	10.0	1.0	5.6	0.08	2.2	322	7
Fattening cattle	80	14.0	4.0	9.0	1.7	2.4	0.08	2.0	395	7
Hog	75	10.0	2.8	7.6	2.7	11.4	0.56	1.6	399	9
Horse	60	13.8	2.0	12.0	1.4	15.7	0.27	2.8	386	6
Sheep	65	28.0	4.2	20.0	1.8	11.7	0.32	3.7	567	14

Source: E.J. Benne *et al.,* "Animal Manures – What Are They Worth Today?" *Agric. Exp. Stn., Mich. State Univ. Bull.,* 1961. With permission.

4.2-39   CATTLE WASTE GENERATION RATES

Waste Characteristics	Generation, lb/day per head			Generation, lb/day per 1,000 lb live weight		
	Dairy bull[a]	Dairy cow	Beef cattle	Dairy bull[a]	Dairy cow	Beef cattle
Total wastes						
Volume (gal/day)	23.4[b]	8.6[c]	1.45[c]	11.0	5.8	2.64
5-day BOD	1.59	1.93	0.56	0.76	1.32	1.02
COD	8.80	8.64	1.79	4.19	5.78	3.26
Total solids	8.86	10.15	1.99	4.21	6.80	3.62
Volatile solids	6.86	8.49	1.74	3.26	5.68	3.17
Fixed solids	2.00	1.66	0.25	0.95	1.12	0.45
NH-Nitrogen	0.32	0.34	0.06	0.15	0.23	0.11
Organic-Nitrogen	0.19	0.21	0.08	0.09	0.14	0.15
Suspended solids[d]						
5-day BOD	0.92	1.02	0.41	0.44	0.70	0.75
COD	6.34	6.87	1.36	3.02	4.59	2.47
Total solids	4.07	6.62	1.26	1.94	4.42	2.29
Volatile solids	3.63	5.91	1.15	1.73	3.95	2.09
Fixed solids	0.44	0.71	0.11	0.21	0.47	0.20
Settleable solids[e]						
5-day BOD	0.19	0.53	0.23	0.09	0.36	0.42
COD	3.11	4.30	0.81	1.48	2.88	1.47
Total solids	6.18	8.38	1.59	2.94	5.60	2.89
Volatile solids	5.40	7.44	1.47	2.57	4.97	2.67
Fixed solids	0.78	0.94	0.12	0.37	0.63	0.22
Human population with equivalent generation						
5-day BOD basis[f]	8.6	11.1	3.3	4.0	7.7	6.0
Suspended solids basis[g]	19.4	31.5	6.0	9.2	21.0	10.9
Volume basis[h]	0.234	0.086	0.0145	0.11	0.058	0.0264

[a] Waste screened through No. 4 mesh.
[b] Liquid manure, including contained wash water.
[c] Raw waste only.
[d] Filtered through Whatman No. 1 filter paper.
[e] Based on 1-hr quiescent settling.
[f] Based on 0.17 lb BOD/capita day for domestic sewage.
[g] Based on 0.21 lb suspended solids/capita day for domestic sewage.
[h] Based on 100 gal/capita day for domestic sewage.

Source: S.A. Witzel *et al*. "Physical, Chemical, and Bacteriological Properties of Farm Wastes (Bovine Animals)," *Management of Farm Animal Wastes*, Am. Soc. Agric. Eng., *Proc. Natl. Symp. Anim. Waste Manage.*, Michigan State University, May 5–7, 1966.

4.2-40   POULTRY WASTE CHARACTERISTICS

Parameter	Value,[a] % total solids
Total nitrogen	1.8–5.9
Ammonia nitrogen, in water slurry	5,400 (mg/l N)
Alkalinity, in water slurry	21,000–52,000 (mg/l $CaCO_3$)
pH	6.4–7.0
P_2O_5	1.0–6.6
K_2O	0.8–3.3
Fe_2O_3	1.1
SO_3	0.4–1.2
CaO	4–12
MgO	0.4–1.2
ZnO	0.8
Total solids, %	10–50
Volatile solids, % dry solids	70–80

[a] Except as noted.

Source: E.A. Cassell and A. Anthonisen, *Studies on Chicken Manure Disposal: Part I, Laboratory Studies,* New York State Department of Health, 1966.

4.3   EFFECTS OF SOLID WASTES

4.3.1   HEALTH

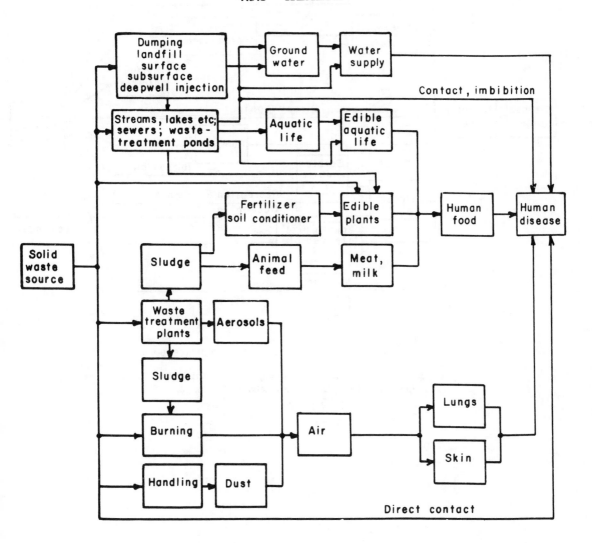

4.3-1   Chemical waste/human disease pathways (postulated).

Source: *Solid Waste/Disease Relationships,* U.S. Department of Health, Education, and Welfare, Public Health Service, 1967, p. 16.

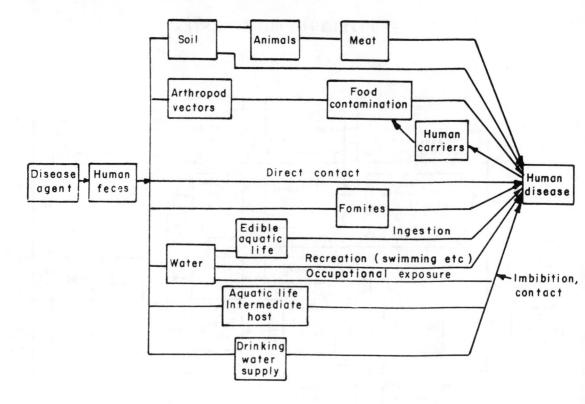

4.3-2   Human fecal waste/human disease pathways (postulated).

Source: *Solid Waste/Disease Relationships,* U.S. Department of Health, Education, and Welfare, Public Health Service, 1967, p. 52.

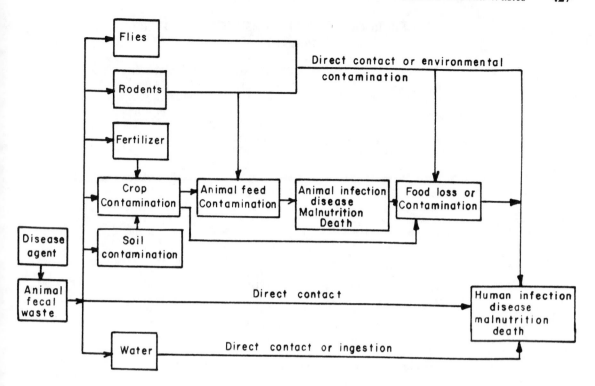

4.3-3   Animal fecal waste/disposal relationships (postulated).

Source: *Solid Waste/Disease Relationships,* U.S. Department of Health, Education, and Welfare, Public Health Service, 1967, p. 78.

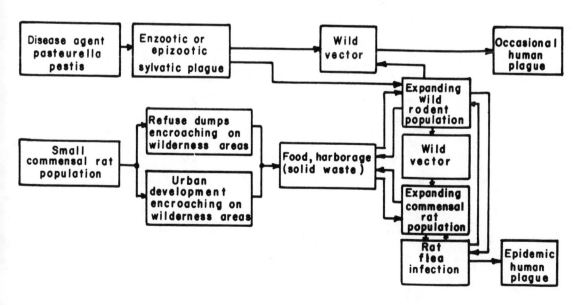

4.3-4   Solid waste/plague pathways (postulated).

Source: *Solid Waste/Disease Relationships,* U.S. Department of Health, Education, and Welfare, Public Health Service, 1967, p. 87.

4.3-5   MICROORGANISMS IN RESOURCE RECOVERY PLANTS

Organisms	Solid waste (CFU/g)	Aerosols (CFU/m^3)
Total plate count	10^7—10^8	10^3—10^7
Fecal coliforms	10^4	10^2—10^3
Staphylococcus aureus	10^4	10^2
Klebsiella pneumoniae	10^4—10^5	10^3
K. oxytoca	10^4—10^5	10^3
Salmonella	ND[a]	ND
Shigella	ND	ND
Legionella	ND	ND
Mycobacterium tuberculosis	ND	ND
Mycobacterium sp.	40	2×10^2
Streptomyces	10^5—10^6	10^3—10^4
Nocardia sp.	10^3	10^2—10^3
N. asteroides	ND	ND
N. brasiliensis	ND	ND
Aspergillis fumigatus	10^3	10^3—10^4
A. flavus	ND	10^3
Viruses	Low[b]	ND

[a] Not detected.
[b] One poliovirus 3 sample was positive, but too low to quantify.

Source: H.R. Pahren, *CRC Crit. Rev. Environ. Control, 17*:187, 1987.

REFERENCE

1. S.Z. Mansdorf, M.A. Golembieski, and M.W. Fletcher, Industrial Hygiene Characterization and Aerobiology of Resource Recovery Systems, Final Report on Contract No. 210-79-0013, National Institute for Occupational Safety and Health, Morgantown, W. Va., 1982.

4.3-6   STORAGE LIMITS FOR SOLID WASTES

Waste	Maximum period of source storage, days	Controlling environmental effects
Municipal Wastes		
Garbage	4	Flies
Residential rubbish	7	Land pollution, flies
Mixed garbage	4	Flies
Street refuse	7	Unsightliness
Dead animals	1	Flies, animal disease, land pollution
Abandoned vehicles	7	Unsightliness, land pollution
Demolition wastes	7	Land pollution
Construction wastes	7	Land pollution
Special wastes	1	Human disease
Sewage treatment residue	7	Flies
Water treatment residue	14	Water quality, air pollution
Ashes	14	Air pollution, unsightliness
Human fecal matter	1	Flies, human disease
Agricultural Wastes		
Barley	7	Plant disease, rodents
Beans, dry	7	Plant disease, rodents
Corn	7	Plant disease, rodents
Cotton lint	7	Plant disease, rodents
Cotton seed	7	Plant disease, rodents
Hay	7	Plant disease, rodents
Oats	7	Plant disease, rodents
Alfalfa	7	Plant disease, rodents
Rice	7	Plant disease, rodents
Safflower	7	Plant disease, rodents
Sorghum	7	Plant disease, rodents
Sugar beets	7	Plant disease, rodents
Wheat	7	Plant disease, rodents
Beans	7	Plant disease, flies
Cabbage	7	Plant disease, flies
Chinese vegetables	7	Plant disease, flies
Sweet corn	7	Plant disease, flies
Cucumbers	7	Plant disease, flies
Melons	7	Plant disease, flies
Onions	7	Plant disease, flies
Peppers	7	Plant disease, flies
Radishes	7	Plant disease, flies
Romaine	7	Plant disease, flies
Squash	7	Plant disease, flies
Sweet potatoes	7	Plant disease, flies
Tomatoes	7	Plant disease, flies
Turnips	7	Plant disease, flies
Almonds	7	Land pollution, flies
Apricots	7	Land pollution, flies
Bushberries	7	Land pollution, flies
Figs	7	Land pollution, flies
Grapefruit	7	Land pollution, flies
Grapes	7	Land pollution, flies
Lemons	7	Land pollution, flies
Nectarines	7	Land pollution, flies

4.3-6   STORAGE LIMITS FOR SOLID WASTES (continued)

Waste	Maximum period of source storage, days	Controlling environmental effects
Agricultural Wastes (*Continued*)		
Olives	7	Land pollution, flies
Oranges	7	Land pollution, flies
Peaches	7	Land pollution, flies
Persimmons	7	Land pollution, flies
Plums	7	Land pollution, flies
Pomegranates	7	Land pollution, flies
Strawberries	7	Land pollution, flies
Walnuts	7	Land pollution, flies
Beef cattle	7	Flies
Dairy cattle	7	Flies
Sheep	7	Animal disease
Hogs	7	Flies
Horses and mules	7	Flies
Chickens	7	Flies
Turkeys	7	Flies
Pigeons	7	Human disease
Rabbits	7	Flies
Livestock feed	7	Rodents, flies
Industrial Wastes		
Cotton trash	7	Land pollution
Fruit and vegetables	1	Land pollution, rodents, flies
Poultry	1	Animal disease, flies
Animal	1	Animal disease, land pollution
Milk solids	1	Land pollution
Wine and spirits	1	Rodents, flies
Vegetable oils	1	Land pollution
Tallow	1	Animal disease, land pollution
Cotton, wool, silk	7	Unsightliness
Lumber and wood products	14	Rodents, unsightliness
Chemicals	7	Safety hazards, toxicity
Petroleum	7	Safety hazards, toxicity
Plastics	14	Unsightliness, rodents
Masonry wastes	7	Land pollution
Metals	14	Land pollution, rodents
Seeds	7	Rodents, safety hazards
Tires	14	Rodents, other insects

Source: *A Systems Study of Solid Waste Management in the Fresno Area*, U.S. Department of Health, Education, and Welfare, Public Health Service, 1969, p. VI-51.

4.4   HANDLING AND DISPOSAL

4.4-1   SOLID WASTE DISPOSAL METHODS

Municipalities of Over 25,000,[a] 1966

Method	No. of cities
Composting	5
Incineration	190[b]
Landfill	738[c]

[a] In 1966 there were approximately 933 cities with a population of over 25,000.

[b] Some cities have more than one incinerator. Some incinerators handle refuse from several cities. There are a total of 250 incinerator plants in the U.S.

[c] Estimate.

Source: *Technical and Economic Study of Solid Waste Disposal Needs and Practices, Municipal Inventory (Volume 1),* U.S. Department of Health, Education, and Welfare, Public Health Service, 1969, p. 15.

4.4-2   SOLID WASTE DISPOSAL

1966 and 1968

Disposal method	Waste, 10^6 ton/yr	
	1966	1968
Municipal incineration	16	19
On-site incineration	57	55
Sanitary landfills	10	29
Open dumps	227	218
Burned	77	82
Nonburned	151	136
Wigwam burners	27	27
Hog feeding	1	1
Composting, treatment plants, etc	19	18

Source: *Nationwide Inventory of Air Pollutant Emissions, 1968,* U.S. Department of Health, Education, and Welfare, Public Health Service, 1970, p. 31.

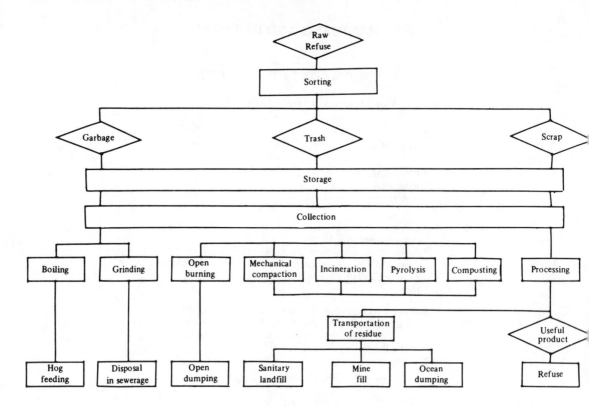

4.4-3   Waste disposal routes.

Source: A. Hershaft, "Solid Waste Treatment," *Science and Technology, 90:*34, June 1969. With permission.

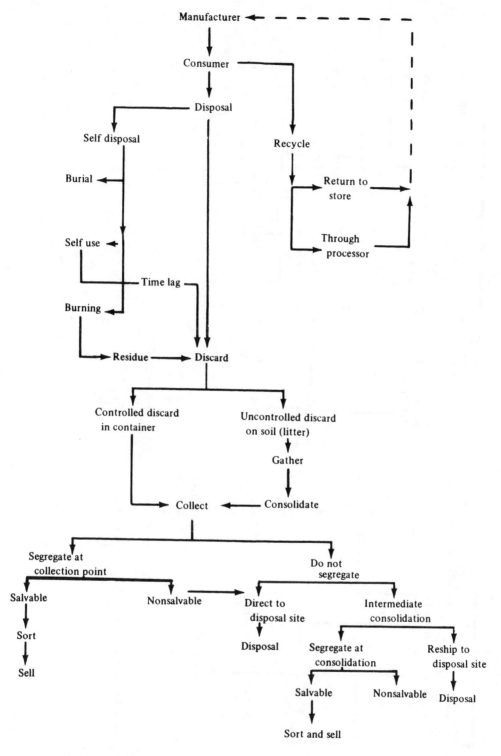

4.4-4   Solid waste flow from consumer to disposal site or recycling.

Source: Midwest Research Institute, Kansas City, Mo. With permission.

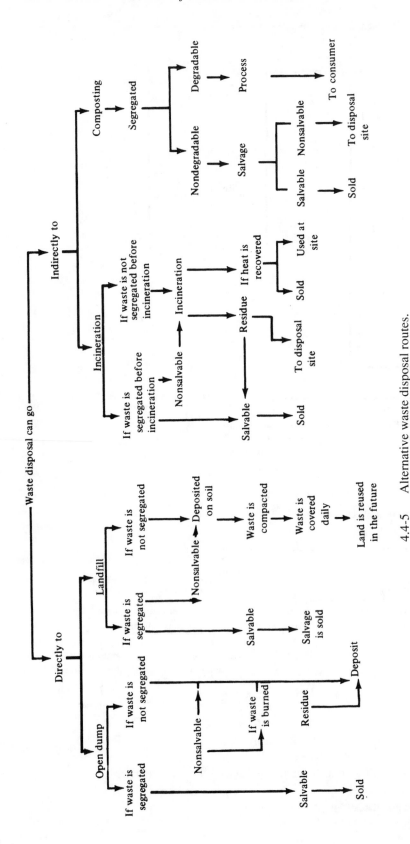

4.4-5   Alternative waste disposal routes.

Source: Midwest Research Institute, Kansas City, Mo. With permission.

4.4.1 RECYCLING AND RECOVERY

4.4-6 ESTIMATED LIFE IN YEARS OF HOUSEHOLD APPLIANCES BEFORE DISCARD

	Primary useful life[1]	Service Life Expectancy[2]		Estimated age at discard
		New when acquired	Used when acquired	
Room air conditioners	15			10
Dehumidifiers				20
Disposers				10
Kitchen ranges	15	16	8	20
Freezers		15	11	20
Refrigerators	12	15	8	20
Dishwashers				10
Washers	5	10	5	8
Dryers		14		15
Water heaters				10

Source: *Disposal of Major Appliances*, National Industrial Pollution Control Council, June 1971.

REFERENCES

1. Battelle Memorial Institute, *Final Report on a Survey and Analysis of the Supply and Availability of Obsolete Iron and Steel Scrap*, January 15, 1957.
2. J.L. Pennock and C.M. Jaeger, "The Household Service Life of Durable Goods," *J. Home Econ.*, January 1964.

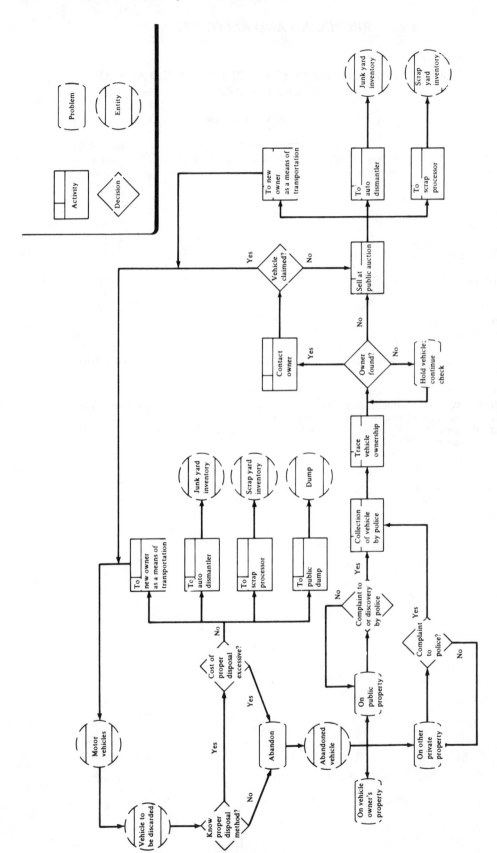

4.4-7   Vehicle disposal current flow.

Source: *Automobile Scrapping Processes and Needs for Maryland*, U.S. Department of Health, Education, and Welfare, Public Health Service, 1970, p. 61.

4.4.2   INCINERATION

4.4-8   QUANTITIES OF MATERIALS FOR INCINERATION

1967 as measured	Volume[c]		Weight[c]	
	yd³/wk	yd³/yr	lb/wk	ton/yr
1. Total wastes[a,b]	39,000	2,026,000	17,720,000	461,000
2. Wastes to incinerator[a]	30,000	1,577,000	9,363,000	243,000
3. Total special tree wastes[b]		423,000		94,000
4. Incineration potential, @ 58%		246,000		
@ 35%				33,000
5. New total wastes (Line 1 less line 3)		1,603,000		367,000
6. New waste to incinerator (Line 2 less line 4)		1,331,000		210,000
7. New waste to sanitary landfill		272,000		156,000
8. New per cent of waste to incinerator		83%		57%
9. 1967 total waste		2,480,000[b]		562,000
10. Less total special tree waste		440,000[b]		103,500
11. 1967 total less special tree waste		2,040,000		458,500
12. Quantity to incineration 1967				
83% x line 11		1,690,000		
57% x line 11				260,000
13. Quantity to incinerate, 20th yr				423,000
57% x 741,000				
14. Quantity to incinerate, avg yr				
1987 accumulation 13,563,000 tons				
1967 accumulation 562,000				
Special tree waste 870,000				
20 yr accumulation 12,131,000 tons				
Average yr 12,131,000 ÷ 20 x .57				346,000

[a] From computer program.
[b] Compacted in place quantity adjusted to "as received." 2.45 "as received" = 1 yd³ compacted.
[c] Adjusted for seasonal variation.

Source: *Collection and Disposal of Solid Waste for the Des Moines Metropolitan Area,* U.S. Department of Health, Education, and Welfare, Public Health Service, 1968, p. 4–5.

4.4-9   VOLUME REDUCTION BY INCINERATION

Starting with 2,000 lb of refuse, the comparable volumes are indicated below:

	As collected at source	Raw refuse landfilled	Incinerated and residue landfilled
Yd³	13.3	2.22	0.194
Vol ratio	68.5	11.5	1.0

Source: *Surgeon General's Conference on Solid Waste Management, Proceedings for Metropolitan Washington,* U.S. Department of Health, Education, and Welfare, Public Health Service, 1967, p. 99.

4.4-10   APPROXIMATE COMBUSTION CHARACTERISTICS OF VARIOUS KINDS OF MATERIALS AND AMOUNTS OF AIR NEEDED FOR COMBUSTION

Material	High heat value,[a] Btu per lb of MAF[b] waste	Air needed for complete combustion, lb per lb of MAF waste[c]
Paper	7,900	5.9
Wood	8,400	6.3
Leaves and grass	8,600	6.5
Rags, wool	8,900	6.7
Rags, cotton	7,200	5.4
Garbage	7,300	5.5
Rubber	12,500	9.4
Suet	16,200	12.1

[a] Values are necessarily approximate, since the ultimate composition of the combustible part of the materials varies, depending upon sources. The heating value of the material as it is received is obtained by multiplying the moisture-free and ash-free Btu value of the materials by $1 - (\% \text{ moisture} + \% \text{ ash})/100$. For example, garbage with an MAF value of 7,300 and containing 35 per cent moisture and 5 per cent ash or other noncombustible material will have an "as-fired" heating value of 4,380 Btu per pound.

[b] MAF means moisture-free and ash-free if ash refers to total noncombustible materials.

[c] These values are also approximate and are based on 0.75 pounds of air per 1,000 Btu for complete combustion. For various percentages of excess air, multiply these values by (100 plus per cent of excess air)/100. For example, if paper is burned with 100 per cent excess air (5.9) 200/100 = 11.8 pounds of air per pound of moisture-free and ash-free paper will be required.

Source: *Municipal Refuse Disposal*, Institute for Solid Wastes, Am. Publ. Works Assn., 1970, p. 174. With permission.

4.4-11   PRODUCTS OF INCINERATION

Stack gases	lb/ton	Volume, ft³	Dry vol, %
Carbon dioxide	1,738	14,856	6.05
Sulfur dioxide	1	6	22 ppm
Carbon monoxide	10	135	0.06
Oxygen	2,980	35,209	14.32
Nitrogen oxides	3	23	93 ppm
Nitrogen	14,557	195,690	79.57
Total dry gas	19,289	245,919	100.00
Water vapor	1,400	29,424	
Total	20,689	275,343	
Solids, dry basis			
Grate residue	471		
Collected fly ash	17		
Emitted fly ash	3		
Grand total	21,180		

Source: *Surgeon General's Conference on Solid Waste Management, Proceedings for Metropolitan Washington, D.C.,* U.S. Department of Health, Education, and Welfare, Public Health Service, 1967.

4.4-12　INCINERATOR RESIDUES

Material	Washington, D.C.,[1] metro-average, grate-type municipal incinerators, % dry weight	Rotary-kiln incinerators,[2] % dry weight
Tin cans	17.2	19.3 + 6.5 (non-metallics)
Mill scale and small iron	6.8	10.7
Iron wire	0.7	0.5
Massive iron	3.5	1.9
Nonferrous metals	1.4	0.1
Stones and bricks	1.3	
Ceramics	0.9	0.2
Unburned paper and charcoal	8.3	3.4 (charcoal)
Partially burned organics	0.7	
Ash	15.4	57.0
Glass	44.1	

REFERENCES

1. C.B. Kenahan and P.M. Sullivan, "Let's Not Overlook Salvage," *APWA Rep., 34*(3):5, 1967.
2. G. Rampacek, "Reclaiming and Recycling Metals and Minerals Found in Municipal Incinerator Residues," Proc. Mineral Waste Utilization Symp., March 27–28, 1968, IIT Research Institute, p. 129.

4.4-13　PARTICLE SIZE OF FLY ASH EMITTED FROM INCINERATORS

Per Cent by Weight

Size, μm	Incinerator guidelines[1]	Gansevoort incinerator, New York City[2]	South Shore incinerator, New York City[3]
<2	13.5		
<4	16.0		
<5		12.0	
<6	19.0		
<8	21.0		
<10	23.0	17.8	
<15	25.0	39.0	
<20	27.5	42.4	
<30	30.0	44.3	
<40		56.8	
<44			28.1
<60		70.0	
<74			66.5
<90		87.7	
<120		94.2	
>120		5.8	
<149			86.7
<250			95.5
<841			99.6
>841			0.4

REFERENCES

1. *Incinerator Guidelines,* U.S. Department of Health, Education, and Welfare, Public Health Service, 1969, p. 52.
2. *Municipal Refuse Disposal,* Institute for Solid Wastes, Am. Publ. Works Assn., 1970, p. 201.
3. *Municipal Refuse Disposal,* Institute for Solid Waste. Am. Publ. Works Assn., 1970, p. 202.

4.4-14   CHEMICAL ANALYSIS OF INCINERATOR FLY ASH

Component	Gansevoort incinerator, New York City,[1] % by weight	South Shore incinerator, New York City,[2] % by weight	Arlington, Va.,[3] %	Jens-Rehm study,[4] %	Kaiser study,[5] %
Carbon					
Organic	14.5	10.4	11.62		
Inorganic	85.5	89.6			
Silicon as SiO_2	36.0	36.1			36.3
Si			18.64	5+	
Aluminum as Al_2O_3	27.7	22.4			25.7
Al			10.79	1–10	
Iron as Fe_2O_3	10.0	4.2			7.1
Fe			2.13	0.5–5.0	
Sulfur as SO_3	9.7	7.6	small or trace		8.0
S					
Calcium as CaO	8.5	8.6			8.8
Ca			4.70	1.0+	
Magnesium as MgO	3.4	2.1			2.8
Mg			0.98	1–10	
Titanium as TiO_2					0.9
Ti			2.24	0.5–5.0	
Ni			small or trace	1–10	
Na			small or trace	1–10	
Zn			small or trace	1–10	
Ba			small or trace	0.1–1.0	
Cr			small or trace	0.1–1.0	
Cu			small or trace	0.1–1.0	
Mn			small or trace	0.1–1.0	
Sn			small or trace	0.05–0.5	
B				0.01–0.1	
Pb			small or trace	0.01–0.1	
Be				0.001–0.01	
Ag			small or trace	0.001–0.01	
V				0.001–0.01	
Sodium and potassium oxides	4.7	19.0			
Na as Na_2O					10.4
K as K_2O					
Na			small or trace		
K			small or trace		
Ga			small or trace		
Hg			small or trace		

4.4-14   CHEMICAL ANALYSIS OF INCINERATOR FLY ASH (continued)

Component	Gansevoort incinerator, New York City,[1] % by weight	South Shore incinerator, New York City,[2] % by weight	Arlington, Va.[3] %	Jens-Rehm study,[4] %	Kaiser study,[5] %
Mo			small or trace		
Ta			small or trace		
Apparent specific gravity	2.58				
Ignition loss			14.45		

REFERENCES

1. *Municipal Refuse Disposal,* Institute for Solid Wastes, Am. Publ. Works Assn., 1970.
2. *Municipal Refuse Disposal,* Institute for Solid Wastes, Am. Publ. Works Assn., 1970.
3. *Municipal Refuse Disposal,* Institute for Solid Wastes, Am. Publ. Works Assn., 1970.
4. W. Jens and F.R. Rehm, "Municipal Incineration and Air Pollution Control," *Proc. Natl. Incin. Conf.,* Am. Soc. Mech. Eng., 1966, p. 74.
5. E.R. Kaiser, "Refuse Composition and Flue-Gas Analyses from Municipal Incinerators," *Proc. Natl. Incin. Conf.,* Am. Soc. Mech. Eng., 1964, p. 35.

4.4-15   CHEMICAL ANALYSIS OF INCINERATOR SLAGS

Compound		Wet chemical analysis, %[1]	X-ray spectro-graphic analysis, %[2]	Average spectrochemical analysis[3] 25 samples from NY, NJ, Conn area, %	3 samples from New York area, %
Silicon dioxide	(SiO_2)	43.01–49.91	40–52	44.73	46.3
Aluminum oxide	(Al_2O_3)	8.73–24.85	8–25	17.44	18.35
Calcium oxide	(CaO)	9.28–11.03	9–11.5	10.52	9.94
Iron oxide	(Fe_2O_3)	6.00–12.78	5.5–8.5	9.26	8.85
Sodium oxide	(Na_2O)	3.16–3.31		6.09	3.25
Titanium oxide	(TiO_2)	2.40–3.31	2.2–3.5	2.92	2.90
Magnesium oxide	(MgO)	2.47–2.65		2.1	2.54
Phosphorus oxide	(P_2O_5)	2.00–2.40	2.0–2.5	1.52	2.15
Potassium oxide	(K_2O)	0.73–2.27	0.5–2.5	1.99	1.36
Zinc oxide	(ZnO)	0.46–2.49	0.25–2.75	1.54 (avg of 6)	1.37
Barium oxide	(BaO)	0.45–0.66		Trace	0.58
Manganese oxide	(MnO_2)			0.29	
Lithium oxide	(Li_2O)			0.06	
	(CuO)			Trace	
Lead oxide	(PbO)			Trace	
	(SO_3)			3.69 (avg of 6)	0.96 (avg of 2)
Equiv.	S			1.48	0.38

REFERENCES
1. D.B. Herbert, "The Nature of Incineration Slags," *Proc. Natl. Incin. Conf.*, Am. Soc. Mech. Eng., 1966, p. 191.
2. A.J. Regis, "X-ray Spectrographic Analysis of Incinerator Slags," *Proc. Natl. Incin. Conf.*, Am. Soc. Mech. Eng., 1966, p. 195.
3. E.R. Kaiser, *J. Air Pollut. Contr. Assn.*, *18*(3):171, 1968.

4.4-16   AMOUNT OF POLYNUCLEAR HYDROCARBONS, INCINERATOR EFFLUENT GAS[a]

	mg/g particulate
Benzo(*a*)pyrene	0.016
Pyrene	1.9
Benzo(*e*)pyrene	0.08
Coronene	0.06
Fluoranthene	2.2
Benzo(*a*)anthracene	0.09

[a] Data on a 250 ton/day municipal incinerator (breeching before settling chamber) when burning rubber tires.

Source: R.P. Hauzebrouck *et al.,* "Emissions of Polynuclear Hydrocarbons and Other Pollutants from Heat-Generation and Incineration Processes," *J. Air Pollut. Contr. Assn., 14*(7), 267, July 1964. With permission.

4.4-17 ANTICIPATED AIR POLLUTANT EMISSIONS IN INCINERATOR EFFLUENT GAS

Kind	Wood, wood products, plants and grass, food waste	Rubber	Plastics	Vary with excess air	Amounts reported
Organic acids					
Formic	x				25–133 ppm[3]
Acetic	x				
Palmitic	x				0.6 lb (as acetic acid)/ton refuse burned[2]
Stearic	x				
Oleic	x				
Palmitoleic	x				
Esters					
Methyl acetate	x				5–137 ppm[3]
Ethyl acetate	x				
Ethyl stearate	x				
Aldehydes					
Acetaldehyde	x		x	x	$1.7–3.9 \times 10^{-4}$ lb/ton;[1] 23.6×10^{-4}
Formaldehyde	x		x	x	1.1 lb/ton refuse burned[2]
CO	x		x	x	35–400 ppm;[3] 0.51 lb/ton
CO_2	x		x	x	
Hydrocarbons	x	x	x		1.4 lb/ton refuse[a]; 0.8 lb/ton with food waste[1]
Polynuclear hydrocarbons	x				
Halogenated hydrocarbons[a]			x		
Phosgene			x		
NH_3	x				0.44–10 ppm[3] 0.3 lb/ton refuse burned[2]
NO_2 (nitrogen dioxide)	x		x	x	0.15–1.5 ppm[3] 0.2–0.33 lb/1,000 lb dry flue gas[1] 2.1 lb/ton refuse burned;[2] 2.7 lb/ton with food waste[1]
SO_2 (sulfur dioxide) and SO_3 (sulfur trioxide)	x				SO_3 –3 and 8 per cent by weight collected and emitted respectively[4] SO^2 –0.25–1.2 ppm[3] SO_2 –1.9 lb/ton refuse burned[2]
Cl_2, F_2, HCN			x		Not reported

[a] Also from pressurized can chemicals.

Source: *Special Studies for Incinerators for the Government of the District of Columbia,* U.S. Department of Health, Education, and Welfare, Public Health Service, 1968, p. 7.

REFERENCES
1. R.L. Stenburg *et al.,* "Field Evaluation of Combustion Air Effects on Atmospheric Emissions from Municipal Incinerators," *J. Air. Pollut. Contr. Assn., 12*(2):83, February 1962.
2. *The Smog Problem in Los Angeles County,* Stanford Res. Inst., Western Oil and Gas Assn., 1954.
3. M.B. Jacobs *et al., Sampling and Analysis of Incinerator Flue Gases,* Proc. Air Pollut. Contr. Assn., May 25–39, 1968.
4. E.R. Kaiser, "Refuse Composition and Flue-Gas Analyses from Municipal Incinerators," *Proc. Natl. Incin. Conf.,* Am. Soc. Mech. Eng., May 18–20, 1964, p. 35.

4.4-18   ESTIMATED EMISSION RATES FROM INCINERATORS

	Pounds of contaminants per ton of fuel burnt	
Contaminant	Household	Municipal
Solids	46.3[a]	24[a]
Sulphur oxides as SO_2	2	2
Nitrogen oxides as NO_2	10.6	2
Ammonia	2.0	0.4
Acids as CH_3COOH	27.4	0.6
Aldehydes as HCHO	5.1	1.4
Other organics	274	1.2

[a] Ether soluble and insoluble aerosols.

REFERENCES
1. *The Smog Problem in Los Angeles County,* Stanford Research Institute, 1950.
2. G.P. Larson, G.I. Fischer, and W.J. Hamming, *Ind. Eng. Chem., 45:*1070, 1953.
3. P.L. Magell and R.W. Benolen, *Ind. Eng. Chem., 44:*1347, 1052.

Source: R. Rickles, *Pollution Control,* Noyes Data Corporation, 1965, p. 177. With permission.

4.4.3   *LANDFILLS*

4.4-19   SEASONAL AND COMPACTION FACTORS AND UNIT WEIGHTS FOR LANDFILLS

Materials	Seasonal[a] factor	Compaction[b] factor	Unit wt, lb/ft^3	Unit wt, lb/yd^3
Demolition–mixed, noncombustible	0.6	0.8	90	2,400
Demolition–mixed, combustible	0.6	0.7	22	600
Dirt, sand or gravel	0.6	0.8	90	2,430
Rock	0.6	0.6	90	2,430
Broken pavement or sidewalk	0.6	0.7	95	2,560
Construction mixed	0.6	0.7	60	1,620
Street sweepings	0.7	0.9	85	2,300
Logs and stumps 10″ diameter and greater	0.9	0.5	25	675
Logs and stumps less than 10″ diameter	0.9	0.5	25	675
Limbs and leaves chipped	0.6	0.4	12	320
Limbs and leaves not chipped	0.6	0.3	10	270
Brush	0.6	0.1	2	54
Grass and garden	0.6	0.2	5	135
Paunch manure	1.0	0.8	64	1,730
Pen sweepings	1.5	0.6	40	1,090
Other meat packing wastes	1.0	0.8	64	1,730
Poultry processing wastes	1.0	0.8	64	1,730
Tires and rubber products – new	1.0	0.2	15	400
Tires and rubber products – used	1.0	0.2	15	400
Rubber manufacturing wastes	1.0	0.3	55	1,500
Oil, tars and asphalts (liquid)	.8	0.1	60	1,620
Bean or grain wastes	1.0	0.8	48	1,300
Potato processing wastes	1.0	0.8	42	1,130
Other food processing waste	1.0	0.6	20	540
Fruits and vegetables	0.8	0.5	35	950
Ashes and cinders	2.0	0.9	45	1,220
Fly ash	1.0	1.0	80	2,160
Cement industry waste	1.0	1.0	90	2,400
Other fine particles	1.0	0.9	60	1,620
Garbage and kitchen waste, domestic	0.9	0.3	6.2	167
Garbage and kitchen waste, commercial	0.9	0.3	6.2	167
Mixed trash and refuse (incl. garbage)	0.9	0.2	5.2	140
Mixed trash and refuse (no garbage)	0.9	0.2	5.2	140
Incinerator residue domestic	1.0	0.7	30	810
Incinerator residue, commercial and industrial	1.0	0.7	30	810
Paper and cardboard	1.0	0.2	4.5	120
Cans	1.0	0.2	6	160
Furniture combustible	1.0	0.2	3	80
Furniture noncombustible	1.0	0.2	3	80
Major appliances	1.0	0.2	11	300
Heavy metal scrap	1.0	0.9	150	4,050
Light metal scrap	1.0	0.3	50	1,350
Wood crates	1.0	0.5	11	300
Glass and bottles	1.0	0.9	26	700
Battery cases and automotive	1.0	0.9	45	1,200
Automobile bodies	1.0	0.3	8	216
Wire	1.0	0.2	20	540
Chemical waste, dry	1.0	0.9	40	1,080
Chemical waste, liquid or wet	1.0	9	60	1,620
Sewage sludge solids	1.0	0.1	65	1,750
Sewage grit	1.0	0.9	80	2,200
Sewage screenings	1.0	0.9	60	1,600
Sewage grease skimmings	1.0	0.9	60	1,600

4.4-19 SEASONAL AND COMPACTION FACTORS AND UNIT WEIGHTS FOR LANDFILLS (continued)

[a] *Seasonal Factor Adjustment.* A factor was assigned to each material which would adjust the amount of material received during the day or week to the average daily or 7-day rate which would be experienced over the entire year.
[b] *Compaction Factor Adjustment.* A factor was assigned to each material which would adjust the material received during the day or week to reduce its volume "as received" to the volume which it would occupy when mixed with other materials and compacted into a landfill. Cover dirt is not included.

Source: *Collection and Disposal of Solid Waste for the Des Moines Metropolitan Area*, U.S. Department of Health, Education, and Welfare, Public Health Service, 1968, p. 2–30.

4.4-20 SANITARY LANDFILL REQUIREMENTS

The American Public Works Association lists the following requirements for a sanitary landfill:
1. Vector breeding and sustenance must be prevented.
2. Air pollution by dust, smoke, and odor must be controlled.
3. Fire hazards must be avoided.
4. Pollution of surface and ground water must be precluded.
5. All nuisances must be controlled.

Source: R.R. Fleming, "Solid Waste Disposal: Part I — Sanitary Landfills," *The American City*, 66:101, January 1966. Reprinted from the *American City Magazine*, January 1966. With permission.

4.4-21 ESTIMATION OF NECESSARY LANDFILL CAPACITY

An American Public Works Association formula provides a means for estimating the necessary landfill capacity:

$$V = \frac{FR}{D} (1 - \frac{P}{100})$$

where

V = landfill volume per capita per year in yd^3;
F = a factor incorporating cover material; averaging 17 per cent for deep fills and 33 per cent for shallow fills, with corresponding F values of 1.17 and 1.33;
R = amount of refuse in lb/capita yr;
D = average density of refuse in lb/yd^3;
P = per cent reduction of refuse volume in the landfill (0–90 per cent).

Completed landfills are most suitable for parks and recreation areas. Structures must incorporate provisions for settlement and methane gas disposal.

Source: R.R. Fleming, "Solid Waste Disposal: Part I — Sanitary Landfills," *The American City*, 66:101, January 1966. Reprinted from the *American City Magazine*, January 1966. With permission.

4.4-22   RESULTS OF SOIL ANALYSES FOR SELECTED LANDFILL SITES

Material description	Unified soil classification	Wet density,[a] lb/ft³	Moisture content, %	Dry density, lb/ft³	Porosity	Void ratio
Gravelly sand	SW	131.0	7.7	122.5	0.272	0.374
Silty shale	–	101.2	22.6	94.2	–	–
Silty clay	CL	128.5	11.1	115.5	0.294	0.416
Sandy silt	ML	110.5	14.8	101.5	0.364	0.573
Silty sand	SM	118.5	7.3	110.5	0.332	0.497
Silty clay (diatomaceous)	CL	51.5	16.0	44.4	0.704	2.38
Silty clay (diatomaceous)	CL	106.2	6.2	100.0	0.359	0.560
Sandy silt	ML	137.5	19.6	115.0	0.300	0.429
Silty clay	CL	99.0	14.5	86.3	0.429	0.750
Gravelly sand	SW-SM	111.5	7.8	103.5	0.382	0.618
Gravelly sand with silt	SM	114.5	4.4	109.5	0.346	0.530
Gravel-sand and silt mixture	GW-GM	101.5	3.2	98.2	0.418	0.707

[a] In place.

Source: *Development of Construction and Use Criteria for Sanitary Landfills,* County of Los Angeles, U.S. Department of Health, Education, and Welfare, Public Health Service, 1960, p. III-6.

4.4-23   MATERIALS LEACHED FROM REFUSE AND ASH

Based on Weight of Refuse as Received

Materials leached	Per cent leached[a]					
	1[a]	1[b]	2[c]	3[d]	3[e]	3[f]
Permanganate value						
30 min	0.039					
4 hr	0.060	0.037				
Chloride	0.105	0.127		0.11	0.087	
Ammoniacal nitrogen	0.055	0.037		0.036		
Biologic oxygen demand	0.515	0.249		1.27		
Organic carbon	0.285	0.163				
Sulfate	0.130	0.084 (as SO_4)		0.011	0.22	0.30
Sulfide	0.011					
Albuminoid nitrogen	0.005					
Alkalinity (as $CaCO_3$)				0.39	0.042	
Calcium				0.08	0.021	2.57
Magnesium				0.015	0.014	0.24
Sodium			0.260	0.075	0.078	0.29
Potassium			0.135	0.09	0.049	0.38
Total iron				0.01		
Inorganic phosphate				0.0007		
Nitrate					0.0025	
Organic nitrogen	0.0075	0.0072		0.016		

Conditions of leaching:
[a] Analysis of leachate from domestic refuse deposited in standing water.
[b] Analyses of leachate from domestic refuse deposited in unsaturated environment and leached only by natural precipitation.
[c] Refuse from Long Beach, California. Material leached in laboratory before and after ignition.
[d] Domestic refuse in Riverside, California, leached by water in a test bin.
[e] Leaching of California incinerator ash in a test bin by water.
[f] Leaching of southern California incinerator ash in a test bin by acid.

Source: G.M. Hughes, *Selection of Refuse Disposal Sites in Northeastern Illinois,* Ill. State Geological Survey, 1967, p. 6. With permission.

REFERENCES
1. *Pollution of Water by Tipped Refuse,* Ministry of Housing and Local Government, Her Majesty's Stationery Office, 1961.
2. J.M. Montgomery and R.D. Pomeroy, *Report of Investigation for City of Long Beach Regarding Probable Effects of Proposed Cut-and-Cover Trash Disposal,* Montgomery and Pomeroy, Engineers-Chemists, 1949.
3. *Effects of Refuse Dumps on Groundwater Quality,* California Resources Agency, 1961.

4.4-24 ANALYSIS OF LEACHATES FROM LANDFILL REFUSE

Determination, ppm	Source of sample			
	Test bin[a1]	Site no. 2 test fill[a2]	Site no. 8 test well[b]	Site no. 11 test well[c]
pH	5.6	5.9	8.3	7.2
Total hardness as $CaCO_3$	8,120	3,260	537	1,540
Calcium	2,570	905	72	302
Magnesium	280	254	87	192
Iron, total	305	33.6	219	2.0
Sodium	1,805	350	600	69
Potassium	1,860	655	NR	13
Boron	NR	7.13	Nil	1.15
Sulfate	630	1,220	99	615
Chloride	2,240	NR	300	NR
Nitrate	NR	5	18	NR
Alkalinity as $CaCO_3$	8,100	1,710	1,290	974
Ammonia nitrogen	845	141	NR	0.3
Organic nitrogen	550	152	NR	2.1
COD	NR	7,130	NR	36
BOD	32,400	7,050	NR	Nil
Total dissolved solids	NR	9,190	2,000	1,960

NR – No result.
[a] Sample from sumps at bottom of test fills.
[b] Sample of water found in refuse drilling of test well.
[c] Sample from test well located 50 feet from large refuse fill.

Source: *Development of Construction and Use Criteria for Sanitary Landfills,* U.S. Department of Health, Education, and Welfare, Public Health Service, 1969.

REFERENCES

1. "Investigation of Leaching of a Sanitary Landfill," California State Water Pollution Control Board, 1954.
2. "In-Situ Investigation of Movement of Gases Produced from Decomposing Refuse," California State Water Quality Control Board, November 1965.

4.4.4   COMPOSTING

4.4-25   TIME-TEMPERATURES REQUIRED FOR ORGANISM DESTRUCTION

Pathogen destruction during the composting process occurs primarily as a result of thermal kill and kill by antibiotic action or by the decomposing organisms and their products. The following table gives a list of key pathogens and their thermal death points:

Organism	Destruction temperature, °F	Time at temperature, min.	Destruction temperature, °F	Time at temperature, min
Salmonella typhosa	131–140	30	140	20
Salmonella sp.	131	60	140	15–20
Shigella sp.	131	60	–	–
Ent. hystolica Cysts	113	Few	131	Few seconds
Taenia saginata	131	Few	–	–
Trichinella spiralis Larvae	131	Quickly	140	Instantly
Brucella abortis, Br. suis	144–145	3	131	60
Micrococcus pyogenes var. aureus	122	10	–	–
Streptococcus pyogenes	129	10	–	–
Mycobacterium tuberculosis var. hominis	151	15–20	152.6	Momentarily
Corynebacterium diptheriae	131	45	–	–
Necator americanus	113	50	–	–
Ascoris lumbricoides Eggs	122	60	–	–
E. coli	131	60	140	15–20

Source: J.S. Wiley, "Pathogen Survival in Composting Municipal Wastes," *J. Water Pollut. Contr. Fed., 34:*80, 1962. With permission.

4.4-26   ANALYSIS OF 42-DAY-OLD COMPOST

Johnson City, Tennessee

Element	Per cent dry weight (average)		Range (all samples)
	Containing sludge, 3–5%	Without sludge	
Carbon	33.07	32.89	26.23–37.53
Nitrogen	0.94	0.91	0.85–1.07
Potassium	0.28	0.33	0.25–0.40
Sodium	0.42	0.41	0.36–0.51
Calcium	1.41	1.91	0.75–3.11
Phosphorus	0.28	0.22	0.20–0.34
Magnesium	1.56	1.92	0.83–2.52
Iron	1.07	1.10	0.55–1.68
Aluminum	1.19	1.15	0.32–2.67
Copper	<0.05	<0.03	
Manganese	<0.05	<0.05	
Nickel	<0.01	<0.01	
Zinc	<0.005	<0.005	
Boron	<0.0005	<0.0005	
Mercury	not detected	not detected	
Lead	not detected	not detected	

Source: A.W. Breidenbach *et al., Composting of Municipal Solid Wastes in the United States,* U.S. Environmental Protection Agency, 1971, p. 47.

4.4.5 PYROLYSIS

4.4-27 PYROLYSIS PRODUCT YIELD

Per Cent by Weight

Temperature, °F	Gases	Pyroligneous acids and tars	Char	Mass accounted for
900	12.33	61.08	24.71	98.12
1200	18.64	59.18	21.80	99.62
1500	23.69	59.67	17.24	100.59
1700	24.36	58.70	17.67	100.73

Source: Reprinted with permission from D. Hoffman, *Environ. Sci. Technol.*, 2(11):1025, November 1968. Copyright 1968, American Chemical Society.

4.4-28 GASES EVOLVED BY PYROLYSIS

Per Cent by Volume

Constituent	900°F	1200°F	1500°F	1700°F
H_2	5.56	16.58	28.55	32.48
CH_4	12.43	15.91	13.73	10.45
CO	33.50	30.49	34.12	35.25
CO_2	44.77	31.78	20.59	18.31
C_2H_4	0.45	2.18	2.24	2.43
C_2H_6	3.03	3.06	0.77	1.07
Accountability	99.74	100.00	100.00	99.99

Source: Reprinted with permission from D. Hoffman, *Environ. Sci. Technol.*, 2(11):1025, November 1968. Copyright 1968, American Chemical Society.

4.4.6 MARINE DISPOSAL

4.4-29   ESTIMATED POLLUTED DREDGE SPOILS

Area	Total spoils, tons	Estimated per cent of total polluted spoils	Total polluted spoils, tons
Atlantic	30,880,000	45	13,896,000
Gulf	13,000,000	31	4,030,000
Pacific	8,320,000	19	1,580,800
Totals	52,200,000	34	19,506,800

Source: Council on Environmental Quality, *Ocean Dumping; A National Policy,* U.S. Government Printing Office, 1970, p. 3. Table revised and updated by James L. Verber, Food and Drug Administration.

4.4-30   MAJOR INDUSTRIAL WASTES BARGED TO SEA

Types of industrial waste	Per cent of total industrial wastes	
	Tonnage	Cost
Waste acid	58	$59
Refinery wastes	12	7
Pesticide wastes	7	15
Paper mill wastes	3	7
Others	20	12

Sources: *J. Sanit. Eng. Div.,* Am. Soc. Civ. Eng., *96:* 1387, 1970 and *Ocean Disposal of Barge-Delivered Liquid and Solid Wastes from U.S. Coastal Cities,* U.S. Environmental Protection Agency, 1971, p. 22.

Section 5

Institutions

5.1   MICROBIOLOGICAL CONSIDERATIONS

5.1.1   *ENVIRONMENTAL MICROBIOLOGY*

5.1-1   SUMMARY OF AIRBORNE CONTAMINATION LEVELS IN VARIOUS HOSPITAL LOCATIONS

Location	No. of samples	Mean no. colonies/ft^3	Kind of microbes found, % of total					
			a	b	c	d	e	f
Chute closets and other waste storage areas	246	72.4	42.9	27.0	11.5	5.7	1.0	5.5
Laundry room	217	27.5	24.8	32.5	10.2	17.2	3.1	4.0
Corridors	891	20.6	32.5	18.9	26.8	15.4	0.8	2.3
Patient rooms	271	20	41.1	12.9	13.6	26.7	0.3	0.5
Substerilizing rooms	94	19.5						
Incinerator room	111	18.6	35.2	37.1	11.1	6.2	2.5	6.2
Utility room	710	18.1	49.6	14.3	5.9	11.8	2.9	3.4
Induction room	123	13.0						
Operating room	795	10.5	66.7	14.3	6.7	6.5	1.0	1.1
Obstetrical delivery room	90	4.5	44.4	24.9	13.3	9.1	1.4	0.3
Outside air (summer only)	138	29.8						

Note: a, Gram + cocci; b, Gram + rods, c, Gram − rods; d, molds, e, yeasts; f, actinomycetes.

REFERENCES

1. W.V. Greene *et al., Appl. Microbiol., 10*:561, 1962.
2. V.W. Greene *et al., Appl. Microbiol., 10*:567, 1962.

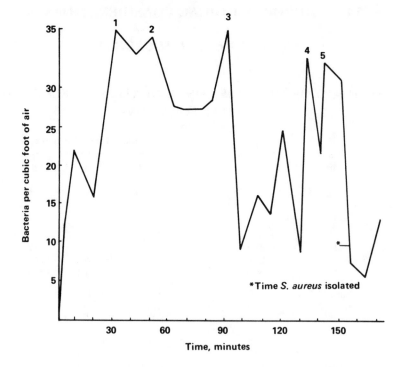

5.1-2   Results of continuous air sampling in Operating Room 2 during cholecystectomy Nov. 11, 1960, showing average number of bacteria per cubic foot of air.

Notes: (1) Patient brought in and operating team enters. (2) Circulating nurse leaves and reenters room. (3) Nurse enters with sterile packs. (4) Patient is removed and operating team leaves. (5) Cleanup period begins.

Source: Reprinted by permission from J.G. Schaffer and J. McDade, *Hospitals*, 38, 40, 1964. Copyright © 1964, American Hospital Publishing, Inc., Chicago.

5.1-3   COMPARISON OF MICROBIAL CHARACTERISTICS OF A LAMINAR CROSSFLOW OPERATING ROOM AND CONVENTIONALLY VENTILATED OPERATING ROOMS

Operating room	Air samples		Surgical instruments		Floor samples taken before surgery
	At wound	In room	Used	Unused	
	Organisms per 100 ft^3		Percent	Positive	Organisms per square inch
Laminar flow room	2.4	86	10	0	11.7
New surgical wing	60.0	120	50	38	2.3
Old surgical area	189.0	174	21	10	4.9

Source: Reprinted by permission from D.G. Fox and M. Baldwin, *Hospitals*, 42, 108, 1968. Copyright © 1964, American Hospital Publishing, Inc., Chicago.

5.1-4 PERCENTAGE REDUCTION OF BACTERIAL AND PARTICLE COUNT AT DIFFERENT VELOCITIES OF LAMINAR FLOW COMPARED WITH CONVENTIONAL VENTILATION

	Air velocity, meters per second					
	0.1	0.2	0.3	0.4	0.5	0.6
Bacteria						
Downflow	79.5	90.0	97.1	98.9	98.8	–
Crossflow	39.0	79.4	90.2	90.5	94.6	93.9
Particles $\geqslant 0.5\ \mu m$						
Downflow	88.6	91.7	94.8	96.9	98.0	–
Crossflow	74.8	87.1	90.1	91.1	92.0	94.5

Source: W. Whyte et al., *J. Hyg.*, 71, 559, 1973. With permission of Cambridge University Press.

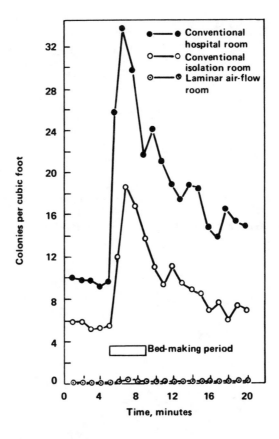

5.1-5   Air contamination in laminar airflow room and conventional isolation and hospital rooms measured by slit-samplers placed 3 ft inside the room entrances. Curve for laminar airflow room is mean of four experiments, others are mean of two experiments.

Source: C.O. Solberg et al., *Appl. Microbiol.*, 21, 209, 1971. With permission.

5.1-6 IDENTIFICATION OF ENVIRONMENTAL CONTAMINANTS FROM LAMINAR FLOW ROOM AND CONVENTIONAL ISOLATION AND HOSPITAL ROOMS

	Number of colonies					
	Laminar airflow room		Conventional isolation rooms		Conventional hospital rooms	
Organisms	Blankets and pillows	Floors, walls, furniture, and air	Blankets and pillows	Floors, walls, furniture, and air	Blankets and pillows	Floors, walls, furniture, and air
Bacillus sp.		16	14	81	10	63
Candida sp.			2	4	1	5
Clostridium sp.			1			
Diphtheroids	2	14	3	28	4	12
Enterobacter			2	2		1
Enterococcus	3	3	2	2	3	3
Escherichia coli	4	2	1	2	4	6
Hereliea sp.			1	1		1
Klebsiella sp.	4	1	2		2	9
Lactobacillus sp.			1	2	1	
Neisseria sp.		9	2	16	2	21
Proteus mirabilis	1	1	2	4	1	3
Pseudomonas aeruginosa	3	2	1	8	3	2
Saccharomyces				1		2
Staphylococcus aureus			2	9	4	11
S. epidermidis	33	68	20	61	22	79
Viridans group streptococcus				3		2
Unidentified molds		2	4	16	3	20
Total	50	118	60	240	60	240

Source: C.O. Solberg et al., *Appl. Microbiol.,* 21, 209, 1971. With permission.

5.1-7   EFFECT OF VARIOUS TREATMENT METHODS ON SURFACE AREAS IN HOSPITALS

Surface area	Treatment	Year	No. hospitals	Mean colony count/Rodac[a]	Range of colony counts
Overbed tables	Before cleaning	1964	9	100	44—193
	Before cleaning	1965	14	75	22—230
	After cleaning	1965	14	32	10—90
Patient room floors	Before cleaning	1964	9	210	158—322
	Before cleaning	1965	14	235	54—682
	After cleaning	1965	14	99	17—443
	Before mopping		4	161[b]	108—283
	After mopping		4	76[b]	38—231
	Before wet vacuum		8	178[b]	64—345]
	After wet vacuum		8	39[b]	6—162
	Before wet vacuum[c]	1967	13	156[b]	24—310
	After wet vacuum[c]	1967	13	46[b]	2—312
	Before wet vacuum[d]	1967	12	169[b]	87—425
	After wet vacuum[d]	1967	12	23[b]	5—85
	Before wet vacuum[e]	1968	12	93[b]	49—179
	After wet vacuum[e]	1968	12	19[b]	4—48
	Before wet vacuum[f]	1968	12	110[b]	17—211
	After wet vacuum[f]	1968	12	29[b]	3—111
	Before wet vacuum[g]	1968	12	94[b]	22—156
	After wet vacuum[g]	1968	12	17[b]	4—43

[a]   Rodac® plate approximately 4 in^2 in area.
[b]   Per room
[c]   Short contact time, average 3 min.
[d]   Long contact time, average 13 min.
[e]   Phenolic product.
[f]   All-purpose cleaner.
[g]   Quaternary compound.

REFERENCES

1. A.K. Pryor *et al., Health Lab. Sci., 4*:153, 1967.
2. D. Vesley *et al., Health Lab. Sci., 7*:256, 1970.

5.1-8   DISINFECTION AND BACTERIOLOGICAL CONTROL OF 73 SALINE-BATH TREATMENTS

	Number of times the listed organisms were grown		
Organisms	Before bath	During bath	After disinfection[a]
Pseudomonas aeruginosa	1	44	2[b]
Proteus spp.	0	30	3
Staphylococcus aureus	1	25	0
Coliforms	0	15	1
S. albus	5	0	1
Enterococci	1	11	0
Bacillus anthracoides	0	1	0
Streptococcus pyogenes	0	0	0
Streptomyces viridans	0	0	0
Diphtheroids	0	0	0

[a] Tego® (MHG) disinfectant used.
[b] Untrained staff used incorrect disinfection procedure on these two occasions.

Source: F. Dexter, *J. Hyg.,* 69, 179, 1971. With permission of Cambridge University Press.

5.1-9   BACTERIAL CONTAMINATION OF TROLLEY WHEELS

Hospital	Number of samples	Mean total organisms per plate	Mean *Staphylococcus aureus* per plate	% of plates showing *S. aureus*	Mean *Clostridium welchii* per plate
1 Theatre trolleys	24	550	18.6	12.5	3.13 ± 0.47
1 Hospital trolleys	24	834.5	18.6	41.7	67.90 ± 7.68
2 Hospital and theatre trolleys	20	287.6	6.0	35	45.10 ± 3.90

Source: G.A.J. Ayliffe et al., *J. Hyg.,* 67, 417, 1969. With permission of Cambridge University Press.

5.1-10 BACTERIAL CONTAMINATION OF TROLLEYS

Hospital	Site of sampling	Number of samples	Mean total organisms per plate	Mean *Staphylococcus aureus* per plate	Mean *Clostridium welchii* per plate
1 Theatre	Top	12	48.2	0.42	0.75
trolleys	Bars	12	89.3	2	–
	Handles	6	12.3	0	–
1 Hospital	Top	12	36.8	5.6	0.92
trolleys	Bars	12	88.7	0.17	–
	Handles	6	12.6	0	–
2 Hospital	Top	10	35.2	0.33	0.5
and	Bars	10	9	0.1	–
theatre	Handles	5	11.4	0	–
trolleys					

Source: G.A.J. Ayliffe et al., *J. Hyg.,* 67, 417, 1969. With permission of Cambridge University Press.

5.1-11 GEOMETRIC MEANS (LOG_{10}) OF SURVIVORS PER SQUARE CENTIMETER ON POLYESTER-COTTON SHEETING BEFORE WASHING, AFTER WASHING, AND AFTER WASHING AND DRYING[a]

Time of count	Wash temperature, °C	T3 phage	*Serratia marcescens*	*Staphylococcus aureus*	*Bacillus stearothermophilus*
Before washing	–	44.1	5.19	5.52	4.68
After washing	24	1.83	3.84	3.77	3.95
	35	1.69	2.41	3.96	4.01
	46	1.34	1.01	2.45	3.34
	57	0.69	0.0	1.03	3.17
	68	0.0	0.0	1.20	2.97
After drying	24	0.24	0.0	0.54	3.17
	35	0.0	0.0	1.12	3.14
	46	0.0	0.0	0.84	2.74
	57	0.0	0.0	0.45	2.73
	68	0.0	0.0	0.63	2.45

[a] A regular detergent was used with the permanent-press wash cycle, and a low (40°C) drying temperature was used.

Source: J.C. Wiksell et al., *Appl. Microbiol.,* 25, 431, 1973. With permission.

5.1-12   ARITHMETIC MEANS (LOG$_{10}$) OF MICROORGANISMS RECOVERED PER MILLILITER OF WASH WATER AT THE END OF THE WASH CYCLE AND OF RINSE WATER AT THE END OF THE RINSE CYCLE[a]

Sample	Wash temperature, °C	T3 phage	*Serratia marcescens*	*Staphylococcus aureus*	*Bacillus stearothermophilus*
Wash water	24	3.66	2.32	1.95	3.62
	35	3.08	1.48	1.54	3.36
	46	3.34	1.36	0.0	3.43
	57	1.30	0.0	0.70	3.54
	68	0.0	0.0	0.70	3.08
Rinse water	24	1.60	0.48	1.00	2.45
	35	0.90	0.0	1.36	2.46
	46	0.0	0.0	1.30	2.40
	57	0.0	0.0	1.36	2.43
	68	0.0	0.0	0.0	2.20

[a] Experimental conditions were the same as those listed for Table 2.1–105.

Source: J.C. Wiksell et al., *Appl. Microbiol.*, 25, 431, 1973. With permission.

5.1–13   AIRBORNE MICROBIAL COUNTS IN DISHWASHING FACILITIES

	Open alcove	Separate room	Physical separation
"Clean" area			
Number of samples	456	464	260
Mean number of colonies (ft^{-3})	12.4	13.9	16.2
Median of means	8.0	13.7	6.7
"Soiled" area			
Number of samples	457	465	263
Mean number of colonies (ft^{-3})	15.9	11.7	70.8[a]
Median of means	15.6	12.7	20.2
In operation			
Number of samples	913	929	523
Mean number of colonies (ft^{-3})	14.2	12.6	43.6
Median of means	12.8	13.7	9.6
Not in operation			
Number of samples	363	318	256
Mean number of colonies (ft^{-3})	13.7	14.6	14.4
Median of means	10.5	13.5	8.1

[a] One hospital had an extremely high count.

Adapted from: W.H. Jopke and D.R. Haas, *Hospitals, 44:*4, 126, 1970.

5.1–14 CHARACTERIZATION OF AIRBORNE CONTAMINATION IN DISHWASHING FACILITIES

Contaminant	Open alcove	Separate room	Physical separation	In operation	Not in operation
Number of isolates	5,727	5,973	2,998	10, 268	4,386
Gram-positive cocci (%)	33.8	34.2	29.1	36.7	24.6
Gram-positive rods (%)	4.1	4.7	2.5	3.8	4.4
Gram-positive, spore-forming rods (%)	0.5	0.7	0.5	0.7	0.6
Gram-negative rods (%)	2.1	2.7	3.1	2.4	2.6
Diphtheroids (%)	22.6	23.4	37.2	26.9	21.0
Actinomycetes (%)	1.1	1.8	1.3	1.8	3.1
Yeasts (%)	3.5	4.8	2.0	3.1	4.7
Molds (%)	32.3	27.8	24.2	24.6	38.9

Adapted from: W. H. Jopke and D.R. Haas, *Hospitals, 44:*4, 126, 1970.

5.1–15 MICROBIAL CONTAMINATION ON HOSPITAL TABLEWARE

Type of tableware[a]	Mean count after washing	Mean count during storage	Mean count before use
Flatware	3.9—24.2	5.5—15.8	3.4—14.6
Silverware	7.6—17.5	30.3—42.4	34.1—109.5

[a] Expressed as colonies per utensil for the flatware and colonies per Rodac® plate for the silverware.

REFERENCE

1. W.H. Jopke *et al., Hosp. Prog.*, *53:*31, 1972.

5.1-16 AVERAGE TEST RESULTS. CHUTES RANKED IN ORDER OF BACTERIAL SURFACE CONTAMINATION (CLEANLINESS)

Hospital	Chute No.	Number of floors served[a]	Department served	Approximate number of patients	Method of disposal	Chute vented[b]	Average counts on the chute surfaces			Average floor count	% of Staphylococcus aureus isolated from chute
							Test 1	Test 2	Mean		
G	11	1	Operating theatres	–	Linen bags	Yes	3.3	1.0	2.1	5.3	0
G	10	1	Operating theatres	–	Linen bags	Yes	5.1	1.8	3.4	12.3	2.0
C	9	3	Neurosurgical wards	60	Linen bags	Yes	6.3	2.7	4.5	107	0.4
C	4	4	Children's wards and operating theatres	100	Linen bags	Yes	3.8	11.7	7.7		
F	8	6	General wards and operating theatres	300	Polythene bags	Yes	6.8	10.6	8.7	125	0.6
E	7	1	Children's ward (infectious diseases)	25	Linen and polythene bags	No	7.2	11.8	9.5	88	0
D	6	5	Medical wards	270	Linen bags	Yes	8.1	19.2	13.6	93	1.1
B	2	2	Wards	95	Linen bags	No	13.7	21.4	17.5	109	0
A	1	6	Maternity wards and operating theatres	180	Loose	No	28.6 (9.6)[d]	26.9 (14.3)	27.7 (12.4)	–	0.1
B	3	2	Wards	80	Linen bags	No	12.3	47.8	30.1	169	0.2
D	5	5	Surgical wards	300	Linen bags	Yes	39.2	32.9	36.1	164	4.0
H[c]	12	3	General wards	66	Loose	Yes	41.5 (19.7)	—	41.5 (19.7)	422 163[e]	0.5
	13	3	General wards		Loose	Yes	53.8 (41.2)	—	53.8 (41.2)	385 126[e]	0.5

Note: All counts are given as average numbers of bacterial colonies per Rodac® plate.

a   The number of floors served does not include the basement or exit floor.
b   Vented by an opening at the top, either to outside or into roof space; none of the chutes had mechanical extract ventilation.
c   Both chutes at Hospital H served the same area, No. 12 being used for pre-rinsed soiled linen, No. 13 for dry dirty linen.
d   Figures in parentheses give the average chute counts with the results from the sloping entry connections omitted.
e   Average floor counts with the results from the basement floor omitted.

Source: W. Whyte et al., *J. Hyg.*, 67, 427, 1969. With permission of Cambridge University Press.

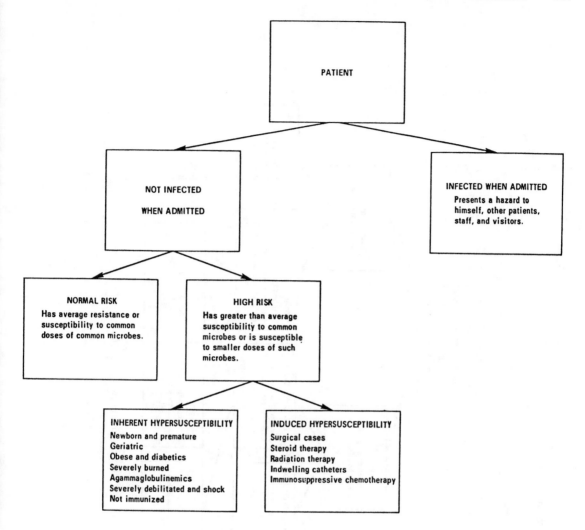

5.1-17   Classification of infection risks in hospitals.

Source: Reprinted by permission from V.W. Greene, *Hospitals, 44*:124, 1970. Copyright © 1964, American Hospital Publishing, Inc., Chicago.

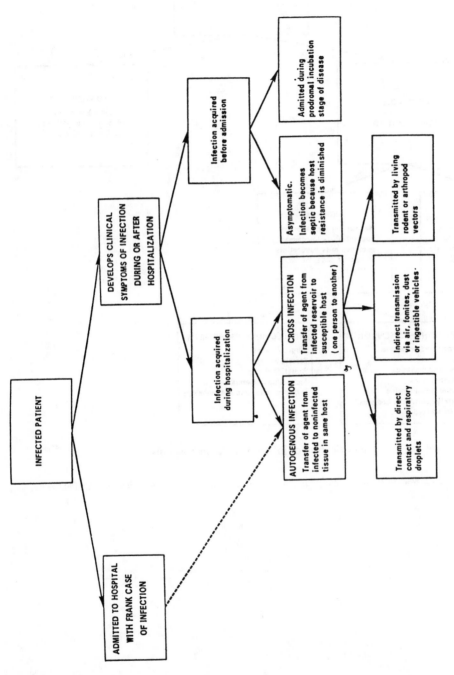

5.1-18   Classification of hospital infection situations according to source.

Source: Reprinted by permission from V. W. Greene, *Hospitals, 44*:124, 1970. Copyright © 1964, American Hospital Publishing, Inc., Chicago.

5.1-19 AIRBORNE BACTERIA INSIDE AND OUTSIDE PLASTIC VENTILATED ISOLATORS

	Settle plate counts[a]		Andersen sampler counts (total per cubic foot of air)		
	Mean counts per plate	Number of observations	Experiment 1 (quiet ward)	Experiment 2 (busy ward)	Experiment 3 (quiet ward)
Isolator with filters	0.1 (range 0–0.4)	5	0.2	<0.01	0.03
Isolator with coarse filter only	–	–	–	0.5	0.22
Isolator with no filter	1.0 (range 0–2.2)	10	1.2	–	0.13
Open ward	46.7 (range 11.0–82.8)	15	2.4	7.3	2.3

[a] Mean counts of colonies on 3½-in. (8.8-cm) plates exposed for 6 hr. Each observation represents a sampling with a number of settle plates on 1 day.

Source: E.J.L. Lowbury et al., *J. Hyg.*, 69, 529, 1971. With permission of Cambridge University Press.

5.1-20 CONTAMINATION OF VARIOUS ITEMS ISSUED TO PATIENTS

		Number of samples contaminated with			
		Staphylococcus aureus[a]		Gram-negative bacilli	
Items	Number of samples	+[b]	CM[c]	+	CM
Books, papers, etc.	20	–	–	2	1[d]
Washing bowls	16	1	3	–	3
Crockery, glassware	55	3	–	1	3
Cutlery	8	–	2	–	–
Clean pillows	7	1	–	–	1
Disposable bedpan supports	8	6	–	2	1[e]
Urine bottles	2	1	–	–	1
Toys	8	1	–	–	–
Receiving bowls	3	–	1	–	1
Foods (various)	45	1	1	4	5
Total	172	14	7	9	16

[a] All strains were found resistant to two or more antibiotics except those from food, which were not tested.

[b] + = growth on solid medium.

[c] CM = growth only in fluid medium (cooked meat).

[d] *Proteus* sp.

[e] *Pseudomonas aeruginosa.*

Source: E.J.L. Lowbury et al., *J. Hyg.*, 69, 529, 1971. With permission of Cambridge University Press.

5.1.2   *STERILIZATION, DISINFECTION, CLEANING TECHNIQUES*

5.1-21   COMPARISON OF VARIOUS PATIENT ROOM FLOOR CLEANING PROCEDURES WITH ARBITRARY GUIDELINES OF A.P.H.A. COMMITTEE ON MICROBIAL CONTAMINATION OF SURFACES

Method	Type of product	Number of participating hospitals	Percent of participating hospitals achieving result		
			Good (<25 colonies per Rodac® after cleaning)	Fair (26–50 colonies per Rodac after cleaning)	Poor (>50 colonies per Rodac after cleaning)
Unstandardized mopping (1965)	Variable	14	7	29	64
Double bucket and mop (1966)	Phenolic	4	0	25	75
Wet-vacuum pick-up – test area only (1966)	Phenolic	8	63	0	38
Wet-vacuum pick-up – entire room – 13-min contact time (1967)	Phenolic	12	67	25	8
Wet-vacuum pick-up – entire room – 3-min contact time (1967)	Phenolic	13	54	31	15
Wet-vacuum pick-up – entire room – 5-min contact time (1968)	Phenolic	12	75	25	0
	Quaternary ammonium	12	83	17	0
	All-purpose cleaner (no disinfectant)	12	50	42	8

Source: D. Vesley et al., *Health Lab. Sci.*, 7, 256, 1970. With permission.

5.1-22   THE EFFECT OF ULTRAVIOLET LIGHT BARRIERS IN AN AIRLOCK

Contact Plates from Floors of Plenum-ventilated Units
Mean bacterial count per plate

Site of sampling	UV on			UV off		
	Number of plates	Total organisms	Total *Staphylococcus aureus*	Number of plates	Total organisms	Total *S. aureus*
Cubicle	71	106.8 ± 18.8	3.6	68	269.9 ± 28.1	3.5
Airlock	81	12.8 ± 2.2	0.02	75	190.8 ± 19.1	3.6

Source: G.A.J. Ayliffe et al., *J. Hyg.,* 69, 511, 1971. With permission of Cambridge University Press.

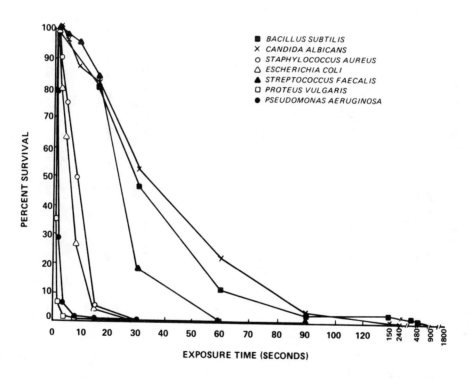

5.1-23   Germicidal efficiencies of the ultraviolet irradiated pass-through chambers of the patient isolator as determined for a select group of microbial species. The exposure distance of monocontaminated surfaces from two ceiling-mounted G-8T5[a] germicidal lamps was 12 in.; effective light transmission of the protective guards enclosing the lamps was 60%.

[a] Lamp Division, General Electric Co.

Source: S. Shadomy et al., *Arch. Environ. Health,* 11, 183, 1965. With permission. Copyright 1965, American Medical Association.

5.1-24 SUPPORT SERVICES FOR LAMINAR FLOW ROOMS

Service	Additional staff	Additional hours per patient	Training	Special supplies	Space	Turnover	Problem areas
Nursing	2 RNs and 1 LPN per 4 rooms, or 1 RN per 2 rooms; 1 RN per critically ill patient	—	5 days to 2 months (depends on sterile techniques background)	—	Twice as much room space per patient	Some immediately because of complicated protocols	Maintaining supply levels; staffing for critically ill patients
Central supply	2 per 4 rooms (with disposables); 4 per 4 rooms (no disposables); 1 1/3 per 2 rooms (with disposables); 20 hr per week per 2 beds for wrapping	—	2 days if personnel have experience	Disposables; gas sterilizer; aeration chamber	20 × 20 ft storage room area for wrapping	Low	Storage; projecting needs; maintaining stock
Pharmacy	None	1–5 hr per patient week	None	Special antibiotics; i.v. tubing	No additional	Low	None
Dietary	2 technicians, ½ dietitian (4 rooms, sterile diet); 1 technician, ¼ dietitian (2 rooms, cooked food diet)	—	3–4 weeks	Bags for sterilizing; individual portions of canned goods	Sterile diet kitchen or room for preparation and storage	Low	Storage, palatability of food
Housekeeping	Only for cleaning after patient discharge	—	4 hr instruction, plus supervised practice	Cleaning solutions, sterile mops, etc.	Storage closet for supplies	Low	Waste generation
Maintenance	None	1 hr per week (4 rooms)	1 week	Replacement parts	Parts storage	Low	Crisis maintenance must be inhouse; maintaining parts inventory

5.1-24 SUPPORT SERVICES FOR LAMINAR FLOW ROOMS (continued)

Service	Additional staff	Additional hours per patient	Training	Special supplies	Space	Turnover	Problem areas
Bacteriology	1 technician per 4 rooms; biologist or bacteriologist	–	Minimal if personnel have experience	–	–	Low	Careful monitoring of standing water required
Social and psychiatric services	Usually none	2–4 hr per week with patients and staff	None	–	–	Low	Providing recreational activities for patients; high referral rate of patients due to close staff observation

Source: Reprinted from E.C. Drazen and A.S. Levine, *Hospitals*, 48:89, 1974. Copyright © 1964, American Hospital Publishing, Inc., Chicago.

5.2 ENVIRONMENTAL HYGIENE AND RADIOLOGICAL HEALTH

5.2.1 VENTILATION AND AIR CONDITIONING

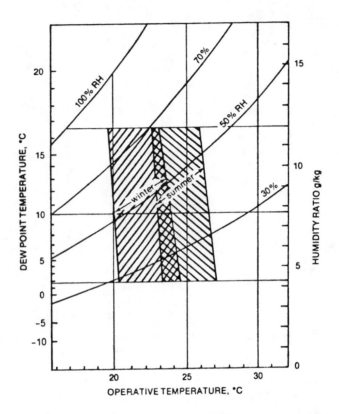

5.2-1   Acceptable ranges of operative temperature and humidity for persons clothed in typical summer and winter clothing, at light, mainly sedentary, activity.

Source: Reprinted by permission from the *ASHRAE Handbook — 1985 Fundamentals* Volume.

5.2-2   EVALUATION OF HEAT-STRESS INDEX

Index of heat stress	Physiologic and hygienic implications of 8-hr exposures to various heat stresses
−20 −10	Mild cold strain. This condition frequently exists in areas where men recover from exposure to heat.
0	No thermal strain.
+10 20 30	Mild to moderate heat strain. Where a job involves higher intellectual functions, dexterity, or alertness, subtle to substantial decrements in performance may be expected. In performance of heavy physical work, little decrement expected unless ability of individuals to perform such work under no thermal stress is marginal.
40 50 60	Severe heat strain, involving a threat to health unless men are physically fit. Break-in period required for men not previously acclimatized. Some decrement in performance of physical work is to be expected. Medical selection of personnel desirable because these conditions are unsuitable for those with cardiovascular or respiratory impairment or with chronic dermatitis. These working conditions are also unsuitable for activities requiring sustained mental effort.
70 80 90	Very severe heat strain. Only a small percentage of the population may be expected to qualify for this work. Personnel should be selected (1) by medical examination and (2) by trial on the job (after acclimatization). Special measures are needed to assure adequate water and salt intake. Amelioration of working conditions by any feasible means is highly desirable, and may be expected to decrease the health hazard while increasing efficiency on the job. Slight indisposition which in most jobs would be insufficient to affect performance may render workers unfit for this exposure.
100	The maximum strain tolerated daily by fit, acclimatized young men.

Source: American Society of Heating, Refrigerating, and Air Conditioning Engineers, *ASHRAE Guide and Data Book: Handbook of Fundamentals,* ASHRAE, New York, 1967, 120. With permission.

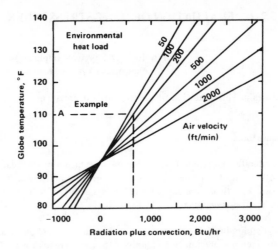

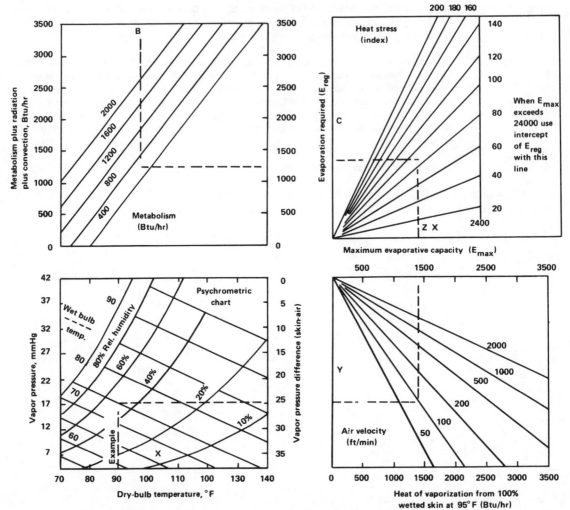

5.2-3   Charts for determining heat stress.

Note: Data are for an 8-hr day exposure, whereas a 24-hr exposure will be the norm for institutional patients.

Source: American Society of Heating, Refrigerating, and Air Conditioning Engineers, *ASHRAE Guide and Data Book: Handbook of Fundamentals,* ASHRAE, New York, 1967. With permission.

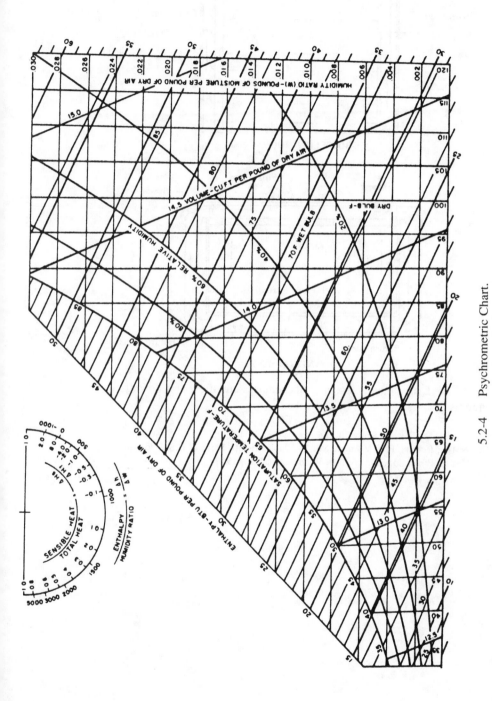

5.2-4   Psychrometric Chart.

Source: Reprinted by permission from the *ASHRAE Handbook* — 1985 *Fundamentals* Volume.

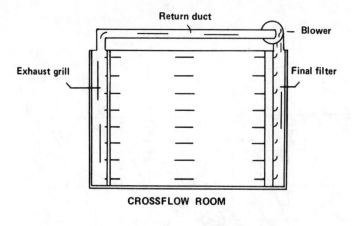

CROSSFLOW ROOM

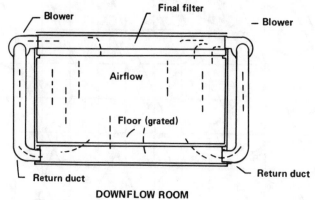

DOWNFLOW ROOM

5.2-5   Diagrammatic drawing of laminar crossflow and downflow rooms.

Source: D.G. Fox, A Study of the Application of Laminar Flow Ventilation to Operating Rooms, PHS Monogr. No. 78, Public Health Service, U.S. Department of Health, Education, and Welfare, Washington, D.C., 1969.

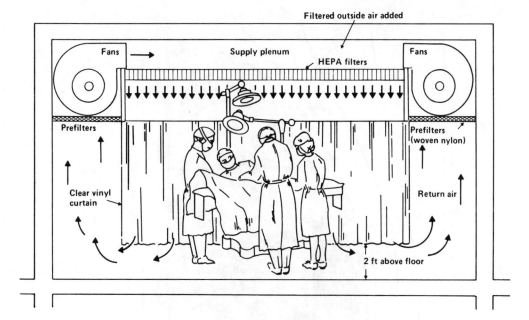

5.2-6   Schematic of vertical laminar-flow shroud area.

Note: HEPA = high-efficiency particulate air.

Source: W. Viesmann, *Heat. Piping Air Cond., 40*:61, 1968. Copyright 1968 by Penton Publishing Company, a division of Pittway Corporation.

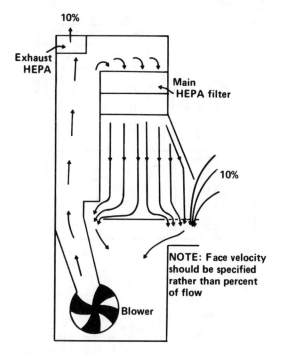

5.2-7   A biohazard hood.

Source: L.L. Coriell and G.J. McGarrity, *Appl. Microbiol.,*
16, 1895, 1968. With permission.

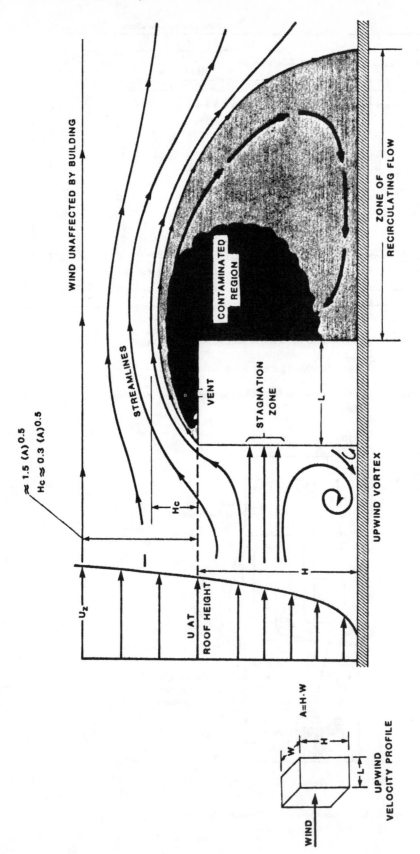

5.2-8   Centerline flow patterns around rectangular building.

Source: Reprinted by permission from the *ASHRAE Handbook* — 1985 *Fundamentals* Volume.

5.2-9 PRESSURE RELATIONSHIPS AND VENTILATION OF CERTAIN HOSPITAL AREAS

Area designation	Pressure relationship to adjacent areas	All supply air from outdoors	Minimum air changes of outdoor air per hour	Minimum total air changes per hour	All air exhausted directly to outdoors	Recirculated within room
Operating room	+	–	5	12	–	No
Emergency operating room	+	–	5	12	–	No
Delivery room	+	–	5	12	–	No
Nursery	+	–	5	12	–	No
Recovery	0	–	2	6	Yes	No
Intensive care	+	–	2	6	–	–
Patient room	0	–	2	2	–	–
Patient area corridor	0	–	2	4	–	No
Isolation room	0	–	2	6	Yes	No
Isolation anteroom	0	–	2	6	Yes	No
Treatment room	0	–	2	6	–	No
X-ray, fluoroscopy room	–	–	2	6	Yes	No
X-ray, treatment room	0	–	2	6	–	–
Physical therapy and hydrotherapy	–	–	2	6	–	No
Soiled workroom	–	–	2	4	–	–
Clean workroom	+	–	2	4	–	No
Autopsy and darkroom	–	–	2	12	Yes	No
Toilet room	–	–	–	10	Yes	No
Bedpan room	–	–	–	10	Yes	No
Bathroom	–	–	–	10	Yes	No
Janitor's closet	–	–	–	10	Yes	No
Sterilizer equipment room	–	–	–	10	Yes	No
Linen and trash chute rooms	–	–	–	10	Yes	No
Laboratory, general[a]	–	–	2	6	–	–
Laboratory, media transfer[b]	+	–	2	4	–	No
Food preparation centers[c]	0	–	2	10	Yes	No
Dishwashing room	–	–	–	10	Yes	No
Dietary day storage	0	–	–	2	–	No
Laundry, general	0	–	2	10	Yes	No
Soiled linen sorting and storage	–	–	–	10	Yes	No

Note: + = positive, – = negative, 0 = equal, – = optional.

5.2-9 PRESSURE RELATIONSHIPS AND VENTILATION OF CERTAIN HOSPITAL AREAS (continued)

Area designation	Pressure relationship to adjacent areas	All supply air from outdoors	Minimum air changes of outdoor air per hour	Minimum total air changes per hour	All air exhausted directly to outdoors	Recirculated within room
Clean linen storage	+	—	2	2	—	—
Anesthesia storage[d]	0	—	—	8	Yes	No
Central medical and surgical supply						
Soiled or decontamination room	−	—	2	4	—	No
Clean workroom	+	—	2	4	—	—
Unsterile supply storage	0	—	2	2	—	—

Note: + = positive, − = negative, 0 = equal, − = optional.

[a]See sec. 8-23D2n and sec. 8-23D2o for additional requirements.
[b]See sec. 8-23D2n for additional requirements.
[c]See sec. 8-23D2q for exceptions.
[d] See sec. 8-23D2s for additional requirements.

Source: Department of Health, Education, and Welfare, General Standards of Construction and Equipment for Hospital and Medical Facilities, PHS Pub. No. 930-A-7, Public Health Service, U.S. Department of Health, Education, and Welfare, Washington, D.C., 1969.

5.2-10 PRESSURE RELATIONSHIPS AND VENTILATION OF CERTAIN NURSING HOME AREAS

Area designation	Pressure relationship to adjacent areas	All supply air from outdoors	Minimum air changes of outdoor air per hour	Minimum total air changes per hour	All air exhausted directly to outdoors	Recirculated within room
Patient room	0	–	2	2	–	–
Patient area corridor	0	–	2	4	–	–
Special purpose room	0	–	2	6	Yes	No
Physical therapy and hydrotherapy	–	–	2	6	–	–
Soiled workroom	–	–	2	4	–	No
Clean workroom	+	–	2	4	–	–
Toilet room	–	–	–	10	Yes	No
Bedpan room	–	–	–	10	Yes	No
Bathroom	–	–	–	10	Yes	No
Janitor's closet	–	–	–	10	Yes	No
Sterilizer equipment room	–	–	–	10	Yes	No
Linen and trash chute rooms	–	–	–	10	Yes	No
Food preparation center	0	–	2	10	Yes	No
Dishwashing room	–	–	–	10	Yes	No
Dietary day storage	0	–	–	2	–	No
Laundry, general	0	–	2	10	Yes	No
Soiled linen sorting and storage	–	–	–	10	Yes	No
Clean linen storage	+	–	2	2	–	–

Note: + = positive, – = negative, 0 = equal, – = optional.

Source: Department of Health, Education, and Welfare, General Standards of Construction and Equipment for Hospital and Medical Facilities, PHS Pub. No. 930-A-7, Public Health Service, U.S. Department of Health, Education, and Welfare, Washington, D.C., 1969.

5.2.2 TOXIC AGENTS

5.2-11 TYPES OF AIRBORNE CONTAMINANTS

Dusts

    Toxic mineral dusts
        Pneumoconiosis
            Asbestosis – repair of asbestos insulation
            Silicosis
    Toxic metallic dusts
        Lead – soldering, grinding, polishing brazing
        Cadmium – soldering, cutting, welding
        Galvanized iron – brazing, welding
        Mercury – see Vapors
    Nuisance dusts
        Mineral dusts – road dust, cement
        Organic dusts – grain, wood, pollen

Fumes

    Toxic metallic dusts – see above

Vapors

    Organic solvents
        Hydrocarbons
            Aromatic – benzene, toluene, xylene
            Aliphatic – petroleum ethers, hexane, naphtha, mineral spirits, gasoline, Stoddard's solvent
            Halogenated – monochloromethane, dichloromethane, chloroform, carbon tetrachloride, methyl chloroform (1,1,1-trichloroethane)

    Alcohols
    Ethers
    Esters } Found in laboratories, maintenance shops, housekeeping departments
    Aldehydes
    Oleofins
    Ketones

    Mercury – broken mercury thermometers, laboratory uses, dental care facilities

Gases

    Chlorine – disinfectant in recreational and therapeutic swimming pools; disinfectant housecleaning activities
    Ammonia – housekeeping activities, refrigerant
    Nitrogen dioxide – produced in maintenance shops in cleaning metals with nitric acid and in electric-arc welding
    Ozone – used for odor control, use of ozone discouraged
    Asphyxiation – maintenance operation
                – carbon monoxide – incomplete combustion

Mists

    Aerosol sprays – insecticides, rodenticides, caustic cleaning compounds, sanitizing agents

Adapted from G. S. Michaelsen, Toxic aspects, in *Environmental Health and Safety in Health-Care Facilities*, R. G. Bond et al., Eds., Macmillan, New York, 1973.

5.2-12 CONTROL OF AIRBORNE CONTAMINANTS

1. Local exhaust ventilation
2. General dilution ventilation
3. Substitution
   a. Less toxic agent substituted for toxic substance
   b. Use of less hazardous procedure
4. Protective devices
   a. Dust respirators
   b. Gas masks
   c. Chemical cartridge respirators
   d. Airline respirators
   e. Hose masks
   f. Self-contained breathing apparatus
5. Good housekeeping

Source: G.S. Michaelsen, Toxic aspects, in *Environmental Health and Safety in Health-Care Facilities,* R.G. Bond et al., Eds., Macmillan, New York, 1973.

5.2-13   REPORTED MERCURY CONCENTRATIONS IN DOCTORS' AND DENTISTS' SUITES AND HOSPITALS

Location	Hg conc (ng/m^3)		Percent above TLV-50,000 ng/m^3
	Range	Median	
Doctors' rooms, Dallas[1]	4,550—5680[a]	4,950	
Dentists' rooms, Dallas[1]	4,770—5,550[b]	5,030	
Dentists' rooms, Dallas[1]	1,135—1295[c]	1,160	
Hospital laboratory, Dallas[1]	307	—	
Hospital ward, Dallas[1]	336		
Dentists' suites			
Air[2,d]	50,000—600,000[e]	100,000	
Rugs[f]	50,000—800,000[e]	200,000	
Dental suites and operations[3]	ND[g]—2,500,000	20,000—32,000	
Utah dental offices[4,h]	14,000—150,000		
Boston, 77 dentists[5]			22
Pennsylvania, dental personnel[6]			20
Minnesota, 128 offices[7]			12.5
Atlanta, 134 offices[8]			5
Maryland[9]			15
South Carolina[9]			16
Texas[9]			10

[a]   Mercury thermometer broken in past.
[b]   Mixing area for Ag amalgam.
[c]   Inactive for previous 4 days (holidays).
[d]   20 lb Hg maliciously spilled on floors, etc., during break in.
[e]   10 days later levels reduced to 20,000 to 40,000 ng/m^3.
[f]   Taken less than 1 in. from floor or rug.
[g]   ND, none detected.
[h]   72 dental offices in Utah.

REFERENCES

1. R.S. Foote, *Science, 177*:513, 1972.
2. L.V. Pagnotto and C.M. Comproni, *J. Am. Dental Assn., 92*:1195, 1976.
3. H. Buchwald, *Am. Ind. Hygiene Assn. J., 33*:492, 1972.
4. D.G. Mantyla and O.D. Wright, *J. Am. Dental Assn., 92*:1189, 1976.
5. C. Cuzacq *et al., J. Mass. Dental Soc., 20*:254, 1971.
6. P.A. Gronka *et al., J. Am. Dental Assn., 81*:923, 1970.
7. Council on Dental Materials and Devices, *J. Am. Dental Assn., 89*:900, 1974.
8. W.B. Eames *et al., J. Am. Dental Assn., 92*:1199, 1976.
9. Council on Dental Materials and Devices, *J. Am. Dental Assn., 92*:1217, 1976.

5.2.3   NOISE PRODUCTION AND CONTROL

5.2-14   MOST PREVALENT SOUNDS IN HOSPITALS, ARRANGED IN ORDER OF ANNOYANCE

Rank position	Sound
1	Radio or television set
2	Staff talk in corridors
3	Other patients in distress and recovery-room sounds
4	Voice paging
5	Talk in other rooms
6	Babies or children crying
7	Telephones
8	Pantry, kitchen, utility room

Source: Noise in Hospitals, PHS Pub. No. 930-D-11, Public Health Service, U.S. Department of Health, Education, and Welfare, Washington, D.C., 1963.

5.2-15   HOSPITAL NOISE—INTERNAL SOURCES

Power plant – boilers, induced draft fans, pumps, combustion noises

Electrical equipment – auxiliary generators, transformers, switch gear

Heating ventilation and air conditioning equipment – air supply and air exhaust openings.

Transportation systems – elevators, dumbwaiters, pneumatic tubes, conveyor systems, linen chutes, rubbish chutes

Mechanical equipment – sterilizers, autoclaves, ultrasonic cleaners, dishwashing machines, ice machines, refrigerators, freezers

Plumbing fixtures – water flowing from faucet, stream of water hitting metal basin, frequent water hammer

Furnishings – cabinet doors and drawers, metal units, chart cases, racks, beds, bed curtains, furniture

Housekeeping facilities – floor-washing machines, floor polishers, vacuum cleaners, metal trash containers

Communication systems – voice paging systems, telephones, carts used for food, medication, linens, supplies, trash collection

Adapted from G. S. Michaelsen, Noise production and control, in *Environmental Health and Safety in Health-Care Facilities,* R. G. Bond et al., Eds., Macmillan, New York,   1973.

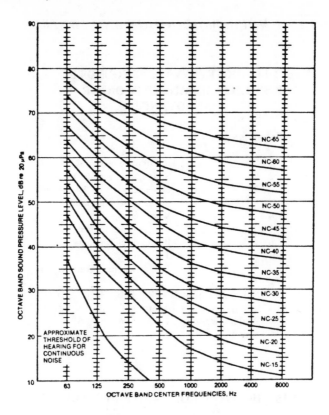

5.2-16   NC (noise criteria) curves for specifying the design level in terms of the maximum permissible sound pressure level for each frequency band.

Source: Reprinted by permission from the *ASHRAE Handbook — 1985 Fundamentals* Volume.

5.2-17   RANGES OF INDOOR GOALS FOR AIR-CONDITIONING SYSTEMS IN HOSPITALS AND CLINICS

Type of area	Range of NC criteria curves[a]
Private rooms	25—35
Operating rooms, wards	30—40
Laboratories, halls, and corridors	35—45
Lobbies and waiting rooms	35—45
Washrooms and toilets	40—50

[a]   NC curves are shown in Figure 5.2-16.

Source: *ASHRAE Guide and Data Book: Systems,* American Society of Heating, Refrigeration, and Air Conditioning Engineers, New York, 1970, p. 497. With permission.

5.2.4 RADIATION PROTECTION

5.2-18 EXAMPLES OF BIOLOGICAL RESPONSE IN HUMAN ORGANS AFTER EXTERNAL PARTIAL BODY IRRADIATION

Organ	Dose schedule	Single dose or extrapolated equivalent (rads)	Effect in relevant organs
Ovary	–	200	Temporary amenorrhea, sterility
	1500 rads/10 days	800[a]	Permanent menopause, sterility
Testis	–	50	Temporary sterility
	1500 rads/10 days	800[a]	Permanent sterility
Bone marrow	25–75 rads in each of 5–10 days. (Portion of bone marrow segments requires higher doses.)	200[a]	Hematopoiesis inhibited in irradiated volume. Usually compensated by marrow activity in unexposed sites.
Kidney	2000 rads/30 days 3000 rads/40 days	800[a]	Nephritis, hypertension
Stomach	1500 rads/20 days 2500 rads/30 days	1000[a]	Atrophic mucosa, anacidity
Liver	3000 rads/30 days 4000 rads/42 days	1500[a]	Hepatitis
Brain and spinal cord	5000 rads/30 days 6000 rads/42 days	2200[a]	Necrosis atrophy
Lung	4000 rads/30 days 6000 rads/42 days	2200[a]	Pneumonitis, fibrosis
Rectum	8000 rads/56 days	2700[a]	Atrophy. Limit of tolerance, most cases.
Bladder	10,000 rads/56 days	3400[a]	Atrophy. Limit of tolerance, most cases.
Ureter	12,000 rads/56 days	4000[a]	Atrophy. Limit of tolerance, most cases.

[a] Extrapolated equivalent calculated from the empirical relation $D_o = D_0 t^{-0.27}$, where D_0 is the extrapolated equivalent single dose, when an actual dose of D_o is spread over the time t days. This relationship essentially assumes an equal daily dose schedule. Other formulations (e.g., the Ellis formula[1]) give successful empirical results for some erratic fractionation schedules. Such refinements are not needed here; as in Table 3.4–3 these entries are meant to be descriptive, rather than definitive.

Source: Basic Radiation Protection Criteria, NCRP Report No. 39, National Council on Radiation Protection and Measurements, Washington, D.C., January 15, 1971. With permission.

REFERENCE

1. F. Ellis, *Clin. Radiol.*, 20, 1, 1969.

5.2-19   PERMISSIBLE EXPOSURE LEVELS FOR UNLIMITED EXPOSURE TO VARIOUS TYPES OF RADIATION

Radiation	Permissible levels	Ref.
Microwaves	10^{-2} Wcm^{-2}	1
Q-switched Ruby laser[a]	10^{-8} Jcm^{-2}	2
Non-Q-switched Ruby laser[a]	10^{-7} Jcm^{-2}	2
Continuous wave Ruby laser[a]	10^{-6} Wcm^{-2}	3
Ultraviolet light	5×10^{-7} Wcm^{-2}	3
Infrared	10^{-1} Jcm^{-2}	3

[a]   7-mm pupil diameter.

REFERENCES

1. S.F. Cleary, *Crit. Rev. Environ. Control*, 7:121, 1977.
2. A.M. Clarke, *Crit. Rev. Environ. Control*, 1:307, 1970.
3. L.R. Setter *et al.*, Regulations, Standards, and Guides for Microwave, Ultraviolet Radiation, and Radiation from Lasers and Television Receivers — An Annotated Bibliography, PHS Pub. No. 99-RH-35, Public Health Service, U.S. Department of Health, Education and Welfare, Washington, D.C., 1969.

5.3   SAFETY

5.3.1   LABORATORY SAFETY

5.3-1   SAFETY REGULATIONS FOR THE USE OF COMPRESSED-GAS CYLINDERS

1. Cylinders shall have the name of the chemical contents appearing in legible form on the cylinder. A color code is not a satisfactory designation.

2. Cylinders shall be held securely in an upright position. String, wire, or similar makeshift materials are not acceptable.

3. Cylinders shall be located so they are not exposed to direct flame or heat in excess of 125°F.

4. Cylinders not in use shall have the valve protective cap securely in place. (Lecture bottles are an exception.)

5. Cylinders shall be moved only in suitable hand carts.

6. Cylinders containing flammable gas shall be used and stored in a ventilated area (ten air changes per hour minimum). No other gases or chemicals shall be stored in the same area.

7. Cylinders containing toxic or corrosive gases shall be used and stored in well-ventilated areas separated from cylinders containing other gases.

8. Cylinders containing corrosive gas shall be returned to the supplier no later than 6 months from time of first use. Cylinder and regulator valves shall be opened and closed at least weekly during periods of nonuse.

9. Cylinders discharging into liquids or closed systems containing other chemicals shall have a trap, check valve, or vacuum breaker between the cylinder and system or liquid.

10. Systems mixing two or more gases shall be provided with necessary control or check valves to prevent contamination of the separate gas sources.

11. Closed systems, or any arrangement that might accidentally become a closed system to which a cylinder is attached, shall have a safety relief valve set at a relief pressure that will prevent damage to any part of the equipment.

12. Cylinder valves and regulators shall have outlets and inlets, respectively, for the specific gas as designated in the American Standard of Compressed Gas Association Pamphlet V-1 or for flush type cylinder valves according to Compressed Gas Association Pamphlet V-3.

13. Use of adaptors between cylinder valve and regulator should be discouraged, but if used, they should be only a type listed in Appendix of Compressed Gas Association Pamphlet V-1.

14. Emergency plans shall be developed to insure control or safe removal of leaking cylinders. Proper protective clothing and equipment for the type of compressed gas being used should be available in the immediate area to allow safe entry should the gas be accidentally released from the cylinder.

Source: Safety Standard, Division of Environmental Health and Safety, University of Minnesota, Minneapolis.

5.3.2   FIRE SAFETY

5.3-2   FIRE CHARACTERISTICS OF FLAMMABLE LIQUIDS COMMONLY FOUND IN HOSPITALS

Liquid	Flash point °F	Flash point °C	Ignition temperature °F	Ignition temperature °C	Health[a]	Flammability[b]	Reactivity[c]
Acetic acid, glacial	103	39	867	463	2	3	1W
Acetone	−4	−20	869	465	1	3	0
Benzene (benzol)	12	−11	928	498	2	3	0
Ethyl alcohol	55	13	685	383	0	3	0
Ethyl ether	−49	−45	320	160	2	4	1
Formalin (with methanol)	122	50			2	2	
Gasoline	−45	−43	536	280	1	3	0
Isopropyl alcohol	53	12	750	399	1	3	0
Isopropyl ether	−18	−28	830	443	2	3	1
Kerosene fuel oil (#1)	110—162	43—72	410	210	0	2	0
Lubricating oil mineral	300—450	149—232	500—700	260—371	0	1	0
Toluene (toluol)	40	4	896	480	2	3	0
Turpentine	95	35	488	253	1	3	0

[a] 2, Materials hazardous to health; 1, materials only slightly hazardous to health; 0, materials offering no hazard to health.

[b] 4, Very flammable gases or volatile flammable liquids; 3, materials which can be ignited under almost all normal temperature conditions; 2, materials which must be moderately heated before ignition will occur; 1, materials that must be preheated before ignition can occur.

[c] 1, Materials which (in themselves) are normally stable but which become unstable at elevated temperatures and pressures or which may react with water with some release of energy but not violently; 0, materials which (in themselves) are normally stable even under fire exposure conditions and which are not reactive with water; indicates avoid use of water.

Adapted from: *Fire Protection Guide on Hazardous Materials,* 7th ed., National Fire Protection Association, Boston, Mass., 1978.

5.3-3   TYPICAL FUEL CONTENTS OF MATERIALS

Material	Fuel load per pound, Btu	Material	Fuel load per pound, Btu
Woods		Fats and waxes	
Douglas fir, untreated	8,400[1]	Animal fats, mean	17,100[2]
Douglas fir, fire retardant	7,050–8,290[1]	Butter fat	16,800[2]
Fir (dry)	9,060[2]	Lard	17,200[2]
Maple, soft, untreated	7,940[1]	Vegetable and fish oils	
Ash (dry)	8,480[2]	Cottonseed	16,920[2]
Beech, 13% moisture	7,510[2]	Linseed	16,860
Elm (dry)	8,510[2]	Olive	17,020[2]
Hardwood, average	8,120[2]	Cod-liver	16,980[2]
Locust	8,640[2]	Flammable liquids	
Oak, 13% moisture	7,180[2]	Acetone	13,500[3]
Pine, 12.3–10.5% moisture	8,080–8,420[2]	Acetylene	21,600[3]
Soft wood, resinous	8,330[2]	Ethyl alcohol	12,900[3]
Average for soft and hard	8,000[3]	Benzene (benzol)	18,000[3]
woods approximately 12% moisture		Bitumen	15,200[3]
Birch, 12% moisture	7,580[2]	Pentane	20,900[3]
Paper		Kerosene	19,800[2]
Average	7,000[3]	Gasoline	20,000[2]
Ash, 7.0%–1.4%	6,710–7,830[2]	Crude oil	19,000[2]
Building paper (asphalt	13,620[1]	Xylene	18,400[3]
impregnated)		Toluene	18,300[3]
Building paper (rosin sized)	7,650[1]	Diethyl ether	16,500[3]
Fibers		Plastics	
Cotton	7,160[3]	Polystyrene	17,420[1]
Silk (raw)	9,200[3]	Natural rubber	17,000[3]
Wool (raw)	9,800[3]		
Flax	6,500[3]		
Metals			
Structural steel, unpainted	230[1]		
Aluminum	30[1]		
Carbon	60[1]		
Magnesium	10,800[1]		

Source: H.E. Nelson, *J. Am. Soc. Saf. Eng.,* 10, 17, 1965. With permission.

REFERENCES

1. Potential Heat of Building Materials in Building Fires, *NBS Tech. Bull.,* November 1960.
2. *Fire Protection Handbook,* 12th ed., National Fire Protection Association, Boston.
3. Fire Gradings of Buildings, Part 1, Joint Committee of the Building Research Board of the Department of Scientific and Industrial Research and of the Fire Offices' Committee, Her Majesty's Stationery Office, London.

5.4   GENERAL SANITATION

5.4.1   SOLID WASTE HANDLING AND DISPOSAL

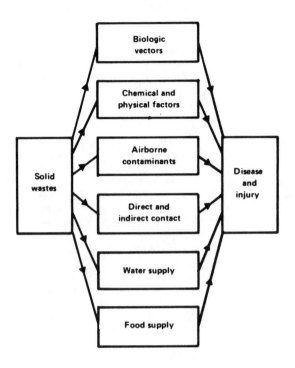

5.4-1 Solid waste human disease relationships (postulated).

Adapted from T.G. Hanks, Solid Waste/Disease Relationships: A Literature Survey, PHS Rep. No. SW-1c, PHS Pub. No. 999-UIH-6, Public Health Service, U.S. Department of Health, Education, and Welfare, Washington, D.C., 1967.

5.4-2 DEFINITIONS APPLICABLE TO WASTES FROM HEALTH-CARE FACILITIES

1. Wastes   Useless, unused, unwanted, or discarded materials
2. Refuse   Includes all solid wastes; in practice this category includes garbage, rubbish, ashes, dead animals.
3. Garbage   Designates putrescible wastes resulting from handling, preparation, cooking, and serving of food.
4. Rubbish   This term includes all nonputrescible refuse except ashes; there are two categories of rubbish, combustible and noncombustible.
   a. Combustible: this material is primarily organic; it includes items such as paper, plastics, cardboard, wood, rubber, bedding.
   b. Noncombustible: this material is primarily inorganic and includes tin cans, metals, glass, ceramics, and other mineral refuse.
5. Ashes   Residue from fires used for cooking, heating, and on-site incineration.
6. Biologic wastes   Wastes resulting directly from patient diagnosis and treatment procedures; includes materials of medical, surgical, autopsy, and laboratory origin.
   a. Medical wastes: these wastes are usually produced in patient rooms, treatment rooms, and nursing stations; the operating room may also be a contributor; items include soiled dressings, bandages, catheters, swabs, plaster casts, receptacles, and masks.
   b. Surgical and autopsy (pathologic wastes): these wastes may be produced in surgical suites or autopsy rooms; items that may be included are placenta, tissues and organs, amputated limbs, and similar material.
   c. Laboratory wastes: these wastes are produced in diagnostic or research laboratories; items that may be included are cultures, spinal-fluid samples, dead animals, and animal bedding.

Sources: D. L. Snow et al., *A Report on Hospital Solid Wastes and Their Handling,* Committee on Hospital Facilities, Engineering and Sanitation Section, American Public Health Association, New York, 1955; American Public Works Association, *Municipal Refuse Disposal,* 2nd ed., Public Administration Service, Chicago, 1966. With permission.

5.4-3 AREAS GENERATING WASTES AND THEIR TYPICAL WASTE PRODUCTS

Area	Waste products
Administration	Paper goods
Obstetrics department, including patient rooms in the department	Soiled dressings; sponges; placentas; waste ampules, including silver nitrate capsules; needles and syringes; disposable masks; disposable drapes; sanitary napkins; disposable blood lancets; disposable catheters; disposable enema units; disposable diapers and underpads; disposable gloves
Emergency and surgical departments, including patient rooms	Soiled dressings; sponges; body tissue, including amputations; waste ampules; disposable masks; needles and syringes; drapes; casts; disposable blood lancets; disposable emesis basins; Levine tubes; catheters; drainage sets; colostomy bags; underpads; surgical gloves
Laboratory, morgue, pathology, and autopsy rooms	Contaminated glassware, including pipettes, petri dishes, specimen containers, and specimen slides; body tissue; organs; bones
Isolation rooms other than regular patient rooms	Paper goods containing nasal and sputum discharges; dressings and bandages; disposable masks; leftover food; disposable salt and pepper shakers
Nurses' stations	Ampules; disposable needles and syringes; paper goods
Service areas	Cartons; crates; packing materials; paper goods; metal containers, including tin cans, drums; bottles, including food containers, solution bottles, and pharmaceutical bottles; wastes from public and patient rooms, including paper goods, flowers; waste food materials from the central and floor kitchens; wastes from X-ray, discarded furniture; rags

Source: Reprinted with permission from V.R. Oviatt, *Hospitals,* 42, 73, 1968. Copyright © 1968, American Hospital Publishing, Inc., Chicago.

5.4-4 CONTRIBUTIONS OF HOSPITAL SOLID WASTE
PER PATIENT PRESENTED IN VARIOUS PUBLICATIONS

Year presented	Contribution of solid waste	Reference
1937	7 lb/day/bed[a] comprising 40 to 50% garbage, and the balance rubbish	1
1949	7.5 lb/day/bed, compared with "government estimates" of 8 to 8.5 lb/day/bed.[a] The latter included 4 lb/day/bed[a] of "kitchen garbage."	2
1952	7 lb/day/patient, average generation	3
1955	Design basis for incinerators of 8 lb/day/bed[a] plus 3 lb/day for each person in an "auxiliary section" (e.g., nurse's residence), plus 1 lb/patient in outpatient department. If garbage were excluded, the first value might be reduced to 5 lb/day/bed. Base incinerator design on 7 hr/day.	4

1956	Mean contributions:	lb/day/patient	ft³/day/patient	5
	Garbage	3.28	0.064	
	Noncombustibles	1.10	0.111	
	Combustibles	2.61	0.521	
	Surgical and autopsy waste	0.14	Omitted	
		7.13	0.696	

Year presented	Contribution of solid waste	Reference
1958	7 lb/day/patient, for general hospitals	6
1961	7 lb/day/patient, or almost 1 ft³/day/patient	7
1961	7 to 8 lb/day/bed	8
1963	9.6 lb/day/bed, or 1.7 ft³/day/bed for a 1000-bed military hospital	9
1964	Estimates of combustible waste generation had risen from 8 lb/day/patient "a few years ago" to 10 lb/day/patient, or more.	10
1965	Recommended values, in absence of better data, were 5 lb/day/bed, or 0.5 ft³/day/bed. A range was reported of 0.1–0.93 ft³/day/bed	11
1965	Estimated increase from approximately 6 lb/day/patient in 1955 to 13 lb/day/patient in 1975	12
1966	19 lb/day/patient at a teaching hospital with 1100 beds (included 8.3 lb/day/patient of garbage and noncombustibles, and 10.7 lb/day/patient of "readily burnable material")	13
1966	Design basis for incinerators of 8 lb/day/bed and 6 hr/day of operation	14
1967	Estimated increase during preceding 10 years from 6–9 lb/day/patient, to 20 lb/day/patient	15
1967	4.3–11.0 lb/day/patient found at hospitals with 200–725 beds.	16
1967	20 lb/day/patient	17
1968	17 lb/day/patient approximately	18
1968	Estimated 12 lb/day/bed	19
1968	Mean of 12.3 lb/day/bed found at 7 hospitals (range of 9–19 lb/day/bed)	20
1968	Estimated increase in preceding 10 years from 6–9 lb/day/patient to 12–19 lb/day/patient	21

5.4-4   CONTRIBUTIONS OF HOSPITAL SOLID WASTE
PER PATIENT PRESENTED IN VARIOUS PUBLICATIONS (continued)

Year presented	Contribution of solid waste	Reference
1969	15 lb/day/bed	22
1969	At least 10 lb/day/bed from "general hospital buildings," excluding food service wastes and loads from auxiliary buildings. Waste contribution was expected to peak at 20 to 25 lb/day/bed.	23

[a] Contribution of solid waste listed on a "lb/day/bed" basis may be converted to "lb/day/patient" by using percent occupancy of the hospital, as follows: lb/day/patient = (100/percent occupancy) (lb/day/bed). Since nongovermental, nonprofit hospitals have an average occupancy of 80% the conversion factor would typically be 1.25.

Adapted from A.F. Iglar, *CRC Crit. Rev. Environ. Control, 1*:507, 1971.

REFERENCES

1. W.J. Overton, *Mod. Hosp.*, 48, 87, 1937.
2. O.E. Olson, *Mod. Hosp.*, 73, 116, 1949.
3. American Hospital Association, *Manual of Hospital Housekeeping*, Pub. No. M 16–52, Committee on Housekeeping in Hospitals, A.H.A., Chicago, 1952.
4. Incinerator Institute of Amercia, *The Selection of Incinerators for Hospital Use*, Bull. H., I.I.A., New York, 1955.
5. American Public Health Association, *Am. J. Public Health*, 46, 357, 1956.
6. Sister M. Clarisse, *Tex. Hosp.*, 14, 30, 1958.
7. O. Vance, *Hosp. Manage.*, 92, 48, 1961.
8. Anonymous, *Institutions*, 49, 127, 1961.
9. J.P. McKenna, A Study of the Requirements for Disposing of Waste Materials in the United States Air Force Hospital (at) Lackland, unpublished report submitted in partial fulfillment of requirements for Master of Hospital Administration degree, Baylor University, Waco, Tex., 1963.
10. R.C. Paul, *Hospitals*, 38, 99, 104, 1964.
11. A.G.R. Farr and H.B. Healy, *Hosp. Eng.*, 19, 65, 1965.
12. J. Falick, *Archit. Eng. News*, 7, 46, 1965.
13. J.A. Holbrook, *Mod. Hosp.*, 107, 126, 1966.
14. Incinerator Institute of America: I.I.A. Incinerator Standards, I.I.A., New York, 1966.
15. A.J.J. Rourke, *Mod. Hosp.*, 109, 132, 1967.
16. V.R. Oviatt, Waste Handling, an Old Problem, Tech. Inform. Rep., Institute of Sanitation Management, Hicksville, N.Y., 1967.
17. Anonymous, *Mod. Hosp.*, 109, 214, 1967.
18. Syska and Hennessey, Inc., Hospital systems. Part II. Materials handling, *Tech. Lett.*, 1968.
19. R.W. Davis, *Mod. Hosp.*, 111, 138, 1968.
20. H.J. Baker, A Summary of Solid Waste Handling in Seven Hospitals within the District of Columbia, unpublished course paper, University of Minnesota School of Public Health, Minneapolis, 1968.
21. R.B. Groce, *Tex. Hosp.*, 24, 32, 1968.
22. T.L. Jacobsen, *Hospitals*, 43, 89, 1969.
23. Syska and Hennessey, Inc., Incineration: an engineering approach to the waste disposal crisis, *Tech. Lett.*, 1968.

5.4–5   BREAKDOWN OF DAILY WASTE PRODUCTION BY TYPES OF WASTES IN POUNDS (SEVEN HOSPITALS)

Types of wastes	Mean	Median	Range
Pathological and surgical	183+	6	Tr—1,000
Sharps, needles, etc.	25	20	3—75
Soiled linens	12,700	5,630	1,120—45,500
Rubbish	4,124	1,722	362—16,200
Reusable patient items	Tr	Tr	Tr
Noncombustibles	454	250	75—1,500
Garbage (nongrindable)[a]	604	475	110—1,800
Food service items	2,990	2,400	600—9,000
Radiological	Tr (4)[b]	Tr (4)	Tr (4)
Ash and residue	22+(6)	20(6)	Tr—50(6)
Animal carcasses	60(5)	23(5)	10—220(5)
Food waste (grindable)	1,030	950	150—2,600
Total production	22,200	12,500	2,452—77,700
Daily production disposable	6,480	4,376	732—23,200
Pounds per bed patient	8.4	7.8	3.6—16.7
Pounds per capita[c]	3.6	3.57	2.08—5.57
Daily production reusable	15,700	8,130	1,720—54,500
Pounds per bed patient	20.6	21.7	11.9—29.6 ·
Pounds per capita[c]	9.12	9.41	7.93—10.20

[a] Predominantly garbage mixed with substantial quantities of paper, plastics, metals, etc.
[b] Number of hospitals reporting
[c] Per capita production based on equivalent 24-hr population

Adapted from: W.E. Small, *Mod. Hosp.*, *117:*100, 1971.

5.4-6 SOURCES OF SOLID WASTE WITHIN THE HOSPITAL

Source	Mean percent of total solid waste weight
Administrative and other offices	1.60
Central supply	0.84
Teaching facilities	0.03
Construction and demolition	0.22
Dietary facilities	49.00
Emergency	0.82
Extended care	0.54
Grounds	0.36
Hospitality shop	0.94
Housekeeping	0.13
Intensive care	0.46
Isolation	0.26
Laboratories	2.10
Laundry	0.46
Maintenance	0.34
Maternity	3.80
Morgue	0.04
Nursing stations, general	20.00
Outpatient department	0.24
Pediatric care	0.85
Pharmacy	0.74
Physical therapy	0.05
Psychiatric care	0.20
Public areas	0.29
Residences	0.76
Storerooms	1.20
Surgery	4.50
X-ray	1.10
Mixed, other and unknown	8.30

Source: A.F. Iglar, Hospital Solid Waste Management, Doctoral thesis, University of Minnesota, Minneapolis, 1970. With permission.

5.4-7 ENUMERATION OF DISPOSABLE ITEMS IN HOSPITALS

Hospital department	Different items[a]
Nursing	126
Dietary	29
Surgery and obstetrics	24
Laboratory	26
Housekeeping	17
Total	222

[a] Examples: canopies, catheters, diapers.

Source: G.S. Michaelsen and D. Vesley, *Hosp. Manage.*, 101, 23, 1966. With permission.

5.4-8 TYPES OF STORAGE CONTAINERS

Type of container	Number of hospitals
No containers	5
Metal cans and similar receptacles	25
Improvised containers	17
Noncompacting bulk receptacles	37
Stationary compactor receiver receptacles	9
Other	7

Source: A.F. Iglar, Hospital Solid Wastes Management, Doctoral thesis, University of Minnesota, Minneapolis, 1970. With permission.

5.4-9   DISPOSITION OF HOSPITAL SOLID WASTE

Method of solid waste disposal	N	Percent by weight for hospitals visited	
		Mean percent[a]	Range of percent[b]
Incineration at the hospital	63[c]	35	3–89
Grinding	63[c]	21	1–47
Municipal incineration	7	2.9	3–100
Sanitary landfill	24	15	3–98
Dumping	49	21	6–100
Hog feeding	13	3.7	4–40
Mixed, other, and unknown	4	1.2	2–45

[a] Average percentage disposed of by this method for all hospitals visited. Means include values of zero.

[b] Ranges include only hospitals for which waste was measured for indicated method of disposal.

[c] Excludes certain instances in which method was used, at least occasionally, but no waste was measured.

Source: A.F. Iglar, Hospital Solid Wastes Management, Doctoral thesis, University of Minnesota, Minneapolis, 1970. With permission.

5.4-10 CONCENTRATIONS OF GASES RELEASED DURING THERMAL DEGRADATION OF POLYVINYL CHLORIDE (PVC)

	Concentration (milligrams per cubic meter)				
		100°C		20°C	40°C
Substance	1 hr	2 hr	3 hr	30 days	30 days
Aldehydes	0.68	1.0	1.32	0	0.64
Dibutyl phthalate	31.2	33.2	35.6	6.4	8.8
Carbon dioxide	14,080	16,000	17,600	3,520	7,920
Fatty acids	Trace	6	10	0	0
Carbon monoxide	38	40	40	30	34
Hydrocarbons	96	168	396	552	720
Vinyl chloride	152	162	164	0	0
Hydrogen chloride	8	12	20	0	0

Source: L.A. Popv and V.D. Yablochkin, *Hyg. Sanit. (USSR),* 32, 114, 1967. (Translations for the U.S. Department of Commerce, Clearinghouse for Federal Scientific and Technical Information.)

5.4-11 GARBAGE GRINDER SIZE

Horsepower	Capacity (lb/hr)	Up to number of persons per meal
½	200	125
¾	400	300
1½	1200	1500
3	2000	1500 plus

Source: R. J. Black, in Environmental Aspects of Hospitals: Vol. II, Supportive Departments, PHS Pub. No. 930-C-16, Public Health Service, U.S. Department of Helath, Education, and Welfare, Washington, D.C., 1967, 20.

5.4.2 LIQUID WASTE COLLECTION AND DISPOSAL

5.4-12 STACK SIZES FOR BEDPAN STEAMERS AND BOILING TYPE STERILIZERS

Stack size, in.	Connection size	
	1½ in.	2 in.
1½[a]	1 or	0
2[a]	2 or	1
2[b]	1 and	1
3[a]	4 or	2
3[b]	2 and	2
4[a]	8 or	4
4[b]	4 and	4

Note: Number of connections of various sizes permitted to various sized sterilizer vent stacks.

[a] Total of each size.
[b] Combination of sizes.

Source: *Report of Public Health Service Committee on Plumbing Standards,* U.S. Dept. of Health, Educ., and Welfare, Public Health Service, Washington, D.C., 1963.

5.4-13 STACK SIZES FOR PRESSURE STERILIZERS

Stack size, in.	Connection size			
	¾ in.	1 in.	1¼ in.	1½ in.
1½[a]	3 or	2 or	1	
1½[b]	2 and	1		
2[a]	6 or	3 or	2 or	1
2[b]	3 and	2		
2[b]	2 and	1 and	1	
2[b]	1 and	1 and		1
3[a]	15 or	7 or	5 or	3
3[b]		1 and	2 and	2
3[b]	1 and	5 and		1

Note: Number of connections of various sizes permitted to various sized vent stacks.

[a] Combination of sizes.
[b] Total of each size.

Source: *Report of Public Health Service Committee on Plumbing Standards,* U.S. Dept. of Health, Educ., and Welfare, Public Health Service, Washington, D.C., 1963.

5.4.3   *WATER SUPPLY, TREATMENT, DISTRIBUTION*

5.4-14   HOSPITAL WATER REQUIREMENTS

St. John's Hospital, St. Louis, Missouri

Item[a]	Water use, gpd/bed	Ratio at designated use to avg
Avg daily	300	1
Max daily	375	1.25
Avg 5 hr maximum	530	1.70
5 hr max	675	2.25
2 hr max	800	2.70

[a] Hospital had 525 beds and a staff of 750 (3 shifts). Estimated 92% of waste returned to sewer.

Source*: Design and Construction of Sanitary and Storm Sewers,* Manual of Practice No. 9, Water Pollut. Control Fed., 1969. With permission.

5.4-15 SUMMARY OF WATER STANDARDS FROM PHARMACOPEIA

Classification of water and water use	Reaction	Heavy metals	Zinc	Foreign volatile matter	Total solids	Bacteriological purity	Pyrogens	pH	Chloride	Antimicrobial agents	Oxidizable substances	Ammonia	Calcium	Carbon dioxide	Sulfate
Water – clear, colorless, odorless liquid	(1)[a]	(2)	(3)	(4) No odor when heated near to boiling point	(5) <0.1%	(6) USPHS Standards for potable water	—	—	—	—	—	—	—	—	—
Water for injection – clear, colorless, odorless liquid prepared by distillation. Intended for use as solvent for the preparation of parenteral solutions.	—	(23)	—	—	(25) <10.0 ppm	(16) Sterility	(7) Pyrogen free	(17) 5.0–7.0	(18)	—	(24)	(20) <0.3 ppm	(21)	(22)	(19)
Bacteriostatic water for injection – clear, colorless liquid, odorless or having odor of antimicrobial substance. Consists of *sterile water for injection* containing one or more antimicrobial	—	(23)	—	—	(9) <40 ppm	(12) Sterility	(11) Pyrogen free	(8) 4.5–7.0	(13) <0.5 ppm	(10)	(14) *Note:* Water for formulation meets requirements for *water for injection.*	(20) *Note:* Water for formulation meets requirements for *water for injection.*	(21)	(22)	(19)

5.4-15 SUMMARY OF WATER STANDARDS FROM PHARMACOPEIA (continued)

Classification of water and water use	Re-action	Heavy metals	Zinc	Foreign volatile matter	Total solids	Bacterio-logical purity	Pyrogens	pH	Chloride	Anti-microbial agents	Oxidiz-able sub-stances	Ammonia	Calcium	Carbon dioxide	Sulfate
Bacteriostatic water for injection – (continued) agents; pre-pared by dis-tillation.															
Sterile water for in-jection – clear, color-less, odorless liquid; *water for injection* suitably sterilized and packaged; prepared by distillation.	–	(23)	–	–	(15) <30 ml: <40 ppm 30–100 ml: <30 ppm >100 ml: <20 ppm	(16) Sterility	(7) Pyrogen free	(17) 5.0–7.0	(13) <0.5 ppm	–	(14)	(20) <0.3 ppm	(21)	(22)	(19)
Purified water – clear, color-less, odorless liquid; ob-tained by distillation or ion-exchange.	–	(23)	–	–	(25) <10.0 ppm	(6) USPHS Drinking Water Standards	–	(17) 5.0–7.0	(18)	–	(24)	(20) <0.3 ppm	(21)	(22)	(19)

Key:

1. Reaction (*Note:* For this test, use indicator test solutions specified for pH determinations.) – Add two drops of methyl red test solution[b] to 10 ml of *water* in a test tube; no pink or red color is produced. To another 10-ml portion of *water* in a test tube add two drops of phenolphthalein test solution: no pink or red color is produced.

2. Heavy metals – Adjust 40 ml of *water* with diluted acetic acid to a pH of 3.0 to 4.0 (using short-range pH indicator paper), add 10 ml of freshly prepared hydrogen sulfide test solution, and allow the liquid to stand for 10 min; the color of the liquid, when viewed downward over a white surface, is not darker than the color of a mixture of 40 ml of the same *water* with the same amount of diluted acetic acid as was added to the sample and 10 ml of *purified water*, matched color-comparison tubes being used for the comparison.

3. Zinc – To 50 ml contained in a test tube add three drops of glacial acetic acid

5.4–15   SUMMARY OF WATER STANDARDS FROM PHARMACOPEIA (continued)

and 0.5 ml of potassium ferrocyanide test solution; the solution shows no more turbidity than that produced by 50 ml of *purified water* in a similar test tube, treated in the same manner, and viewed downward over a dark surface.

4. Foreign volatile matter – When heated nearly to the boiling point and agitated, it evolves no odor.

5. Total solids – Evaporate 100 ml on a steam bath to dryness, and dry the residue at 105°C for 1 hr; not more than 100 mg of residue remains (0.1%).

6. Bacteriological purity – It meets the United States Public Health Service regulations for potable water with respect to bacteriological purity.

7. Pyrogen – When previously rendered isotonic by the addition of 900 mg of pyrogen-free sodium chloride for each 100 ml, it meets the requirements of the Pyrogen Test.

8. pH, between 4.5 and 7.0, determined potentiometrically in a solution prepared by the addition of 0.30 ml of saturated potassium chloride solution to 100 ml of *bacteriostatic water for injection.*

9. Total solids – Evaporate 30 ml on a steam bath to dryness, and dry the residue at 105°C for 1 hr or at such higher temperature and for such period of time as may be required to drive off any volatile added substance. Weigh the residue, and correct for any nonvolatile substances declared on the label; the weight does not exceed 1.3 mg (40 ppm).

10. Antimicrobial agent(s) – It meets the requirements under *Antimicrobial Agents – Effectiveness*, Page 845 of the original source of this table, and meets the labeled claim for content of the antimicrobial agent(s), as determined by the method set forth under *Antimicrobial Agents – Content*, Page 902 of the original source.

11. Pyrogen – When previously rendered isotonic by the addition of 900 mg of pyrogen-free sodium chloride for each 100 ml, it meets the requirements of the *Pyrogen Test*, Page 886 of the original source, the test dose being 5 ml/kg, injected very slowly.

12. Sterility – It meets the requirements under *Sterility Tests*, Page 851 of the original source.

13. Chloride – To 20 ml in a color-comparison tube add five drops of nitric acid and 1 ml of silver nitrate test solution, and gently mix; any turbidity formed within 10 min is not greater than that produced in a similarly treated control consisting of 20 ml of *special distilled water*[c] containing 10 mcg of Cl (0.5 ppm), viewed downward over a dark surface with light entering the tubes from the side.

14. Oxidizable substances – To 100 ml add 10 ml of diluted sulfuric acid and heat to boiling. Add 0.2 ml of 0.1 N potassium permanganate, and boil for 5 min; the pink color does not completely disappear.

15. Total solids – Proceed as directed in the test for *Total Solids* (see Item 25). The following limits apply for *sterile water for injection* in glass containers: up to and including 30 ml size, 40 ppm; from 30 ml up to and including 100-ml size, 30 ppm; and for larger sizes, 20 ppm.

16. Sterility – It meets the requirements under *Sterility Tests*, Page 851 of the original source.

17. pH between 5.0 and 7.0, determined potentiometrically in a solution prepared by the addition of 0.30 ml of saturated potassium chloride solution to 100 ml of *purified water.*

18. Chloride – To 100 ml add five drops of nitric acid and 1 ml of silver nitrate test solution; no opalescence is produced.

19. Sulfate – To 100 ml add 1 ml of barium chloride test solution; turbidity is produced.

20. Ammonia – To 100 ml add 2 ml of alkaline mercuric-potassium iodide test solution; any yellow color produced immediately is not darker than that of a control containing 30 mcg of added NH₃ in *special distilled water*[c] (0.3 ppm).

21. Calcium – To 100 ml add 2 ml of ammonium oxalate test solution; no turbidity is produced.

22. Carbon dioxide – To 25 ml add 25 ml of calcium hydroxide test solution; the mixture remains clear.

23. Heavy metals – Adjust 40 ml of *purified water* with diluted acetic acid to a pH of 3.0 to 4.0 (using short-range pH indicator paper), add 10 ml of freshly prepared hydrogen sulfide test solution, and allow the liquid to stand for 10 min; the color of the liquid, when viewed downward over a white surface, is not darker than the color of a mixture of 50 ml of the same *purified water* with the same amount of diluted acetic acid as was added to the sample, matched color-comparison tubes being used for the comparison.

24. Oxidizable substances – To 100 ml add 10 ml of diluted sulfuric acid and heat to boiling. Add 0.1 ml of 0.1 N potassium permanganate, and boil for 10 min; the pink color does not completely disappear.

25. Total solids – Evaporate 100 ml on a steam bath to dryness, and dry the residue at 105°C for 1 hr; not more than 1 mg of residue remains (10 ppm).

b **Methyl red solution:** Dissolve 24 mg of methyl red sodium in sufficient purified water to make 100 ml. If necessary, neutralize the solution with 0.02 N sodium hydroxide so that the titration of 100 ml of *special distilled water*, containing five drops of indicator, does not require more than 0.02 ml of 0.02 N sodium hydroxide to effect the color change of the indicator, which should occur at a pH of 5.6.

c **Special distilled water:** The water used in these tests has a specific conductivity, determined at 20°C, just prior to use, of not greater than 1 μmho. It meets the requirements of the test for heavy metals under purified water, and is free from copper. The water may be prepared by redistilling once-distilled water from a still of such proportions that the steam flows at low velocity when the still is operating at rated capacity. All parts of the still in contact with the water or steam are of borosilicate glass. Prior to distillation add, for each 1000 ml of water contained in the still, 1 drop of phosphoric acid that previously has been diluted with an equal volume of distilled water and boiled to the point of appearance of dense fumes. Reject the first 10 to 15% of the distillate, and retain the next 75%.

Source: *Pharmacopeia of the United States of America,* 18th rev., United States Pharmacopeial Convention, Washington, D.C., 1970.

5.4-16   AAMI[a] STANDARDS FOR
DIALYSATE WATERS

Contaminant	Suggested maximum level (mg/l)
Calcium	2 (0.1 mEq/l)
Magnesium	4 (0.3 mEq/l)
Sodium	70 (3 mEq/l)
Potassium	8 (0.2 mEq/l)
Fluoride	0.2
Chlorine	0.5
Chloramines	0.1
Nitrate (N)	2
Sulfate	100
Copper	0.1
Barium	0.1
Zinc	0.1
Aluminum	0.01
Arsenic	0.005
Lead	0.005
Silver	0.005
Cadmium	0.001
Chromium	0.014
Selenium	0.09
Mercur	0.002
Viable microbes	<200/ml

[a]   Association for the Advancement of Medical Instrumentation.

Source: W.J. Dorson, Jr., Water problems associated with patients being treated with dialysis, in *Drinking Water and Human Health*, American Medical Association, Chicago, 1974, p. 86.

5.4-17 WATER FOR PARENTERAL SOLUTIONS

Resistance (specific, ohms)	500,000
pH	5.7–6.0
Residue after evaporation (1 hr at 105°C)	1.0 ppm
Chlorides (Cl)	0.1 ppm
Ammonia (NH_3)	0.1 ppm
Heavy metals (as Pb)	0.01 ppm

Source: J.J. Perkins, *Principles and Methods of Sterilization in Health Sciences,* 2nd ed., 1969. Courtesy of Charles C Thomas, Publisher, Springfield, Ill.

5.4-18 MEDICAL CARE WATER USES AT THE UNIVERSITY OF MINNESOTA HOSPITALS

Type of water used	Water uses
Sterile distilled water	Filling of humidity reservoirs in isolettes,[a] dilution of infant formula,[a] ultrasonic nebulizers (tracheotomy),[a] ultrasonic nebulizers,[a] MAI respirators,[a] Pleura-Evac (chest suction device),[a] filling of infant heating pads[a]
Sterile distilled water for normal saline	Rinsing of catheters,[a] filling of trap on Emerson Suction Devices (closed drainage system)[a]
Normal saline (30-cc bottles)	Lavage,[a] irrigation of plugged IV's,[a] nebulizers on intermittent positive pressure breathing (IPPB) machines[a]
Normal saline	Wound irrigation, bladder irrigation, nasal tube irrigation,[a] gastric tube irrigation,[a] irrigation of GI bleeder, burn patient irrigation
Bacteriostatic water or normal saline (30-cc vials)	Dilution of injectable preparations
Nonsterile distilled water or sterile distilled water	Oxygen bubble jets attached to oxygen system,[a] oxygen bubble jets attached to tracheotomy[a]
Nonsterile distilled	Ultrasonic nebulizers (non-tracheotomy),[a] filling of instrument sterilizer on nursing station,[a] lubrication of endotracheal tubes prior to sterilization[a]
Tap water (municipal)	Filling of room humidifiers,[a] filling of water bath for heart-lung machine[a]
Tap water wash followed by tap water rinse in washer-claves	Washing of surgical instruments
a. Tap water stagnant soak b. Tap water running soak c. Distilled water-detergent wash d. Distilled water rinse prior to sterilization	Reprocessing needles and syringes[a]

5.4-18   MEDICAL CARE WATER USES AT THE UNIVERSITY
OF MINNESOTA HOSPITALS (continued)

Type of water used	Water uses
Wash cycle and initial rinse with tap water followed by rinse with distilled water. When machine is used for equipment that does not come into contact with bloodstream, then final rinse is with tap water.	Equipment washer — reusable syringes, surgical instruments, etc. which may come into contact with bloodstream[a]
Pyrogen-free sterile distilled water	Preparation of injectables
Distilled water from pharmacy system[b]	Preparation of irrigating solution, i.e., irrigating solution for flushing body cavities after transplant surgery; nonsterile preparations, i.e., solution used in mist tents for cystic fibrosis patients; final rinse in bottle washer for (a) bottles for nonsterile germicides and antiseptics, (b) bottles (500- and 1000-ml) reused for irrigating solutions, and (c) serum vials for water for injection; rinsing of 100-gal mixing tanks used by pharmacy;[a] filling of bottles of nonsterile distilled water distributed to nursing stations
Purchased water for injection	Eye solution formulation
Water from a reverse osmosis unit	Water for kidney dialysis[a]

[a] No recognized standard for water quality.

[b] A specially designed style for production of pyrogen-free water.

5.4-19   DISINFECTION OF LEGIONELLA IN HOSPITAL WATER SUPPLIES

Source	Disinfectant	Dose (mg/l)	Remarks	Ref.
Tap water (cold)	Free chlorine	1—2	United Kingdom	1
Potable water supply	Free chlorine	4	Columbus, Ohio; not effective; not water tanks	2
Water supply	Free chlorine	3.3—6.6	Kingston, U.K.	3, 4
Water	Free chlorine, 21°C, pH 7.6, 20-min contact	0.1	Laboratory study; 99% kill	5
Water	Free chlorine 4°C, pH 7.6, 80-min contact	0.1	Laboratory study; 99% kill	5
Water tanks, etc.	Free chlorine, 10 days, then	30		6
	Free chlorine	1.5	6	
Hot and cold water	Chlorine	2—4	Los Angeles	1
Water supply	Ozone	0.36	*In vitro* studies, no field confirmation	8
Water	Ozone	1—2	6-hr contact, 5-log kill	9
Water	UV	$1 \ \mu\text{W/cm}^2$	17 min, 50% kill	10
Water	UV	$380 \ \mu\text{W/cm}^2$	50% kill	10
		$920 \ \mu\text{W/cm}^2$	90% kill	10
		$1840—2760 \ \mu\text{W/cm}^2$	99.9% kill	10
Water supply	Heat	60—70°C	Pittsburgh, Penna.	1
Water supply	Heat	55—60°C	Kingston, U.K.	1

Compiled from G. Kaledi and M.A. Shapiro, *Crit. Rev. Environ. Control, 17*:133, 1987.

REFERENCES

1. Anon., *Lancet, 2*:381, 1983.
2. I.M. Baird, W. Potts, J. Smiley, N. Click, S. Schleich, C. Connole, and K. Davidson, *Control of Endemic Nosocomial Legionellosis by Hyperchlorination of Potable Water,* Proc. 2nd Int. Symp., C. Thornsberry, A. Balows, J.C. Feeley, and W. Jakubowski, Eds., AMS, Washington, D.C., 1984, 333.
3. A.P. Dufour and W. Jakubowski, *J. Am. Water Works Assn., 47*:631, 1982.
4. D. Skaliy, T.A. Thompson, G.W. Gorman, G.K. Morris, H.V. McEachem, and D. Mackel, *Appl. Environ. Microbiol., 48*:697, 1980.
5. J.M. Kuchta, S.J. States, A.M. McNamara, R.M. Wadowsky, and R.B. Yee, *Appl. Environ. Microbiol., 46*:1134, 1983.
6. L.E. Witherell, L.A. Orciari, R.W. Duncan, K.N. Stone, and J.M. Lawson, *Disinfection of Hospital Hot Water Systems Containing Legionella pneumophila,* in *Legionella,* Proc. 2nd Int. Symp., C. Thornsberry, A. Balows, J.C. Feeley, and W. Jakubowski, Eds., AMS, Washington, D.C., 1984, 336.
7. European Regional Office (EURO) Reports and Studies, Legionnaires' Disease Report by a WHO Working Group, Copenhagen, 1982.
8. P.H. Edelstein, R.E. Whittaker, R.L. Kreileng, and C.L. Howell, *Appl. Environ. Microbiol., 44*:1330, 1982.
9. P. Muraca, J. Stout, and V.L. Yu, *Appl. Environ. Microbiol., 53*:447, 1987.
10. R.W. Gilpin, Laboratory and Field Applications of UV Light Disinfection of Six Species of *Legionella* and Other Bacteria in Water, in *Legionella,* Proc. 2nd Int. Symp., Thornsberry, C., Balows, A., Feeley, J.C., and Jakubowski, W., Eds., AMS, Washington, D.C., 1984, 337.

Index

INDEX